Wave Scattering by Time-Dependent Perturbations

PRINCETON SERIES IN APPLIED MATHEMATICS

Edited by

The Princeton Series in Applied Mathematics publishes high quality advanced texts and monographs in all areas of applied mathematics. Books include those of a theoretical and general nature as well as those dealing with the mathematics of specific applications areas and real-world situations.

Wave Scattering by Time-Dependent Perturbations

An Introduction

G. F. Roach

PRINCETON UNIVERSITY PRESS
PRINCETON AND OXFORD

Published by Princeton University Press, 41 William Street, Princeton, New Jersey 08540

In the United Kingdom: Princeton University Press, 3 Market Place, Woodstock,
Oxfordshire OX20 1SY

Library of Congress Cataloging-in-Publication Data

Roach, G. F. (Gary Francis)
Wave scattering by time-dependent perturbations: an introduction / G. F. Roach.
p. cm. — (Princeton series in applied mathematics)
Includes bibliographical references and index.
ISBN-13: 978-0-691-11340-1 (alk. paper)
ISBN-10: 0-691-11340-8 (alk. paper)
1. Waves—Mathematics. 2. Scattering (Physics)—Mathematics. 3. Perturbation (Mathematics) I. Title.

QC157.R53 2007
531′.1133—dc22 2006052858

British Library Cataloging-in-Publication Data is available

This book has been composed in LaTeX.

Printed on acid-free paper. ∞

press.princeton.edu

Printed in the United States of America

10 9 8 7 6 5 4 3 2 1

Contents

Preface

The use of various types of wave energy as a probe is an increasingly promising nondestructive means of detecting objects and of diagnosing the properties of quite complicated materials. An analysis of this technique requires a detailed understanding of, first, how signals evolve in the medium of interest in the absence of inhomogeneities and, second, the nature of the scattered or echo field when the original signal is perturbed by inhomogeneities that might exist in the medium. The overall aim of the analysis is to calculate relationships between an unperturbed signal waveform and an associated echo waveform and indicate how these relationships can be used to characterise inhomogeneities in the medium.

When the media involved are stationary with time-independent material properties, the mathematical analysis of the associated propagation and scattering problems is now well developed and a number of efficient techniques are available for constructing solutions to both direct and inverse wave scattering problems. However, investigations of the corresponding propagation and scattering phenomena when the media are moving or have time-dependent material characteristics or both have not yet reached such a well-developed stage. Whilst problems that involve time-dependent perturbations are intriguing from a purely theoretical standpoint, nevertheless, they are also of considerable interest in the applied sciences; they frequently occur, for example, in such areas as radar, sonar, nondestructive testing, geophysical prospecting and ultrasonic medical diagnosis.

An initial aim of this monograph is to give a largely self-contained introductory account of acoustic wave propagation and scattering in the presence of time-dependent perturbations. Later chapters of the book will indicate how the approach adopted here for dealing with acoustic problems can be extended to similar problems in electromagnetism and elasticity.

In this monograph we gather together the principal mathematical topics that can be used when dealing with wave propagation and scattering problems involving time-dependent perturbations. In so doing we will provide a unified and reasonably self-contained introduction to an active research area that has been developing over recent years. We will also indicate how the material can be used to develop constructive methods of solution. The overall intention is to present the material so that is just as persuasive to the theoretician as to the applied scientist who may not have the same mathematical background. This book is meant to be a guide that indicates the technical requirements when investigating wave scattering problems. Throughout, the emphasis will be on concepts and results rather than on the fine detail of proof. The proofs of results that are simply stated in the text, many of which are lengthy and very much of a technical nature, can be found in the references cited in the commentaries in chapter 11.

Many of the results described in this book represent the works of a large number of authors, and an attempt has been made to provide a reasonably comprehensive bibliography. However, particular mention must be made of the pioneering works of Ikebe, Lax and Phillips and of Wilcox when studying time-independent perturbation problems and of Cooper, Kleinman, Mack, Strauss, Tanabe and Vainberg for time-dependent perturbation problems. The influence of the works of these authors has been considerable and is gratefully acknowledged. In particular, a profound debt of gratitude is owed to Rolf Leis and Calvin Wilcox, who have been such an inspiration over the years.

I would also like to express my gratitude to the many colleagues with whom I have had such useful discussions. In particular, I would thank Christos Athanasiadis, Aldo Belleni-Morante, Wilson Lamb and Ioannis Stratis, who have read various parts of the manuscript and offered so many suggestions.

A special word of thanks must also go to Mary Sergeant for the patient way in which she tackled the seemingly endless task of typing and proof corrections.

Finally, I want to express my appreciation to David Ireland and Vickie Kearn of Princeton University Press for their efficient and friendly handling of the publication of this book

GFR
Glasgow
2005

Wave Scattering by Time-Dependent Perturbations

Chapter One

Introduction and Outline of Contents

1.1 INTRODUCTION

The use of various types of wave energy as a probe is an increasingly promising nondestructive means of detecting objects and of diagnosing the properties of quite complicated materials.

An analysis of this technique requires a detailed understanding of, first, how waves evolve in the medium of interest in the absence of any inhomogeneities and, second, the nature of the scattered or echo waves generated when the original wave is perturbed by inhomogeneities that might exist in the medium. The overall aim of the analysis is to calculate the relationships between the unperturbed waveform and the echo waveform and to indicate how these relationships can be used to characterise inhomogeneities in the medium.

The central problem with which we shall be concerned in this monograph can be simply stated as follows.

A system consists of a medium containing a transmitter and a receiver. The transmitter emits a signal that is eventually detected at the receiver, possibly after it has been perturbed, that is, scattered, by some inhomogeneity in the medium. We are interested in the manner in which the emitted signal evolves through the medium and the form that it assumes at the receiver. Properties of the scattered or echo signal are then used to estimate the properties of any inhomogeneity in the medium.

Classifying inhomogeneities in the medium into identifiable classes by means of their echoes is known as the *inverse scattering problem.* An associated problem is that of *waveform design*, which is concerned with the choice of the signal waveform that optimises the echo signal from classes of prescribed inhomogeneities. These problems are of considerable interest and importance in engineering and the applied sciences. However, in order to be able to investigate them, the problem of knowing how to predict the echo signal when the emitted signal and the inhomogeneities are known must be well understood. This is called the *direct scattering problem.*

When the media involved are either stationary or possess time-independent characteristics—these are called *autonomous problems* (APs)—the mathematical analysis of the associated scattering effects is now quite well developed and a number of efficient techniques are available for constructing solutions to both the direct and the inverse problems. However, when the media are either moving or have time-dependent characteristics—these are known as *nonautonomous problems* (NAPs)—the investigations of corresponding scattering phenomena have not reached such a well-developed stage. Nevertheless, there are many significant problems of interest in the applied sciences that are NAPs. For instance, this type of problem can often arise when investigating sonar, radar, nondestructive testing and ultrasonic medical diagnosis methods. Indeed, they occur in any system that is either in motion or has components that either can be switched on or off or can be altered periodically. We shall study some of these systems in later chapters. These NAPs are intriguing both from a theoretical standpoint and from the point of view of developing constructive methods of solution; they certainly present a nontrivial challenge.

In our study here of NAPs we take as a starting point the assumption that all media involved consist of a continuum of interacting infinitesimal elements. Consequently, a disturbance in some small region of a medium induces an associated disturbance in neighbouring regions with the result that some sort of disturbance eventually spreads throughout the medium. We call the progress or evolution of such disturbances *propagation*. Typical examples of this phenomenon include, for instance, waves on water, where the medium is the layer of water close to the surface, the interaction forces are fluid pressure and gravity and the resulting waveform is periodic. Again, acoustic waves in gases, liquids and solids are supported by an elastic interaction and exhibit a variety of waveforms which can be, for example, sinusoidal, periodic, transient pulse or arbitrary. However, in principle any waveform can be set in motion in a given system provided suitable initial or source conditions are imposed.

The above discussion can be conveniently expressed in symbolic form as follows.

Consider first a system that has no inhomogeneities. Let $f_0(x,s)$ be a quantity that characterises the state of the system at some initial time $t=s$ and let $u_0(x,t)$ be a quantity that characterises the state of the system at some later time $t>s$. We shall be concerned with systems for which states can be related by means of an "evolution rule," denoted by $U_0(t-s)$, that determines the *evolution* in time of the system from its initial state $f_0(x,s)$ to a state $u_0(x,t)$ at a later time $t>s$. This being the case, we write

$$u_0(x,t) = U_0(t-s)\, f_0(x,s),$$

where it is understood that $U_0(0) = I =$ the identity.

In a similar manner, when inhomogeneities are present in the system, we will assume that we can express the evolution of the system from an initial state $f_1(x,s)$ to a state $u_1(x,t)$ at a later time $t>s$ in the form

$$u_1(x,t) = U_1(t-s)\, f_1(x,s), \qquad U_1(0) = I,$$

where $U_1(t-s)$ denotes an appropriate evolution rule. Thus we see that we are concerned with two classes of problems. When there are no inhomogeneities present in the system, we shall say that we have a *free problem* (FP). When inhomogeneities are present in a system, we shall say that we have a *perturbed problem* (PP). We shall express this situation symbolically in the form

$$u_j(x,t) = U_j(t-s)\, f_j(x,s), \qquad U_j(0) = I, \qquad j = 0, 1,$$

where when $j=0$ we will assume that we have a FP whilst when $j=1$ we have a PP.

The principal aim of this monograph is to make the preceding discussions more precise and, in so doing, indicate means of developing sound, constructive methods of solution from what might be originally thought to be a purely abstract mathematical framework. In this connection we are immediately faced with a number of fundamental questions.

- What are the mathematical equations that define (model) the systems of interest?
- What is meant by a solution of the defining equations?
- Under what conditions do the defining equations have unique solutions?
- When solutions of the defining equations exist, can they be expressed in the form

$$u_j(x,t)=U_j(t-s)f_j(x,s), \qquad U_j(0)=I, \qquad j=0,1,$$

  where $U_j(t-s)$, $j=0, 1$, is an evolution rule?
- How can $U_j(t-s)$ be determined?
- What are the basic properties of $U_j(t-s)$, $j=0, 1$?
- If a given problem is regarded as a PP, then an associated FP can be taken to be a problem that is more easily solved than the PP. We then ask, Is it possible to determine an initial state of the system defined by the FP so that the state of this system at some later time t, denoted $u_0(x,t)$, is equal in some sense to $u_1(x,t)$, the state at time t of the system defined by the PP? If we can do this, then it is readily seen that we will have taken a large step towards creating a firm basis from which to develop robust, constructive methods for determining the required quantity $u_1(x,t)$ in terms of a more readily obtainable quantity $u_0(x,t)$.

In connection with the possible equality of $u_j(x,t)$, $j=0, 1$, we first recognise and make use of the fact that in most experimental procedures, measurements in a system are made far away from any inhomogeneities that might exist in the system. Consequently, we will be mainly concerned here with the nature of the solutions $u_j(x,t)$, $j=0, 1$, and of their differences in the *far field* of any nonhomogeneity. With this in mind we shall see that it will be sufficient for our purposes to ask, What is the behaviour of $u_j(x,t)$, $j=0, 1$, as $t\to\infty$? Once this asymptotic behaviour is known, we can clarify what we mean by the equality of $u_j(x,t)$, $j=0, 1$, and turn to determining the conditions that will actually ensure when, in the far field at least, $u_j(x,t)$, $j=0, 1$, can be considered equal.

Although we are particularly interested in addressing these various questions when dealing with NAPs, we would point out that even for APs a detailed mathematical analysis of such questions can be technically very demanding. However, we would emphasise that this monograph is not a book on such topics as functional analysis, mathematical scattering theory, linear operator theory and semigroup theory but rather is meant as a guide through these various areas with the intention of highlighting their uses in practical problems. Examples will be given whenever it is practical to do so, as it is felt that abstract theories are often best appreciated, at first, by means of examples. The presentation in this monograph will frequently be quite formal, and we rely very much on the often held view expressed by Goldberger and Watson that "any formal manipulations which are not obviously wrong are assumed to be correct" [44]. Nevertheless, references will always be given in the text to where more precise and often quite general details can be found. Furthermore, we emphasise that in this monograph we are not interested in investigating the evolutionary processes mentioned above in full generality but rather shall confine our attention to those systems involving waves.

1.2 SOME ILLUSTRATIONS

As we have already mentioned, our main interest in this monograph will centre on those physical phenomena whose evolution can be described in terms of propagating waves. In the case of APs a simple example of a propagating wave is provided by a physical quantity y, which is defined by

$$y(x,t) = f(x - ct), \qquad (x,t) \in \mathbf{R} \times \mathbf{R}, \tag{1.1}$$

where c is a real constant. We notice that y has the same value at all those x and t for which $(x - ct)$ has the same values. Thus (1.1) represents a wave that moves with constant velocity c along the x-axis without changing shape. If f is assumed to be sufficiently differentiable, then on differentiating (1.1) twice with respect to x and t, we obtain

$$\left\{ \frac{\partial^2}{\partial t^2} - c^2 \frac{\partial^2}{\partial x^2} \right\} y(x,t) = 0, \qquad (x,t) \in \mathbf{R} \times \mathbf{R}. \tag{1.2}$$

Similarly, we notice that a physical quantity w defined by

$$w(x,t) = g(x + ct), \qquad (x,t) \in \mathbf{R} \times \mathbf{R}, \tag{1.3}$$

where c is a real constant, represents a wave moving with constant velocity c along the x-axis without changing shape but moving in the opposite direction to the wave $y(x,t)$ defined in (1.1). Furthermore, we see that $w(x,t)$ also satisfies an equation of the form (1.2). Equation (1.2) is referred to as the *classical wave equation.* The term "classical" will only be used in order to avoid possible confusion with other wave equations that we might consider.

Because the wave equation is linear, the compound wave

$$u(x,t) = y(x,t) + w(x,t) = f(x - ct) + g(x + ct), \tag{1.4}$$

where f, g are arbitrary functions, is also a solution of the wave equation (1.2). This is the celebrated d'Alembert solution of the wave equation. In specific problems the functions f, g are determined in terms of the imposed initial conditions that solutions of (1.2) are required to satisfy [71, 105]. Indeed, if $u(x,t)$ is required to satisfy the initial conditions

$$u(x,0) = \varphi(x), \qquad u_t(x,0) = \psi(x),$$

then (1.4) can be expressed in the form

$$u(x,t) = \frac{1}{2}\{\varphi(x-ct)+\varphi(x+ct)\} + \frac{1}{2c}\int_{x-ct}^{x+ct}\psi(s)\,ds \tag{1.5}$$

(see chapter 2 and [71, 105]). Consequently, if a system is defined by the wave equation (1.2), then (1.5) indicates how the initial state of the system, characterised by the functions φ and ψ, evolves into a state defined by $u(x,t)$, $t > 0$.

As a simple illustration of such an evolutionary process, consider a system consisting of an infinite string in the particular case when $\psi \equiv 0$. The initial state of this system is then completely characterised by the function φ. For simplicity, assume that the initial displacement of the string is of finite amplitude and is only nonzero over a finite length of the string. The function φ defines this profile. The result (1.5) indicates that this initial state of the system evolves into a state consisting of two propagating waves travelling in opposite directions along the string, each having the same profile as the initial state but with only half the initial amplitude.

As an illustration of a scattering process, again consider waves on a string. We assume, as before, that $\psi \equiv 0$, but now we impose the extra requirement that the displacement of the string should be zero at $x = 0$ for all time. Essentially, this means that we are considering waves on a semi-infinite string. Arguing as before, we see that the initial state of the system, which again is completely characterised by the function φ, evolves into a state consisting of two propagating waves travelling in opposite directions and having the same profile as the initial state but with only half the initial amplitude. However, this situation does not persist. After a certain time, say T, which is dependent on the velocity of the waves, one of the propagating waves will strike the "barrier" at $x = 0$. This wave will bounce off the barrier; that is, it will be reflected or scattered by the barrier. Consequently, from time T onwards the initial state of the system evolves into a state consisting of three components. Specifically, it will consist of the two propagating waves mentioned earlier and a reflected or scattered wave. Thus the evolved state can become quite difficult to describe analytically. Quite how difficult matters can become will be demonstrated in chapter 2. Further complications arise when we have to work either in more than one dimension or with defining equations that are more difficult to analyse than the wave equation (1.2). It was largely with these prospects in mind that scattering theory was developed.

We would emphasise at this stage that not all solutions of the wave equation yield propagating waves. For example, if the wave equation is solved using a separation of variables technique, then *stationary wave* solutions can be obtained. They are called stationary waves because they have certain features, such as nodes and antinodes, that retain their positions permanently with respect to time. Such solutions can be related to the *bound states* appearing in quantum mechanics and to the *trapped wave* phenomenon of classical wave theory.

Since the classical wave equation occurs in so many areas of mathematical physics, we shall adopt it as a prototype for much of our present study. We shall, of course, discuss other types of wave equations after we have settled the main features of wave scattering associated with this particular equation.

1.3 TOWARDS GENERALISATIONS

Scattering means different things to different people. Broadly speaking, it can be thought of as the interaction of an evolutionary process with a nonhomogeneous and possibly nonlinear medium. Certainly, the study of scattering phenomena has played a central role in mathematical physics over the years, with perhaps the earliest investigation of them being attributed to Leonardo da Vinci who studied the scattering of light into the shadow of an opaque body. Subsequently, other scattering effects have been discovered and investigated in such diverse fields as acoustics, quantum mechanics, medical diagnosis and many other nondestructive testing procedures.

Scattering phenomena arise as a consequence of some perturbation of a given known system, and they are analysed by developing an associated scattering theory. These scattering theories are concerned, broadly speaking, with the comparison of two systems, one being regarded as a perturbation of the other, and with the study of the manner in which the two systems might eventually become equal in some sense. Since a NAP can often be regarded as a perturbation of an AP, the development of scattering theories involving NAPs and APs seems to offer good prospects for providing a sound basis from which to develop robust approximation methods for obtaining solutions to a NAP in terms of the more readily obtainable solutions to an associated AP.

An initial aim in this monograph is to present an introductory study of acoustic wave propagation and scattering in the presence of *time-dependent perturbations*. Later chapters of the book will indicate how the analysis of the acoustic case can be extended to similar problems in electromagnetism and elasticity. The study will be made from the standpoint of modern spectral analysis and scattering theory, and the emphasis throughout will be on the development of constructive methods.

Scattering processes can be conveniently characterised in terms of either one or the other of two basic problems. In one the governing differential equation is perturbed but the spatial domain of interest is unaffected. This gives rise to *potential scattering*. In the other the governing differential equation is unaltered but the spatial domain of interest is perturbed. This gives rise to *target scattering*. In developing the material that follows we shall often begin with a discussion of APs, in particular with potential scattering problems, where the notion of a perturbation is perhaps rather more clear, as much because the technical requirements are less than those for target scattering as because the rich theory and analytical techniques of quantum scattering can quite readily be used to indicate the way we might proceed in our analysis. Afterwards we turn our attention to target scattering.

For APs there are traditionally two main approaches to the analysis of scattering phenomena. The first is by means of a *time-independent* or *stationary scattering* theory. This consists of separating out the time dependence in the problems and then studying solutions of the resulting spatial equations. This is a *frequency domain* analysis in which a central interest is the asymptotic behaviour of solutions at large distances. The second approach is by means of a *time-dependent scattering* theory in which the time evolution of the states of the systems and their asymptotic behaviour for large time is a dominant interest. This is a *time domain analysis.*

We remark that not all states of a system necessarily lead to scattering events; for example, as mentioned earlier, a system can have bound states. However, those states of a system that do lead to scattering events and which, as $t \to \pm\infty$, are *asymptotically equal* (AE) in some sense to scattering events of some simpler system are said to satisfy an *asymptotic condition.* An investigation of this condition is a principal study of time-dependent scattering theory. The importance of the time-independent theory for APs lies in the fact that the actual calculation of expressions can often be more readily carried out than for similar expressions occurring in the time-dependent theory. However, we shall see that for NAPs *the separation of variables technique is no longer available.* Consequently, an associated time-independent theory is *not* immediately available.

In the quantum mechanics of the scattering of elementary particles by a potential, the wave packets that characterise scattered particles can be shown to be AE, for large time, to the corresponding wave packets for free particles. The correspondence between these two systems, one describing the scattered particles and the other describing the free particles, is effected by means of Møller operators [71]. Wilcox [154, 153] has developed analogous concepts for wave propagation problems associated with the equations of classical physics. Specifically, Wilcox established conditions that ensure that waves propagating in a inhomogeneous medium are AE, for large time, to corresponding waves propagating in a homogeneous medium. The correspondence between the two wave systems is given by an analogue of the Møller operators of quantum scattering, which we shall simply refer to as *wave operators* (WOs). Since wave propagation problems in homogeneous media can often be solved explicitly, a knowledge of an appropriate WO then provides information concerning the asymptotic behaviour of solutions to wave propagation problems in inhomogeneous media. The intention here is to indicate how these ideas of Wilcox might extend to an investigation of NAPs.

As we remarked above, the study of NAPs has been motivated by problems arising in such areas as, for example, radar, sonar, nondestructive testing and ultrasonic medical diagnosis. In all these areas a powerful diagnostic is the dynamical response of the media to the emitted signal. Mathematically, many of these problems can be conveniently modelled in terms of an *initial boundary value problem* (IBVP). To fix ideas, we shall confine our attention initially to acoustic problems and to IBVPs for the classical wave equation. Results for electromagnetic problems and for elastic problems will be discussed in later chapters.

Specifically, for NAPs we shall be interested in IBVPs that have the following typical form.

Determine a quantity $w(x,t)$ that satisfies

$$\left\{\partial_t^2 + L(x,t)\right\} w(x,t) = f(x,t), \qquad (x,t) \in Q, \tag{1.6}$$

$$w(x,s) = w_{0s}(x), \quad w_t(x,s) = w_{1s}(x), \quad x \in \Omega(s), \tag{1.7}$$

$$w(x,t) \in (bc)(t), \quad (x,t) \in \partial\Omega(s) \times \mathbf{R}, \tag{1.8}$$

where

$$L(x,t) = -c^2\Delta + q(x,t), \tag{1.9}$$

with Δ denoting the usual Laplacian in $\mathbf{R}^n$ and

$$Q := \{(x,t) \in \mathbf{R}^n \times \mathbf{R}\},$$
$$\Omega(t) := \{x \in \mathbf{R}^n : (x,t) \in Q\},$$
$$B(t) := \{x \in \mathbf{R}^n : (x,t) \notin Q\}.$$

The region Q is an open set in $(\mathbf{R}^n \times \mathbf{R})$, and $\Omega(t)$ denotes the exterior, at time t, of a scattering target $B(t)$. For each value of t the domain $\Omega(t)$ is open in $\mathbf{R}^n$ and is assumed to have a smooth boundary $\partial\Omega(t)$. The lateral surface of Q, denoted ∂Q, is defined by

$$\partial Q := \bigcup_{t \in I} \partial\Omega(t), \tag{1.10}$$

where, for some fixed $T > 0$, $I := \{t \in \mathbf{R} : 0 \leq t < T\}$.

The quantities f, c, w_{0s}, w_{1s} and q are given *data functions*, and $s \in \mathbf{R}$ denotes a *fixed initial time*. The notation in (1.8) indicates that the solution $w(x,t)$ is required to satisfy certain conditions, denoted (bc), imposed on the boundary that might also depend on t.

For purposes of illustration in this introductory chapter, we will discuss here, in an entirely formal manner, scattering phenomena associated with *initial value problems* (IVPs) of the form

$$\{\partial_t^2 + L(x,t)\}u(x,t) = 0, \quad (x,t) \in \mathbf{R}^n \times \mathbf{R}, \tag{1.11}$$

$$u(x,s) = f_s(x), \quad u_t(x,s) = g_s, \quad x \in \mathbf{R}^n, \tag{1.12}$$

where $s \in \mathbf{R}$ is a fixed initial time, L is defined via (1.9), and f_s and g_s are given data functions. We remark that when investigating (1.6), an appropriate version of Duhamel's principle (see chapter 5) indicates that in the first instance it is sufficient to study an equation of the form (1.11).

Problems of the form (1.6)–(1.8) divide quite naturally into the following two broad classes.

Target scattering problems: (i) $q(x,t) \equiv 0$; (ii) the (bc) depend on t.
Potential scattering problems: (i) $q(x,t) \neq 0$; (ii) the (bc) are independent of t.

Target scattering problems divide quite readily into two important types.

- Exterior problems in which waves are scattered by obstacles of either finite or infinite extent. These arise, for example, in studies of antennae, propellers and layered media.
- Interior problems in which waves are scattered within some cavity.

We shall be particularly interested in this monograph in the NAP versions of these two types of problems, for example, when an obstacle might be moving or a cavity pulsating.

The NAP versions of potential scattering problems are typical of those that can occur as a consequence of linearising about a time-periodic solution of some nonlinear problem. These can arise, for example, when investigating nonlinear materials and also in certain approaches to inverse scattering problems.

Work on these two types of problems has presented a number of interesting features. For instance, for problems posed in a bounded region, that is, for an interior problem, it has been found that solutions can grow exponentially in energy yet remain finite in amplitude. Whether or not this is also true for exterior problems is not yet fully settled. Numerical evidence indicates that growth of solution occurs by virtue of a parametric resonance similar to that found in studies of the Hill and the Mathieu equations. Again, for such NAPs the familiar separation of variables technique and associated Fourier transform methods are not immediately available. Consequently, a thorough time-dependent theory is required for such problems. Towards this end we shall often give relatively simple worked examples of wave motions on strings to illustrate the salient features of the various analytical techniques that will be required when developing more general time-dependent scattering theories for both APs and NAPs.

To illustrate some of the strategies that will be adopted in this monograph when developing scattering theories, let us begin by considering two systems that are governed, respectively, by IVPs of the form

$$\{\partial_t^2 + L_0(x)\}u_0(x,t) = 0, \quad u_0(x,0) = \varphi_0(x), \quad u_{0t}(x,0) = \psi_0(x), \tag{1.13}$$

$$\{\partial_t^2 + L_1(x)\}u_1(x,t) = 0, \quad u_1(x,0) = \varphi_1(x), \quad u_{1t}(x,0) = \psi_1(x), \tag{1.14}$$

where $(x,t) \in \mathbf{R}^n \times \mathbf{R}$ and, recalling (1.11), (1.12), we have set, for ease of presentation and without any loss of generality, $s = 0$.

In (1.14) the differential expression $L_1(x)$ is assumed to be some perturbation of the differential expression $L_0(x) = -\Delta$, the n-dimensional Laplacian. For this reason we shall refer to (1.13) as a free problem and (1.14) as a perturbed problem.

We remark that, as they are written, (1.13) and (1.14) are APs. However, when q and/or (bc) is also time-dependent, (1.13) and (1.14) are NAPs. We shall restrict our attention, for the moment, to the AP case. We emphasise that the approach we adopt in this introductory chapter will be almost entirely formal. The necessary mathematical structure required to support much of the presentation will be provided in chapters 3–5.

An analysis of such problems must begin with a statement that indicates what is meant by solutions of (1.13) and (1.14); that is, a *solution concept* must be introduced. For example, we might want to confine our attention to a *local classical solution* of (1.13) (or (1.14)). This would be an element v with the property that

$$v \in C(\mathbf{R}^n \times [0,T), \mathbf{R}) \cap C^2(\mathbf{R}^n \times [0,T), \mathbf{R}), \tag{1.15}$$

which, for some $T > 0$, satisfies (1.13) (or (1.14)). The notation here means that v is a continuous function, defined on $\mathbf{R}^n \times [0,T)$ with values in $\mathbf{R}$, that is also twice continuously differentiable (see chapter 3 for more details). A local classical solution is called a *global solution* if we can take $T=\infty$. A *p-periodic solution* is a global solution that is p-periodic in $t \in \mathbf{R}$.

Since we shall be concerned with problems that model real-life situations, we shall always require that mathematically the problem be *well posed*; that is, it is expected that

- a solution exists and is unique for the class of data of interest,
- the solution depends continuously on the data.

Even when it is possible in principle, the actual determination of a classical solution is often a very difficult task in practice. The situation can be eased considerably by realising, that is, by interpreting, IVPs (1.13) and (1.14) in some more convenient collection of functions than that indicated in (1.15). For example, we might choose to realise IVPs (1.13) and (1.14) in the collection of functions denoted by

$$H := L_2(\mathbf{R}^n \times \mathbf{R}) \equiv L_2(\mathbf{R}^n \times \mathbf{R}, \mathbf{C}),$$

which consists of all those functions that are square-integrable functions on their domain of definition ($\mathbf{R}^n \times \mathbf{R}$) and which take values in $\mathbf{C}$, the collection of complex numbers. This choice is not as contrived as it might appear at first sight. We shall see that such a selection will enable us to deal quite automatically with important physical quantities such as, for example, energy. We can obtain a realisation of IVPs (1.13) and (1.14) in the collection $H := L_2(\mathbf{R}^n \times \mathbf{R}) \equiv L_2(\mathbf{R}^n \times \mathbf{R}, \mathbf{C})$ by introducing the spatial *mappings* or *operations* A_j, $j=0, 1$, defined by (see chapter 3 for more details)

$$\begin{aligned} &A_j : H \to H, \qquad j=0, 1 \\ &A_j u_j = L_j u_j, \qquad u_j \in D(A_j), \qquad j=0, 1, \\ &D(A_j) = \{u_j \in H : L_j u_j \in H,\ u_j \in (bc)\}. \end{aligned} \tag{1.16}$$

Here $D(A_j)$ denotes the *domain of definition* of the mapping A_j. In the event that we are only dealing with potential scattering problems, the qualifier $u_j \in (bc)$ is omitted from $D(A_j)$.

Notice the following.

- We have assumed, for ease of presentation, that (1.13) and (1.14) can be posed in the same collection H. This will not always be the case.
- The definition of $D(A_j)$ ensures that throughout any mathematical manipulations involving A_j, we always "stay in H."
- When we have more mathematical structure available, we will be able to replace the word "collection" by "space" and the word "operations" by "operators" (see chapter 3).

The *classical IVPs* (1.13) and (1.14) can now be replaced by the *abstract IVPs*

$$\begin{gathered} \{\partial_t^2 + A_j\} u_j(x, t) = 0, \\ u_j(x, 0) = \varphi_j(x), \qquad u_{jt}(x, 0) = \psi_j(x), \qquad j=0, 1. \end{gathered} \tag{1.17}$$

If we now understand u_j to be a "rule" defined by

$$u_j = u_j(.,.) : t \to u_j(.,t) =: u_j(t) \in H \qquad \text{for all } t \in \mathbf{R},$$

then IVPs (1.17) can be written equivalently in the form

$$\{d_t^2 + A_j\}u_j(t) = 0, \qquad u_j(0) = \varphi_j, \qquad u_{jt}(0) = \psi_j, \qquad j = 0, 1, \tag{1.18}$$

which is an *ordinary differential equation* in H. Consequently, with this understanding, problems (1.18) have solutions (in H) that can be written in the form

$$u_j(t) = (\cos tA_j^{1/2})\varphi_j + A_j^{-1/2}(\sin tA_j^{1/2})\psi_j, \qquad j = 0, 1. \tag{1.19}$$

Hence the solutions of (1.17) can be written in the form

$$\begin{aligned} u_j(x,t) &= (\cos tA_j^{1/2})\varphi_j(x) + A_j^{-1/2}(\sin tA_j^{1/2})\psi_j(x), \qquad j = 0, 1 \\ &= \mathrm{Re}\{v_j(x,t)\}, \qquad j = 0, 1, \end{aligned} \tag{1.20}$$

where

$$\begin{aligned} v_j(x,t) &= \exp\{-itA_j^{1/2}\}h_j(x) =: U_j(t)h_j(x), \\ h_j(x) &= \varphi_j(x) + iA_j^{-1/2}\psi_j(x). \end{aligned} \tag{1.21}$$

The quantities $v_j(x,t)$, $j = 0, 1$, in (1.21) define the complex-valued solutions of IVPs (1.17). The quantity $U_j(t)$ is a rule that governs the evolution (in time) of $h_j(x)$, the initial state of the jth system, into $v_j(x,t)$, the state of the jth system at some other time t. For this reason it will be called, eventually, an *evolution operator*.

In order to develop a scattering theory, we must first show that the FP and the PP actually have solutions in some convenient collections of functions. Once this has been done, we will need a means of comparing these solutions; that is, we will need some suitable formula for measuring their distance apart as either $t \to \pm\infty$ or $x \to \infty$. In general, such a formula is called a *norm*. A familiar example of this is the formula for measuring the distance, in $\mathbf{R}^2$, between the origin O and a point P with coordinates (x, y). The length of the line OP, denoted $\|OP\|$, is well known to be

$$\|OP\| := \sqrt{x^2 + y^2}. \tag{1.22}$$

In (1.22) we are working with *numbers*. We shall see (chapter 3) that the notion of a norm can be generalised so that we will be able to measure the separation of two *functions* rather than having to deal with the separation of the *numerical values* of these functions. With all this in mind we shall write

$$v_j(.,t) := v_j(t), \qquad j = 0, 1, \tag{1.23}$$

and compare the solutions of the FP and the PP by considering an expression of the form

$$\|v_1(t) - v_0(t)\|, \tag{1.24}$$

where $\|.\|$ denotes some suitably chosen norm defined in terms of spatial coordinates. We then find that

$$\begin{aligned}\|v_1(t) - v_0(t)\| &= \|U_1(t)h_1 - U_0(t)h_0\| \\ &= \left\|U_0^*(t)U_1(t)h_1 - h_0\right\| \\ &= \|W(t)h_1 - h_0\|\,, \end{aligned} \tag{1.25}$$

where the symbol $*$ denotes that $U_0(t)$ has a modified form (something like an inverse) as a result of moving it from one quantity to another.

The time dependence in the above comparison can be removed by taking the limit as $t \to \pm\infty$. We then obtain

$$\lim_{t\to\pm\infty} \|v_1(t) - v_0(t)\| = \|W_{\pm}h_1 - h_0\|, \tag{1.26}$$

where

$$W_{\pm} := \lim_{t\to\pm\infty} W(t) = \lim_{t\to\pm\infty} U_0^*(t)U_1(t). \tag{1.27}$$

If it can be shown that the two solutions $v_1(t)$ and $v_0(t)$ exist and that the various steps leading to (1.27) can be justified, then it will still remain to be shown that the limits in (1.27), which define the *wave operators* $W_{\pm}$, actually exist.

When all the above has been achieved, we see that if the initial data for the FP and PP are related according to

$$h_0 = W_{\pm}h_1, \tag{1.28}$$

then the limit in (1.26) is zero, thus indicating that the PP is *asymptotically free* as $t \to \pm\infty$. That is, solutions of the PP with initial data (state) h_1 are *asymptotically equal* in time to solutions of a FP with initial data (state) h_0, which is given by (1.28).

Consequently, if solutions of the two systems are known to exist, then, keeping (1.28) in mind, we would expect there to exist elements $h_{\pm}$ such that

$$v_1(t) \sim U_0(t)h_{\pm} \quad \text{as } t \to \pm\infty, \tag{1.29}$$

where $\sim$ denotes *asymptotic equality* and the $\pm$ are used to indicate the possibly different limits as $t \to \pm\infty$. We would emphasise that it is not automatic that both the limits implied by (1.29) should exist. Indeed, a solution such as v_1 could be asymptotically free as $t \to +\infty$ but not as $t \to -\infty$.

If we use (1.29) in conjunction with the definition $v_j = U_j h_j$, $j = 0, 1$, then we have

$$U_0^*(t)U_1(t)h_1 =: W(t)h_1 \sim h_{\pm} \tag{1.30}$$

and we can conclude that

$$W_{\pm} : h_1 \to h_{\pm}. \tag{1.31}$$

The two initial conditions $h_\pm$ for the FP are related. This is illustrated by noticing that the above discussion implies

$$h_+ = W_+ h_1 = W_+ W_-^* h_- =: Sh_-, \tag{1.32}$$

where S is called the *scattering operator* (SO) for the problem.

Thus we see that the existence of the wave operators $W_\pm$ is equivalent to the *asymptotic condition*

$$\lim_{t\to\pm\infty} \|v_1(t) - v_0(t)\| = \|W_\pm h_1 - h_0\| = 0. \tag{1.33}$$

Consequently, the construction of $W_\pm$ will be one of our primary goals.

For the approach outlined above to be of any practical use, we have to deal with two basic tasks.

Task 1: Provide a means of interpreting, in a constructive manner, such terms as $(\cos t A_j^{1/2})$ and $\exp\{-itA_j^{1/2}\}$, $j=0, 1$.

Task 2: Determine the far-field forms of the solutions $v_j(x, t)$, $j=0, 1$, as these are the quantities to be compared.

In addressing these two tasks we shall look first at the FP ($j=0$). Our intention will be to use as much as possible of our treatment of the FP as a pattern for analysis of the associated PP ($j=1$).

With regard to task 1 recall that when working in a finite-dimensional space setting, that is, when we are concerned with matrix equations rather than with differential equations, the corresponding situation is resolved in terms of eigenvector expansions (also called spectral decompositions). We shall see (chapter 4) that in an infinite-dimensional setting the matter can be resolved by means of the celebrated spectral theorem. However, there is an inherent difficulty associated with this particular approach. It centres on the practical, constructive determination of an appropriate collection of *projection operators*, the *spectral family*, associated with A_0. The situation can be eased considerably by noticing that the FP we are considering at the moment is posed in all of $\mathbf{R}^n$. This means that, for the wave equation at least, we can make use of the Plancherel theory of Fourier transforms [144]. This theory indicates that for any $f \in H \equiv H(\mathbf{R}^n)$, the following integrals exist and are convergent.

$$(Ff)(p) = \hat{f}(p) = \int_{\mathbf{R}^n} \overline{w_0(x, p)} f(x)\, dx, \tag{1.34}$$

$$f(x) = (F^* \hat{f})(x) = \int_{\mathbf{R}^n} w_0(x, p)\, \hat{f}(p)\, dp, \tag{1.35}$$

$$((\Phi A_0) f)(x) = \int_{\mathbf{R}^n} w_0(x, p) \Phi(|p|^2)\, \hat{f}(p)\, dp, \tag{1.36}$$

where Φ is a function that is "sufficiently nice" for our purposes and $w_0(x, p)$ is the usual Fourier kernel

$$w_0(x, p) = \frac{1}{(2\pi)^{n/2}} \exp(ix.p), \qquad x, p \in \mathbf{R}^n. \tag{1.37}$$

We notice that $w_0(x, p)$ satisfies the Helmholtz equation

$$(A_0 + |p|^2) w_0(x, p) = 0, \qquad x, p \in \mathbf{R}^n. \tag{1.38}$$

Recalling the results obtained when dealing with matrix equations, it might be thought that w_0 is an eigenfunction of $A_0 = -\Delta$ with associated eigenvalue $|p|^2$ and as such could provide the basis for a spectral decomposition of A_0. However, we shall find that for most cases of interest here, w_0 does *not* belong to H, the underlying collection of functions. Consequently, w_0 will have to be regarded as being in some sense a generalised eigenfunction of A_0. (See chapter 4 for details and also the remarks in Chapter 11.)

We emphasise that the integrals in (1.34)–(1.36) are improper integrals and as such must be interpreted by means of a limiting process intimately connected with the specific problem being considered. This aspect will be discussed in detail in chapter 6.

With these several remarks in mind we see that the Plancherel theory, which has been developed *quite independently* of any scattering phenomena, indicates that (1.34)–(1.36) provide the required spectral decomposition of A_0 and as such will be referred to as a *generalised eigenfunction expansion theorem* for A_0.

The above generalised eigenfunction expansion theorem provides the means for interpreting the various terms in (1.20) and (1.21). Indeed, a straightforward application of (1.36) yields the required solutions in the following forms.

$$u_0(x, t) = \int_{\mathbf{R}^n} w_0(x, p) \left\{ \hat{\varphi}_0(p) \cos t|p| + \hat{\psi}_0(p) \frac{\sin t|p|}{|p|} \right\} dp \tag{1.39}$$

and

$$\begin{aligned} v_0(x, t) &= \int_{\mathbf{R}^n} w_0(x, p) \exp(-it|p|) \hat{h}_0(p)\, dp \\ &= \frac{1}{(2\pi)^{n/2}} \int_{\mathbf{R}^n} \exp\{i(x.p - t|p|)\} \hat{h}_0(p)\, dp. \end{aligned} \tag{1.40}$$

The results (1.39) and (1.40) complete task 1 for the case $j = 0$.

As far as task 2 is concerned, we shall use (1.40) to calculate the asymptotic behaviour of the complex waveform v_0. To this end and for convenience at this stage, we shall assume

$$\operatorname{supp} \hat{h}_0(p) = \{p : 0 < a \le |p| \le b\}.$$

Here "supp" denotes "support." The support of a function is defined as the closure of the set where it is nonvanishing (see chapter 3). In this case (1.40) will read

$$\begin{aligned} v_0(x, t) &= \frac{1}{(2\pi)^{n/2}} \int_{a \le |p| \le b} \exp\{i(x.p - t|p|)\} \hat{h}_0(p)\, dp \\ &= \int_{a \le |p| \le b} w_0(x, p) \exp(-it|p|) \hat{h}_0(p)\, dp. \end{aligned} \tag{1.41}$$

If we now introduce spherical polar coordinates for p in (1.41) according to

$$p=\rho\omega, \qquad \rho\geqslant 0, \qquad dp=\rho^{n-1}\,d\rho\,d\omega, \qquad \omega\in S^{n-1}:=\{x\in\mathbf{R}^n : |x|=1\}, \tag{1.42}$$

then (1.41) becomes

$$v_0(x,t)=\frac{1}{(2\pi)^{n/2}}\int_a^b\int_{S^{n-1}}\exp\{i\rho(x.\omega-t)\}\,\hat{h}_0(\rho\omega)\rho^{n-1}\,d\omega\,d\rho \tag{1.43}$$

$$=\frac{1}{(2\pi)^{n/2}}\int_a^b\exp(-i\rho t)V_0(x,\rho)\rho^{n-1}\,d\rho, \tag{1.44}$$

where

$$V_0(x,\rho):=\int_{S^{n-1}}\exp\{i\rho(x.\omega)\}\,\hat{h}_0(\rho\omega)\,d\omega \tag{1.45}$$

and $x=|x|\,\eta$ with $\eta\in S^{n-1}$. Furthermore, it is easily verified that $V_0(x,\rho)$ is a solution of the Helmholtz equation $(\Delta+\rho^2)V_0(x,\rho)=0$ for all $x\in\mathbf{R}^n$. Our aim now is to display the asymptotic behaviour of v_0 as $t\to\infty$. This we do by first calculating the asymptotic behaviour of $V_0(x,\rho)$ as $|x|\to\infty$ and then using this knowledge to determine the required behaviour of $v_0(x,t)$ as $t\to\infty$. To give some indication of how this can be done, we begin by noticing that the integral in (1.45) suggests use of the method of stationary phase. A precise version of this method has been developed for integrals of the form (1.45) by Littman [76] and by Matsumura [78]. Their results for our present case lead to the following representations.

$$v_0(x,t)=|x|^{(1-n)/2}\,G\left(\frac{x}{|x|},|x|-t\right)+|x|^{(1-n)/2}\,G_1\left(\frac{x}{|x|},|x|+t\right)+q_1(x,t), \tag{1.46}$$

where $G(\eta,r)$ and $G_1(\eta,r)$ are functions of $\eta\in S^{n-1}$ and $r\in\mathbf{R}$. Here

$$G(\eta,r)=\frac{1}{(2\pi)^{1/2}}\int_a^b\exp(i\rho r)\,\hat{h}_0(\rho\eta)(-i\rho)^{(n-1)/2}\,d\rho. \tag{1.47}$$

The precise forms of G_1 and q_1 are not really of immediate interest to us in this introductory chapter since in many specific cases of interest it can be proved that they tend to zero as $t\to\infty$; consequently, we shall set them to zero here.

The relation (1.47) has the flavour of a Fourier transform. Indeed, in terms of the usual one-dimensional Fourier transform, denoted here by F_1, we have the following defining relations.

$$\hat{G}_{(1)}(\eta,\rho)=(F_1G)(\eta,\rho)=\lim_{M\to\infty}\frac{1}{(2\pi)^{1/2}}\int_{|r|\leq M}\exp(-i\rho r)G(\eta,r)\,dr, \tag{1.48}$$

$$G(\eta,r)=(F_1^*\hat{G}_{(1)})(\eta,r)=\lim_{M\to\infty}\frac{1}{(2\pi)^{1/2}}\int_{|\rho|\leq M}\exp(i\rho r)\,\hat{G}_{(1)}(\eta,\rho)\,d\rho, \tag{1.49}$$

where the limits are taken in the $L_2(S^{n-1}\times\mathbf{R})$ sense. Furthermore, the Fourier-Plancherel theory [144, 154] indicates that

$$\|G\| = \left\|\hat{G}_{(1)}\right\| \quad \text{for all } G\in L_2(S^{n-1}\times\mathbf{R}),$$

where $\|.\|$ is the "'size" rule, or norm, in $L_2(S^{n-1}\times\mathbf{R})$. We also notice that for any $h\in L_2(\mathbf{R}^n)$, and in particular for h_0 above, the Fourier transforms (1.34)–(1.36) in conjunction with the Parseval equality, using $\|.\|$ to denote the size rule in $L_2(\mathbf{R}^n)$, indicate

$$\begin{aligned}\|h_0\|^2 = \left\|\overset{\wedge}{h_0}\right\|^2 &= \int_{\mathbf{R}^n}\left|\overset{\wedge}{h_0}(p)\right|^2 dp = \int_0^\infty\int_{S^{n-1}}\left|\overset{\wedge}{h_0}(\rho\eta)\right|^2 \rho^{n-1}\,d\rho\,d\eta\\ &= \int_0^\infty\int_{S^{n-1}}\left|(-i\rho)^{(n-1)/2}\overset{\wedge}{h_0}(\rho\eta)\right|^2 d\rho\,d\eta\\ &= \int_0^\infty\int_{S^{n-1}}\left|\hat{G}_{(1)}(\eta,\rho)\right|^2,\end{aligned} \tag{1.50}$$

where

$$\hat{G}_{(1)}(\eta,\rho) := \begin{cases}(-i\rho)^{(n-1)/2}\overset{\wedge}{h_0}(\rho\eta), & \rho\geqslant 0,\\ 0, & \rho<0.\end{cases} \tag{1.51}$$

To see that this notation makes sense, we rewrite the definition (1.47) in a slightly different form as follows.

$$\begin{aligned}G(\eta,r) &= \frac{1}{(2\pi)^{1/2}}\int_a^b \exp(i\rho r)\,\overset{\wedge}{h_0}(\rho\eta)(-i\rho)^{(n-1)/2}\,d\rho\\ &= \frac{1}{(2\pi)^{1/2}}\int_{\mathbf{R}}\chi_{ab}(\rho)\exp(i\rho r)\,\overset{\wedge}{h_0}(\rho\eta)(-i\rho)^{(n-1)/2}\,d\rho,\end{aligned} \tag{1.52}$$

where $\chi_{ab}(\rho)$ is the characteristic function of the interval $[a,b]$.

We now use (1.49) to obtain

$$G(\eta,r) = F_1^*\{\chi_{ab}(.)\,\overset{\wedge}{h_0}((.)\eta)(-i)(.))^{(n-1)/2}\}(r,\eta). \tag{1.53}$$

Hence

$$(F_1G)(\eta,\rho) = \hat{G}_{(1)}(\eta,\rho) = \chi_{ab}(\rho)\,\overset{\wedge}{h_0}(\rho\eta)(-i\rho)^{(n-1)/2}, \tag{1.54}$$

and we conclude that the Fourier transform of $G(\eta,r)$ is

$$\chi_{ab}(\rho)\,\overset{\wedge}{h_0}(\rho\eta)(-i\rho)^{(n-1)/2}.$$

Bearing in mind (1.54), we see that the right-hand side of (1.51) is a Fourier transform in the sense of (1.48) and as such defines a unique function $G\in L_2(S^{n-1}\times\mathbf{R})$, hence the notation. These several remarks suggest the following.

Definition 1.1 *(i) The function G defined by (1.51) and (1.49) is the wave profile associated with each* $h_0 \in L_2(\mathbf{R}^n)$.
(ii) Corresponding to the wave function v_0 *defined by*

$$v_0(x, t) = \exp\{-itA_0^{1/2}\}h_0(x), \qquad h_0 \in L_2(\mathbf{R}^n),$$

there is an asymptotic wave function v_0^∞ *defined by*

$$v_0^\infty(x, t) := |x|^{(1-n)/2} G\left(\frac{x}{|x|}, |x| - t\right), \tag{1.55}$$

where $x \in \mathbf{R}^n \setminus \{0\}$, $t \in \mathbf{R}$, *and G is the wave profile for* h_0.

The manner in which we arrived at definition 1.1 indicates that it will only have practical significance if we can actually *prove* a result of the following form.

Theorem 1.2 *For every* $h_0 \in L_2(\mathbf{R}^n)$

$$\lim_{t\to\infty} \left\| v_0(, .t) - v_0^\infty(., t) \right\| = 0.$$

A proof of such a theorem requires that the solutions involved have certain decay properties. We develop this aspect in later chapters.

Once we reach this stage, we have completed tasks 1 and 2 for the FP. We would now like to have a similar pattern of results for the PP.

When the free system is perturbed either by the presence of an obstacle (target scattering) or by additional terms being attached to L_0 (potential scattering), the FP solution $v_0(x, t)$ will be perturbed to yield $v_1(x, t)$, the solution of the PP. Furthermore, we can expect that the generalised eigenfunction expansion given by (1.34)–(1.36) will also become perturbed. Essentially, this means that in order that we might have a generalised eigenfunction expansion theorem appropriate for the PP and which has the same form as (1.34)–(1.36), we can expect to have to use a kernel $w(x, p)$, which is some perturbation of the kernel $w_0(x, p)$ used for the FP.

Now, $w_0(x, p)$ satisfies the Helmholtz equation (1.38). Furthermore, a separation of variables technique indicates that it is a *steady-state* solution of the wave equation and represents a plane wave propagating in the direction p with a time dependence of the form $\exp\{-it\,|p|\}$. Consequently, in order to obtain a generalised eigenfunction expansion theorem for the mapping A_1 that characterises the PP as a perturbation of that obtained for the mapping A_0 that characterises the FP, we can expect to have to use a perturbed (distorted) kernel $w(x, p)$ satisfying

$$\{A_1 + |p|^2\}w(x, p) = 0, \tag{1.56}$$

where we assume

$$w(x, p) = w_0(x, p) + w'(x, p). \tag{1.57}$$

This will then emphasise that the total wave field associated with the PP will be characterised by the plane wave $w(x, p)$, which consists of a *free* or *incident wave*, characterised by $w_0(x, p)$, and a *scattered wave*, characterised by $w'(x, p)$. For this reason we say that $w(x, p)$ is a *distorted plane wave* [54].

We will find that there are two families of generalised eigenfunction expansions arising from the use of the distorted plane wave $w(x, p)$. For one family the distorted plane wave consists of an incident free wave and a scattered wave that is *outgoing*. For the other family the scattered wave is *incoming*.

The outgoing and incoming distorted plane waves will be denoted by $w_+(x, p)$ and $w_-(x, p)$, respectively. Hence we write

$$w(x, p) \equiv w_\pm(x, p) = w_0(x, p) + w'_\pm(x, p). \tag{1.58}$$

The precise meaning of "outgoing" and "incoming" will be given in chapter 6 and in later chapters as required. For the time being just think of them as waves that either approach or recede from a barrier.

The actual construction of $w'_\pm(x, p)$ is achieved via a study of the Helmholtz equation [62]. We remark that this will rely on a detailed investigation of $\sigma(A_1)$, the *spectrum* of A_1, and on $R_\lambda(A_1) = (A_1 - \lambda I)^{-1}$, the *resolvent* of A_1. This lies in the province of *stationary* (*steady-state*) scattering theory and will be dealt with later.

Once more is known about the distorted plane wave kernels $w_\pm(x, p)$, in many cases of practical interest it is possible to *prove* generalised eigenfunction expansion theorems of the following form.

$$(F_\pm f)(p) := \hat{f}_\pm(p) = \int_{\mathbf{R}^n} \overline{w_\pm(x, p)} f(x)\, dx, \tag{1.59}$$

$$f(x) = (F_\pm^* \hat{f}_\pm)(x) = \int_{\mathbf{R}^n} w_\pm(x, p)\, \hat{f}_\pm(p)\, dp, \tag{1.60}$$

$$(\Phi(A) f)(x) = \int_{\mathbf{R}^n} w_\pm(x, p) \Phi(|p|^2)\, \hat{f}_\pm(p)\, dp, \tag{1.61}$$

where, as before, we emphasise that the integrals in (1.59)–(1.61) are improper integrals and must be interpreted by means of a limiting process intimately connected with the particular problem being studied.

The above theorems, when proved, will then provide outgoing and incoming spectral representations of the complex-valued solutions

$$v_1(x, t) = \exp\{-it A_1^{1/2}\} h_1(x). \tag{1.62}$$

Specifically, we will have the following.

Outgoing representation:

$$v_1(x, t) = \int_{\mathbf{R}^n} w_+(x, p) \exp\{-it\,|p|\}\, \hat{h}_+(p)\, dp. \tag{1.63}$$

Incoming representation:

$$v_1(x,t) = \int_{\mathbf{R}^n} w_-(x,p) \exp\{-it\,|p|\} \hat{h}_-(p)\,dp, \tag{1.64}$$

where

$$\hat{h}_\pm(p) = \int_{\mathbf{R}^n} \overline{w_\pm(x,p)} h_1(x)\,dx = (F_\pm h_1)(p).$$

If we use the incoming representation (1.64) and substitute $w_-(x,p)$ given by (1.58) into (1.64), then we obtain

$$v_1(x,t) = v_0^+(x,t) + v_1^+(x,t), \tag{1.65}$$

where

$$v_0^+(x,t) = \int_{\mathbf{R}^n} w_0(x,p) \exp\{-it\,|p|\} \hat{h}_-(p)\,dp, \tag{1.66}$$

$$v_1^+(x,t) = \int_{\mathbf{R}^n} w'_-(x,p) \exp\{-it\,|p|\} \hat{h}_-(p)\,dp. \tag{1.67}$$

Recalling our discussion of the FP, we see that the first two terms under the integral sign in (1.66) indicate that $v_0^+(x,t)$ is a free wave of the form

$$v_0^+(x,t) = \exp\{-itA_0^{1/2}\}h_0^+(x), \tag{1.68}$$

where

$$h_0^+(x) = v_0^+(x,0). \tag{1.69}$$

Expanding (1.68) by means of (1.34)–(1.36) yields

$$v_0^+(x,t) = \exp\{-itA_0^{1/2}\}h_0^+(x) = \int_{\mathbf{R}^n} w_0(x,p) \exp\{-it\,|p|\} \hat{h}_0^+(p)\,dp. \tag{1.70}$$

For (1.66) and (1.70) to be compatible we must have $\hat{h}_-(p) = \hat{h}_0^+(p)$, that is,

$$h_0^+(x) = (F^* \hat{h}_-)(x) = (F^* F_- h_1)(x). \tag{1.71}$$

Recalling the definitions of AE and the wave operators (see (1.28)), we see that (1.71) suggests that the wave operator that relates $h_0^+(x)$, the initial state of the free system, to $h_1(x)$, the initial state of the perturbed system, is

$$W_+ = F^* F_-. \tag{1.72}$$

Similarly, if we start with the outgoing representation and substitute for w_+ from (1.58), we can obtain

$$W_- = F^* F_+$$

and so conclude that

$$W_{\pm} = F^* F_{\mp}. \tag{1.73}$$

The free wave $v_0^+(x, t)$ defined in (1.68) is asymptotically equal to the asymptotic wave function (see (1.55) and theorem 1.2)

$$v_0^{+,\infty}(x, t) = |x|^{(1-n)/2} G\left(\frac{x}{|x|}, |x| - t\right), \tag{1.74}$$

where (see (1.54))

$$\hat{G}(\eta, \rho) = (-i\rho)^{(n-1)/2} H(\rho)\, \hat{h}_0^+(\rho\eta), \qquad \rho \in \mathbf{R}, \qquad \eta \in S^{n-1}, \tag{1.75}$$

with $H(\rho)$ denoting the Heaviside unit function.

Since by (1.71) we have

$$\hat{h}_0^+(p) = (F h_0^+)(p) = (F_- h_1)(p) = \hat{h}_-(p), \tag{1.76}$$

(1.75) can be written in the form

$$\hat{G}(\eta, \rho) = (-i\rho)^{(n-1)/2} H(\rho)\, \hat{h}_-(\rho\eta), \qquad \rho \in \mathbf{R}, \qquad \eta \in S^{n-1}. \tag{1.77}$$

In most cases of physical interest, and certainly for those investigated in this monograph, it will be possible to obtain results of the following form.

$$\lim_{t\to\infty} \left\| v_1(., t) - v_0^+(., t) \right\| = 0, \tag{1.78}$$

$$\lim_{t\to\infty} \left\| v_0^+(., t) - v_0^{+,\infty}(., t) \right\| = 0, \tag{1.79}$$

$$\lim_{t\to\infty} \left\| v_1(., t) - v_0^{+,\infty}(., t) \right\| = 0. \tag{1.80}$$

These several observations and remarks suggest the following definition.

Definition 1.3 *The asymptotic wave function associated with the wave function v defined by*

$$v(x, t) = \exp\{-itA^{1/2}\} h(x)$$

is

$$v^{\infty}(x, t) = |x|^{(1-n)/2} G\left(\frac{x}{|x|}, |x| - t\right), \qquad x \in \mathbf{R}^n, \qquad t \in \mathbf{R},$$

where G is defined via (1.77).

We notice that G defined through (1.77) has been obtained with AE very much in mind, from (1.75), which is a result for FPs, by using (1.71) and (1.76). Nevertheless, in order for definition 1.3 to have practical significance, we must be able to establish a result of the following form.

Theorem 1.4

$$\lim_{t\to\infty} \left\| v(.,t) - v^{\infty}(.,t) \right\| = 0.$$

This result follows from (1.78)–(1.80) and the fact that when the initial data are related according to (1.71) and (1.76), we have v^{∞} and $v_0^{+,\infty}$ coincident.

When we have *proved* that we can reach this stage, we will have completed tasks 1 and 2 for the PP. In addition we will have shown, by theorem 1.4 and the remarks following it, that the asymptotic wave function for the FP and for the PP can be made to coincide.

An alternative method frequently used when discussing wave motions in the autonomous case is to reduce the given IVP to an equivalent first-order system. We illustrate the method here, briefly and entirely formally, for an AP.

The FP (1.13) and the PP (1.14) can be written in the form

$$\begin{bmatrix} u_j \\ u_{jt} \end{bmatrix}_t (x,t) + \begin{bmatrix} 0 & -I \\ A_j & 0 \end{bmatrix} \begin{bmatrix} u_j \\ u_{jt} \end{bmatrix} (x,t) = \begin{bmatrix} 0 \\ 0 \end{bmatrix}, \qquad j=0,1, \tag{1.81}$$

$$\begin{bmatrix} u_j \\ u_{jt} \end{bmatrix} (x,0) = \begin{bmatrix} \phi_j \\ \psi_j \end{bmatrix} (x), \qquad j=0,1. \tag{1.82}$$

The problem (1.81), (1.82) can be written compactly as

$$\{\partial_t - i\mathbf{M}_j\}\mathbf{\Phi}_j(x,t) = 0, \qquad \mathbf{\Phi}_j(x,0) = \mathbf{\Phi}_{j0}(x), \qquad j=0,1, \tag{1.83}$$

where

$$\mathbf{\Phi}_j(x,t) = \begin{bmatrix} u_j \\ u_{jt} \end{bmatrix} (x,t) =: \langle u_j, u_{jt} \rangle (x,t),$$

$$\mathbf{\Phi}_{j0}(x) = \begin{bmatrix} \phi_j \\ \psi_j \end{bmatrix} (x) =: \langle \varphi_j, \psi_j \rangle (x),$$

$$-i\mathbf{M}_j = \begin{bmatrix} 0 & -I \\ A_j & 0 \end{bmatrix}.$$

If we now decide to analyse the problem (1.83) in some certain collection of functions, denoted by H_E, then we can interpret $\mathbf{\Phi}_j$, $j=0,1$, as H_E-valued functions of t in the sense

$$\mathbf{\Phi}_j \equiv \mathbf{\Phi}_j(.,.) : t \to \mathbf{\Phi}_j(.,t) =: \mathbf{\Phi}_j(t) \in H_E.$$

Implicit in this notation is the requirement that the collection H_E can be decomposed as a "sum" of two collections, H_1 and H_2. We denote this symbolically by setting $H_E = H_1 \oplus H_2$, with the understanding that we have the interpretations

$$u_j = u_j(.,.) : t \to u_j(.,t) =: u_j(t) \in H_1, \qquad j=0,1,$$

$$u_{jt} = u_{jt}(.,.) : t \to u_{jt}(.,t) =: uj(t) \in H_2, \qquad j=0,1.$$

We can now reformulate (1.83) as an IVP for an *ordinary differential equation* in H_E of the form

$$\{d_t - i\mathbf{M}_j\}\boldsymbol{\Phi}_j(t) = 0, \qquad \boldsymbol{\Phi}_j(0) = \boldsymbol{\Phi}_{j0}. \tag{1.84}$$

Thus we see that we have a means of replacing the original partial differential equation that involves numerically valued functions with an ordinary differential equation that involves functions with values in some collection of functions (H_E in the present case). We shall see that a bonus in adopting this approach will turn out to be that we can use analogues of familiar results for scalar ordinary differential equations to solve abstract ordinary differential equations such as (1.84).

If we now make what appears to be the outrageous assumption that the $\mathbf{M}_j$, $j = 0, 1$, in (1.84) are constants, then, using an integrating factor technique, a solution of (1.84) can be obtained in the form

$$\boldsymbol{\Phi}_j(t) = e^{it\mathbf{M}_j}\boldsymbol{\Phi}_{j0} =: \mathbf{U}_j(t)\boldsymbol{\Phi}_{j0}, \qquad j = 0, 1. \tag{1.85}$$

Hence

$$\boldsymbol{\Phi}_j(x, t) = e^{it\mathbf{M}_j}\boldsymbol{\Phi}_{j0}(x) =: \mathbf{U}_j(t)\boldsymbol{\Phi}_{j0}(x), \tag{1.86}$$

and we notice the following features of this solution of the system governed by IVP (1.13).

- The vector function $\boldsymbol{\Phi}_{j0}$ can be considered as defining the *state* of the system of interest at time $t = 0$, that is, the *initial state* of the system.
- The function $\boldsymbol{\Phi}_j$ can be considered as defining the state of the system at some other time t.
- The quantity $\mathbf{U}_j$ can be regarded as an "operator" that controls the manner in which the initial state of the system $\boldsymbol{\Phi}_{j0}(x)$ evolves to the state $\boldsymbol{\Phi}_j(x, t)$, and for this reason it will eventually be called an *evolution operator*.

We remark that the collection H_E will later be identified as an energy space, hence the subscript.

We now notice that if the $\mathbf{U}_j$, $j = 0, 1$, can be determined and interpreted in a meaningful and practical manner, then the required solutions of (1.13) and (1.14) will be given by the first component of $\boldsymbol{\Phi}_j$ in (1.86).

Once the formal nature of the above illustrations has been removed and questions of existence and uniqueness of solution have been properly settled, we can turn our attention to constructing an appropriate scattering theory and to developing methods for actually determining such entities as the wave operators and ultimately the required solutions, albeit often only asymptotically, as $t \to \pm\infty$.

It turns out that a profitable way of investigating the IVPs for NAPs is again to reduce them to first-order systems. This we shall do in essentially the same way as we did for APs.

With (1.11), (1.12) and (1.81), (1.82) in mind and proceeding as for APs, we find that we are led to a consideration of the following NAPs.

$$\{d_t - i\mathbf{N}_j(t)\}\boldsymbol{\Psi}_j(t) = 0, \qquad \boldsymbol{\Psi}_j(s) = \boldsymbol{\Psi}_{js}, \qquad j = 0, 1, \tag{1.87}$$

where for $j=0, 1$,

$$\boldsymbol{\Psi}_j(t) = \langle u_j(t), u_{jt}(t)\rangle, \qquad -i\mathbf{N}_j(t) = \begin{bmatrix} 0 & -I \\ A_j(t) & 0 \end{bmatrix}, \tag{1.88}$$

$$\mathbf{N}_j(t) : H_E(\mathbf{R}^n) \supseteq D(\mathbf{N}_j(t)) \to H_E(\mathbf{R}^n), \tag{1.89}$$

$$D(\mathbf{N}_j(t)) := \{\boldsymbol{\xi} = \langle \xi_1, \xi_2\rangle \in H_E(R^n) : A_j(t)\xi_1 \in H_2(R^n), \xi_2 \in H_1(R^n)\}.$$

$$A_j(t) = L_j(x, t),$$

$$\boldsymbol{\Psi}_j(s) = \langle u_j, u_{jt}\rangle(s) = \langle \varphi_j, \psi_j\rangle = \boldsymbol{\Psi}_{js},$$

where $0 < s \in \mathbf{R}$ is a fixed initial time and, as before, $H_E \equiv H_E(\mathbf{R}^n) = H_1(\mathbf{R}^n) \oplus H_2(\mathbf{R}^n)$.

In contrast to the AP case the matrix operator in (1.88) is now a function of t. Nevertheless, (1.87) can still be solved by an integrating factor technique. In this case we obtain the required solution in the following form.

$$\boldsymbol{\Psi}_j(t) = \exp\{i \int_s^t \mathbf{N}_j(\eta)\, d\eta\}\boldsymbol{\Psi}_{js} =: \mathbf{U}_j(t, s)\boldsymbol{\Psi}_{js}, \tag{1.90}$$

where the entities $\mathbf{U}(t, s)$, $j=0, 1$, are known as the *propagators* for the NAP (1.87).

A wide range of physically significant problems in such fields as acoustics, electromagnetics and elasticity are in fact NAPs. When discussing these problems in this monograph, we shall have in mind the following three main aims.

- To provide results concerning the existence and uniqueness of solutions to problems of the type (1.87).
- To develop constructive methods for determining the propagators $\mathbf{U}_j(t, s)$ and hence the required solutions.
- To develop an appropriate and constructive scattering theory.

We shall first discuss NAPs in an abstract setting and then illustrate the practical relevance of the results obtained by indicating their use when dealing with some particular problems.

We remark that in a series of papers Cooper and Strauss have provided an elegant extension to NAPs of the Lax-Phillips theory, which was developed originally for scattering phenomena associated with APs (see chapter 11 and the cited references). Here we adopt a different approach that is more in keeping with the strategies and theory developed for APs by Wilcox [154] and his colleagues. This, we shall see, leads in quite a natural manner to constructive methods for solving (1.11), (1.12) and also to the development of an associated scattering theory.

1.4 CHAPTER SUMMARIES

Since this book is meant to be an introductory text for some if not all readers, it is felt that it should be as self-contained as possible. The intention is first to give, at a leisurely pace, a reasonably comprehensive overview of the various concepts, techniques and strategies associated with the development of scattering theories for APs and then to show how these various ideas have to be adjusted and added to in order to deal with corresponding NAPs. With this in mind the book has the following structure.

In chapter 2 we give examples, each in one space variable, that illustrate relatively simply some of the notations, methods and techniques that will be used later when studying more complicated systems than acoustic waves on a string. Free problems are considered first, and a number of solution methods are indicated. This is followed by a discussion of associated perturbed problems of both the AP and NAP types. The APs are used to illustrate such features as reflected and transmitted waves, a scattering matrix and resonances. The NAPs are introduced in turn to illustrate such aspects as lack of energy conservation, the nonavailability of the separation of variables technique and scattering frequencies. The final section deals with a scattering problem generated by a moving bead on a semi-infinite string. This gives us an opportunity to introduce, amongst other things, the quasi-stationary approximation method.

We introduce in chapter 3 a number of concepts and results from analysis that are used regularly in subsequent chapters. The presentation is mainly restricted to giving basic definitions and formulating theorems. Few, if any, proofs will be given, but all the results will be referenced and, in chapter 11, a commentary chapter, indications for further reading will be provided. More advanced topics in analysis than those appearing here will be introduced as required. This will have the virtue of emphasising their particular roles in the development of scattering theories.

In finite-dimensional spaces a linear operator can be represented in terms of a matrix with respect to a basis of associated eigenvectors. A generalisation of this notion to an infinite-dimensional space setting is complicated by the fact that the spectrum of an operator on such a space can now consist of more than just eigenvalues. To show how this situation can be eased, we introduce, in chapter 4, the spectral theorem and indicate how it can provide the required generalisations of the notions of spectral representation of an operator and the spectral decomposition of the underlying function space, with respect to an operator, that were available in a finite-dimensional space setting.

In chapter 5 we give a number of results from the theory of semigroups and also provide an overview of results and techniques from the abstract theory of Volterra integral equations. We indicate how these various results can be used to settle questions of existence and uniqueness of solutions to equations governing scattering processes. The final section deals with determination of the propagator (evolution operator) for NAPs and related approximation methods.

In chapter 6 we recall salient features of scattering theories that have been developed for APs. Some of these have already been alluded to in chapter 1. However, here we provide a rather more precise account. Topics to be covered include propagation aspects, solution decay, scattering states, solutions with finite energy, representations of solutions, expansion theorems and construction of solutions. The comparison of solutions for large time is discussed, as are the evolution operator for a wave equation and the asymptotic equality of solutions. Results are recalled concerning the existence, uniqueness and completeness of wave and scattering operators. Mention is also made of the principles of limiting absorption and limiting amplitude. Furthermore, a method is outlined for the construction of wave and scattering operators. In the final section of chapter 6 indications are given of how the philosophy used and the results obtained when dealing with APs can be adapted when dealing with NAPs.

The material presented in chapters 7–9 deals almost entirely with NAPs and relies very much on the preparation offered in chapters 2–6.

Chapter 7 is concerned with determination of the echo wave field. In chapter 8 the influence of time-periodic perturbations on the scattered field is examined. In chapter 9 mention is made of inverse scattering problems in a nonautonomous setting. A method that offers good prospects for dealing with this class of NAPs is introduced and discussed.

In chapter 10 indications are given of the manner in which methods for dealing with nonautonomous acoustic problems can be extended to similar problems in electromagnetism and elasticity.

All references are collected together in a much extended bibliography in chapter 11. Furthermore, a commentary on the material presented in this monograph can be found in this final chapter. This will provide more historical background, additional references and a guide to further reading and more specialised texts.

Chapter Two

Some Aspects of Waves on Strings

The main purpose of this chapter is to introduce some of the methods used when investigating two classes of problem we refer to as *free problems* and *perturbed problems*, respectively. We illustrate these methods by taking as prototype problems those that arise when studying waves on strings. The FP with which we shall be concerned in this chapter deals with wave motions on an infinite homogeneous string. Associated with this FP we consider a number of naturally occurring PPs that arise when such aspects as, for example, variable coefficients or boundary conditions, are introduced into the discussion. Of particular interest are those PPs for which the perturbations are time-dependent. We shall indicate a number of independently established methods for solving FPs. However, not all of these methods can be extended either easily or indeed at all in some cases to deal with associated PPs when time-dependent perturbations are involved. Nevertheless, we shall indicate an approach that does offer good prospects for providing the required extensions and in so doing also lays the foundations for dealing with problems more complicated than waves on a string.

The approach adopted in this chapter is almost entirely formal.

2.1 INTRODUCTION

Wave scattering phenomena involve three ingredients, an incoming or incident wave, an interaction zone in which the incident wave is perturbed in some manner and an outgoing wave that arises as a consequence of the perturbation. The waves in the interaction zone have, almost always, a very complicated structure. We shall see that this particular difficulty can be avoided to a large extent if we concentrate on the *consequences* of the interaction rather than on the interaction itself. This we shall do by developing relationships between the incoming and outgoing processes; that is, we shall construct a scattering theory. To do this, we first need to know how waves propagate in the absence of perturbations; that is, we need to study the FP. When details of the solutions to the FP are well understood, we can then turn to an investigation of the more demanding PP, which can embrace such features as boundary conditions, forcing terms, variable coefficients and so on.

In the following sections we work through, in a number of different ways, specific problems associated with waves on an infinite string. This we do, on the one hand, to enable us to recall familiar results obtained by classical treatments and, on the other hand, to illustrate how these same results might be obtained using rather more abstract methods. Unlike their classical counterparts, these abstract methods often have the virtue of offering good prospects for dealing, elegantly and efficiently, with more complicated problems than those associated with waves on a string.

2.2 A FREE PROBLEM

It is well known that the small-amplitude transverse wave motion of a string is governed by an equation of the form ([11, 138])

$$\{\partial_t^2 - c^2\partial_x^2\}u(x,t) = f(x,t), \qquad (x,t) \in \mathbf{R}\times\mathbf{R}, \tag{2.1}$$

where f characterises a force applied to the string, $u(x,t)$ denotes the transverse displacement of the string at a point x at time t, and c represents the velocity of a wave that might have been generated in the string. Throughout, we use the notation ∂_t^n to denote the nth partial derivative with respect to the variable t, and similarly for other variables. Also, subscript notation will be used to denote the differentiation of dependent variables.

Our aims when analysing an equation of the form (2.1) will be to determine the displacement of the string at any point x and time t and to prescribe how the displacement evolves in time.

We shall see later that it is quite sufficient for most of our purposes to study, in the first instance, the homogeneous form of (2.1). Specifically, we shall see that if we can solve the homogeneous equation, then we will also be able to solve the inhomogeneous equation by using Green's function techniques and Duhamel's principle. Consequently, in this chapter we take as a prototype equation

$$\{\partial_t^2 - c^2\partial_x^2\}u(x,t) = 0, \qquad (x,t) \in \mathbf{R}\times\mathbf{R}. \tag{2.2}$$

Solutions $u(x,t)$ of this equation will in general be required to satisfy appropriate initial conditions to control the variations with respect to the time variable t and, similarly, certain boundary conditions to control the variations in the displacements with respect to the space variable x. However, in most cases of practical interest, at all those points x that are a long way away from any boundary, the effect of the boundary will be minimal since it could take quite a time, depending on the wave velocity c, for the boundary influences to have any substantial effect at these points x. Thus, for large values of t the wave motion is largely unaffected by the boundaries; that is, the waves are (virtually) free of the boundary influence.

In this chapter we shall take as our prototype free problem the IVP

$$\{\partial_t^2 - c^2\partial_x^2\}u(x,t) = 0, \qquad (x,t) \in \mathbf{R}\times\mathbf{R}, \tag{2.3}$$

$$u(x,0) = \varphi(x), \qquad u_t(x,0) = \psi(x), \qquad x \in \mathbf{R}. \tag{2.4}$$

Associated with this FP are a variety of perturbed problems that could involve, for example, forcing terms, variable coefficients, boundary conditions or combinations of these.

We now see that in any study of waves and their echoes there are initially three main tasks.

Task 1: Determine the general form of solutions to the equation governing the wave motion.
Task 2: Investigate initial value problems associated with the equation governing the motion and develop constructive methods for obtaining solutions. This is taken as the underlying FP.
Task 3: Investigate PP associated with the above FP and develop constructive methods of solution.

Once these three tasks have been satisfactorily addressed, we will be well placed actually to compare solutions of FPs and PPs and so develop a scattering theory. By means of such a scattering theory, which at first sight will appear to have a very abstract structure, we will show that we will be able to analyse, in an efficient and thoroughly constructive manner, the echo signals arising from perturbations of an otherwise free system.

In this chapter we concentrate on methods for dealing with these tasks and confine our attention to systems for which the FP has the form of the IVP (2.3), (2.4). We shall recall, in the next few sections, a number of results obtained from a classical analysis of waves on infinite strings. We shall also take the opportunity to illustrate some perhaps rather less familiar techniques for discussing these problems. However, as we mentioned earlier, these techniques offer good prospects for dealing with more demanding problems than waves on strings. If in some parts the account might become, for the sake of illustration, somewhat formal, then reference will be made to where more precise details are to be found either in this book or elsewhere.

2.3 ON SOLUTIONS OF THE WAVE EQUATION

In this section we obtain the general form of solutions of the one-dimensional wave equation

$$\{\partial_t^2 - c^2\partial_x^2\}u(x,t) = 0, \qquad (x,t) \in \mathbf{R} \times \mathbf{R}. \tag{2.5}$$

To this end we introduce new variables, the *characteristic coordinates*

$$\xi = x - ct \quad \text{and} \quad \eta = x + ct. \tag{2.6}$$

We remark that the lines $\xi =$ constant and $\eta =$ constant are called *characteristic lines* for (2.5) [149]. Transforming (2.5) to the new variables ξ, η, we have

$$2x = \eta + \xi \quad \text{and} \quad 2ct = \eta - \xi, \tag{2.7}$$

and we write

$$u(x,t) = v(\xi, \eta). \tag{2.8}$$

Consequently, using the chain rule, we obtain

$$2cv_\xi = cu_x - u_t, \tag{2.9}$$

$$4c^2 v_{\xi\eta} = c^2 u_{xx} + cu_{xt} - cu_{tx} - u_{tt} = c^2 u_{xx} - u_{tt}.$$

Thus the wave equation (2.5) transforms under (2.6) into

$$v_{\xi\eta}(\xi, \eta) = 0. \tag{2.10}$$

The equation (2.10) has a general solution of the form

$$v(\xi, \eta) = f(\xi) + g(\eta), \tag{2.11}$$

where f, g are arbitrary but sufficiently differentiable functions that are determined in terms of imposed conditions to be satisfied by a solution of interest.

Returning to the original variables, we have

$$u(x, t) = f(x - ct) + g(x + ct). \tag{2.12}$$

In order that (2.12) be a solution of (2.5) in the classical sense, the functions f and g must be twice continuously differentiable functions of their arguments. We shall see later that we can relax this requirement.

In (2.12) the function f characterises a wave travelling to the right, unchanging in shape and moving with a velocity $c > 0$. To see this, consider the f-wave when it is at position x_0 at time $t = 0$. Then in this case the wave has a shape (profile) given by $f(x_0)$. At some future time $t \neq 0$ the wave will reach a point $x = x_0 + ct$. Consequently, since this yields $x_0 = x - ct$, we have

$$f(x - ct) = f(x_0),$$

which indicates that the shape of the wave is the same at the point (x, t) as it is at the point $(x_0, 0)$. Clearly, since $x > x_0$, we see that $f(x - ct)$ represents a wave travelling to the right with velocity c and which is unchanging in shape.

Similarly, $g(x + ct)$ represents a wave travelling to the left with velocity c and which is unchanging in shape.

We notice that since

$$\{c\partial_x + \partial_t\} f(x - ct) = 0, \qquad \{c\partial_x - \partial_t\} g(x + ct) = 0, \tag{2.13}$$

both $f(x - ct)$ and $g(x + ct)$ individually satisfy the wave equation

$$\{\partial_t^2 - c^2 \partial_x^2\} w(x, t) = 0. \tag{2.14}$$

It will be convenient at this stage to introduce some notation. This we can do by considering the following particular solution of (2.14).

$$w(x, t) = a\cos(kx - \omega t - \varepsilon), \qquad a > 0, \qquad \omega > 0. \tag{2.15}$$

This is a *harmonic wave* defined in terms of the quantities

$k =$ *wave (propagation) number*
$\omega =$ *angular frequency*
$a =$ *amplitude.*

A number of perhaps more familiar wave features can be defined in terms of these quantities, specifically,

$c = \omega/k =$ *wave velocity*. (The wave is travelling in the positive direction if k>0.)
$\lambda = 2\pi/k =$ *wavelength*
$f = \omega/2\pi =$ *frequency* of the (harmonic) oscillation
$T = f^{-1} = 2\pi/\omega =$ *period* of the oscillation
$\theta(x, t) = kx - \omega t - \varepsilon =$ *phase* of the wave.

It will often be convenient to define the corresponding *complex harmonic wave*

$$\psi(x, t) = a \exp\{i\theta(x, t)\} = C \exp\{i(kx - \omega t)\}, \tag{2.16}$$

where $w(x, t) = \text{Re}(\psi(x, t))$, with $a = |C|$ and $\varepsilon = -\arg C$.

More generally, if a depends on either x or t, then w is referred to as an *amplitude-modulated wave.*

If $\theta(x, t)$ is nonlinear in either x or t, then w is referred to as a *phase-modulated wave.*

A wave of the form

$$w(x, t) = e^{-pt} \cos(kx - \omega t - \varepsilon), \qquad p > 0, \tag{2.17}$$

is a *damped harmonic wave.*

A wave of the form

$$w(x, t) = e^{-qx} \cos(kx - \omega t - \varepsilon), \qquad p > 0, \tag{2.18}$$

is an *attenuated harmonic wave.*

Solutions of (2.14) that have the specific form

$$w(x, t) = X(x)T(t) \tag{2.19}$$

are known as *separable solutions*. A typical example of such a solution is

$$w(x, t) = \sin \pi x \cos \pi ct. \tag{2.20}$$

Direct substitution of (2.20) into (2.14) readily shows that (2.20) is indeed a solution of the wave equation. A general feature of waves such as (2.20) is that they are constant in time; that is, they are *stationary* or *nonpropagating waves*. To see this, notice that the *nodes* of (2.20), that is, those points x at which $w(x, t) = 0$, and the *antinodes* of (2.20), that is, those points x at which $w_x(x,t)=0$, maintain permanent positions, $x=\ldots,-1,0,1,2,\ldots$ and $x=\ldots,-\frac{1}{2},\frac{1}{2},\frac{3}{2},\ldots$, respectively, for all time t. We notice that (2.20) can be written in the form

$$w(x,t)=\sin\pi x\cos\pi ct=\frac{1}{2}\sin\pi(x-ct)+\frac{1}{2}\sin\pi(x+ct). \qquad (2.21)$$

Thus the *stationary wave* (2.20) is seen to be a superposition of two travelling waves, travelling in opposite directions with velocity c.

We have seen that solutions of (2.14) are of the general form

$$w(x,t)=f(x-ct)+g(x+ct). \qquad (2.22)$$

We will often be interested in solutions (2.22) that have a particular time dependence. This can mean that we might look for solutions that have, for example, the following specific and separated form.

$$w(x,t)=X(x)\exp\{-i\omega t\}. \qquad (2.23)$$

Substituting (2.23) into (2.14), we find that

$$\{\partial_x^2+k^2\}X(x)=0, \qquad k=\omega/c. \qquad (2.24)$$

This equation has the solution

$$X(x)=Ae^{ikx}+Be^{-ikx}=:Au_+(x)+Bu_-(x), \qquad (2.25)$$

and we thus obtain, using (2.23) and (2.25),

$$w(x,t)=A\exp\{i(kx-\omega t)\}+B\exp\{-i(kx+\omega t)\}. \qquad (2.26)$$

Thus, comparing (2.22) and (2.26), we arrive at the following sign convention. For waves with a time dependence $\exp\{-i\omega t\}$,

$\exp\{ikx\}$ characterises a wave travelling to the right (increasing x),
$\exp\{-ikx\}$ characterises a wave travelling to the left (decreasing x).

2.4 ON SOLUTIONS OF INITIAL VALUE PROBLEMS FOR THE WAVE EQUATION

In this section we turn our attention to the second of our three basic tasks. Specifically, we study IVP of the form (2.3), (2.4) with $|x|<\infty$ and $t>0$.

We have seen that the general solution of (2.3) has the form

$$u(x,t)=f(x-ct)+g(x+ct). \qquad (2.27)$$

Substituting the initial conditions (2.4) into (2.27), we obtain

$$\varphi(x) = f(x) + g(x), \tag{2.28}$$

$$\psi(x) = -cf'(x) + cg'(x). \tag{2.29}$$

We notice that since f and g are assumed at this stage to be twice continuously differentiable, it follows that the initial conditions must be such that φ is twice continuously differentiable and that ψ is once continuously differentiable.

Differentiate (2.28) and obtain

$$\varphi'(x) = f'(x) + g'(x). \tag{2.30}$$

Solving (2.29) and (2.30) for f' and g' and integrating the results, we obtain

$$f(x) = \frac{1}{2}\varphi(x) - \frac{1}{2c}\int_{x_0}^{x} \psi(s)\,ds + A, \tag{2.31}$$

$$g(x) = \frac{1}{2}\varphi(x) + \frac{1}{2c}\int_{x_0}^{x} \psi(s)\,ds + B, \tag{2.32}$$

where x_0 is an arbitrary constant, whilst A and B are integration constants. We see that (2.28), (2.31) and (2.32), taken together, imply that

$$A + B = 0.$$

Substituting (2.31), (2.32) into (2.27), we obtain

$$u(x,t) = \frac{1}{2}\{\varphi(x-ct) + \varphi(x+ct)\} + \frac{1}{2c}\int_{x-ct}^{x+ct} \psi(s)\,ds. \tag{2.33}$$

This is the celebrated d'Alembert solution of the one-dimensional wave equation. The interval $[x_1 - ct_1, x_1 + ct_1]$, $t_1 > 0$, of the x-axis is the *domain of dependence* of the point (x_1, t_1). The reason for this name is that (2.33) indicates that $u(x_1, t_1)$ depends only on the values of φ taken at the ends of this interval and the values of ψ at all points in this interval. The region D, for which $t > 0$, $x - ct \leq x_1$ and $x + ct \geq x_1$, is known as the *domain of influence* of the point x_1. This is because the value of φ at x_1 influences the solution $u(x,t)$ on the boundary of D, whilst the value of ψ at x_1 influences $u(x,t)$ throughout D.

We notice that if we introduce

$$u_+(x,t) := \begin{cases} \frac{1}{2}\varphi(x-ct) + \frac{1}{2c}\int_{x-ct}^{\infty} \psi(s)\,ds, & x > 0, \\ \frac{1}{2}\varphi(x+ct) + \frac{1}{2c}\int_{-\infty}^{x+ct} \psi(s)\,ds, & x < 0, \end{cases} \tag{2.34}$$

$$u_-(x,t) := \begin{cases} \frac{1}{2}\varphi(x+ct)\} - \frac{1}{2c}\int_{x+ct}^{\infty} \psi(s)\,ds, & x > 0, \\ \frac{1}{2}\varphi(x-ct) - \frac{1}{2c}\int_{-\infty}^{x-ct} \psi(s)\,ds, & x < 0, \end{cases} \tag{2.35}$$

then (2.33) can be written in the alternative form

$$u(x,t) = u_+(x,t) + u_-(x,t). \tag{2.36}$$

This form will be useful when we try to learn more about the behaviour of the solution $u(x, t)$ for large t. We notice that $u_+(x, t)$ is a function of $(x - ct)$ for $x > 0$ and a function of $(x + ct)$ for $x < 0$. Consequently, we can write more compactly

$$u_+(x,t) = f_+(|x| - ct), \qquad x \in R. \tag{2.37}$$

Similarly, we have

$$u_-(x,t) = f_-(|x| + ct), \qquad x \in R. \tag{2.38}$$

We also notice that in the region $x > 0$ there are two "waves," $u_+(x, t)$ travelling to the right (increasing x) and $u_-(x, t)$ travelling to the left (decreasing x). Thus, with respect to the origin $x = 0$ the wave $u_+(x, t)$ is *outgoing*, whilst the wave $u_-(x, t)$ is *incoming*.

Similarly, in the region $x < 0$ the wave $u_+(x, t)$ is *outgoing* (increasing negative x) with respect to the origin, whilst $u_-(x, t)$ is *incoming* (decreasing negative x).

The concepts of incoming and outgoing waves are of crucial importance in scattering theory. They will be discussed in more detail in chapters 6 and 11.

2.5 INTEGRAL TRANSFORM METHODS

In this and the following section we introduce some alternative methods of constructing solutions to FPs. These methods have the virtue that they generalise quite readily when we need to deal with more complicated and demanding problems than waves on strings. Furthermore, we shall see that they also provide an efficient means for developing robust constructive methods for solving quite difficult problems.

An explicit method for constructing solutions to IVPs for the wave equation is provided by the Plancherel theory of the Fourier transform [105, 144, 154]. Specifically, we have the following basic formulae in R^n.

$$(Ff)(p) =: \hat{f}(p) = \lim_{M\to\infty} \frac{1}{(2\pi)^{n/2}} \int_{|x|\le M} \exp(-ix.p) f(x)\, dx, \tag{2.39}$$

$$f(x) = (F^* \hat{f}) = \lim_{M\to\infty} \frac{1}{(2\pi)^{n/2}} \int_{|p|\le M} \exp(ix.p)\, \hat{f}(p)\, dp, \tag{2.40}$$

where $x = (x_1, x_2, \ldots, x_n)$, $p = (p_1, p_2, \ldots, p_n)$, and $x.p = \sum_{j=1}^{n} x_j p_j$. Here F^* denotes the inverse of the transform F. We would emphasise that the integrals in (2.39), (2.40) are improper integrals, and care must be taken when interpreting the limits in (2.39), (2.40). We return to these points in detail in chapter 6. With this understanding, we shall refer to (2.39), (2.40) as a *Fourier inversion theorem*.

This inversion theorem can be used to provide a representation, a *spectral representation*, of differential expressions with constant coefficients. Such a representation will often reduce the complexities and inherent difficulties of a given problem. This is a consequence of the relation

$$(F(D_j f))(p) = ip_j(Ff)(p), \tag{2.41}$$

where $D_j = \partial/\partial x_j$, $j = 1, 2, \ldots, n$. For example, if we write, for convenience,

$$A := -\Delta = -\sum_{j=1}^{n} \frac{\partial^2}{\partial x^2}$$

and if Φ is a "sufficiently nice" function, then using (2.41) we can obtain the representation

$$(\Phi(A)f)(x) = \lim_{M\to\infty} \frac{1}{(2\pi)^{n/2}} \int_{|p|\le M} \exp(ixp)\Phi(|p|^2)\,\hat{f}(p)\,dp. \tag{2.42}$$

In later chapters we shall refer to the three results (2.39), (2.40), (2.42) collectively either as a (*generalised*) *eigenfunction expansion theorem* or as a *spectral representation theorem* (with respect to A).

To illustrate the use of the above Fourier transforms, we consider again the following IVP governing waves on a string.

$$\{\partial_t^2 - c^2\partial_x^2\}u(x,t) = 0, \qquad (x,t) \in \mathbf{R}\times\mathbf{R}, \tag{2.43}$$

$$u(x,0) = \varphi(x), \qquad u_t(x,0) = \psi(x), \qquad x \in \mathbf{R}. \tag{2.44}$$

We now only need consider the case when $n = 1$, for which the inversion theorem (2.39), (2.40) can be conveniently written in the form

$$(Ff)(p) =: \hat{f}(p) = \frac{1}{(2\pi)^{1/2}} \int_{\mathbf{R}} \exp(-ixp) f(x)\,dx = \int_{\mathbf{R}} \overline{w(x,p)} f(x)\,dx, \tag{2.45}$$

$$f(x) = (F^* \hat{f}) = \frac{1}{(2\pi)^{1/2}} \int_{\mathbf{R}} \exp(ixp)\,\hat{f}(p)\,dp = \int_{\mathbf{R}} w(x,p)\,\hat{f}(p)\,dp, \tag{2.46}$$

where again it is understood that the improper integrals appearing in (2.45), (2.46) are interpreted as limits as indicated above. We notice that the Fourier kernel

$$w(x,p) = \frac{1}{(2\pi)^{1/2}} \exp(ixp) \tag{2.47}$$

satisfies

$$\{\partial_x^2 + p^2\}w(x,p) = 0. \tag{2.48}$$

If we take the Fourier transform (2.45) of the IVP for the partial differential equation (2.43), we obtain the following IVP for an ordinary differential equation.

$$\{d_t^2 + c^2 p^2\} \hat{u}(p, t) = 0, \tag{2.49}$$

$$\hat{u}(p, 0) = \hat{\varphi}(p), \qquad \hat{u}_t(p, 0) = \hat{\psi}(p). \tag{2.50}$$

This IVP is easier to solve than that for the partial differential equation (2.43). Indeed, we see immediately that the solution is

$$\hat{u}(p, t) = (\cos ptc)\, \hat{\varphi}(p) + \frac{1}{pc}(\sin ptc)\, \hat{\psi}(p). \tag{2.51}$$

If we now apply the inverse Fourier transform (2.46) to (2.51), then we obtain the required solution of the IVP (2.43), (2.44) in the form

$$u(x, t) = \int_{\mathbf{R}} w(x, p)\{(\cos pct)\, \hat{\varphi}(p) + \frac{1}{pc}(\sin ptc)\, \hat{\psi}(p)\}\, dp = \{F^* \hat{u}(., t)\}(x). \tag{2.52}$$

To see how this solution form relates to that obtained earlier, we first expand $\cos pct$ in the form

$$\cos pct = \frac{1}{2}\{e^{ipct} + e^{-ipct}\}$$

and use the result [65, 101 (vol. 2)]

$$(F(f(x - L)))(p) = e^{-ipL} \hat{f}(p). \tag{2.53}$$

It is then a straightforward matter to show that

$$F^*(\hat{\varphi}(p) \cos pct) = \frac{1}{2}\{\varphi(x + ct) + \varphi(x - ct)\}. \tag{2.54}$$

Similarly,

$$\begin{aligned} F^*\left(\frac{1}{pc}\hat{\psi}(p) \sin pct\right) &= \frac{1}{2i\sqrt{2\pi}} \int_R \hat{\psi}(p) \frac{1}{pc}\{e^{ip(x+ct)} - e^{ip(x-ct)}\}\, dp \\ &= \frac{1}{2c\sqrt{2\pi}} \int_R \hat{\psi}(p) \left\{\int_{x-ct}^{x+ct} e^{ips} ds\right\} dp \\ &= \frac{1}{2c} \int_{x-ct}^{x+ct} \psi(s)\, ds. \end{aligned} \tag{2.55}$$

Combining (2.52), (2.54) and (2.55), we obtain

$$u(x, t) = \frac{1}{2}\{\varphi(x + ct) + \varphi(x - ct)\} + \frac{1}{2c} \int_{x-ct}^{x+ct} \psi(s)\, ds, \tag{2.56}$$

which is the familiar d'Alembert solution obtained earlier.

We remark again that the Fourier transform of the given IVP for a partial differential equation yields, in this instance, an IVP for an ordinary diferential equation. Whilst the ordinary differential equation is more readily solved than the partial differential equation, there will remain the matter of inversion of the Fourier transform. Thus three questions will always have to be addressed if we choose to adopt the integral transform approach. First, what is the most appropriate integral transform for use in reducing the given partial differential equation to an equivalent ordinary differential equation? Second, is there an inversion theorem of the form (2.39) available for use in dealing with the given IVP? Third, is there available a (spectral) representation theorem of the form (2.39), (2.40), (2.42) for use in dealing with the given IVP?

We emphasise that in dealing with our present FP we have been very lucky because if we use the Fourier integral transform, then the Fourier-Plancherel theory is available and we can answer the last two questions above in the affirmative. However, for a perturbation of this FP and indeed for more general problems than waves on a string, we must always *prove* the availability of a representation theorem of the form (2.42). We return to this matter in more detail in chapter 6 and in subsequent chapters.

Finally, in this section we remark that we could have obtained (2.52) and hence (2.56) another way. It turns out that this approach will offer potentially powerful means of addressing a wide range of physically realistic problems.

Essentially, the method rests on how the partial differential equation for the problem of interest is cast into the form of an equivalent ordinary differential equation. Again for the purpose of illustration we consider the IVP (2.43), (2.44). We start by setting $A = -c^2\partial_x^2$ and then make what seems to be an outrageous assumption, namely, that for our immediate purposes A can be treated as a constant! (This will be justified later). This being done, we arrive at the following IVP for an ordinary differential equation.

$$\{d_t^2 + A\}u(x,t) = 0, \qquad u(x,0) = \varphi(x), \qquad u_t(x,0) = \psi(x). \tag{2.57}$$

This IVP has a solution that can be written in the form

$$u(x,t) = (\cos tA^{1/2})\varphi(x) + A^{-1/2}(\sin tA^{1/2})\psi(x). \tag{2.58}$$

It now remains to interpret such quantities as $\cos tA^{1/2}$.

From the standard theory of Fourier transforms [105],

$$(F(Af))(p) = (F(-c^2\partial_x^2 f))(p) = c^2p^2 \hat{f}(p). \tag{2.59}$$

It then follows, because of the particularly simple form of A that we are using here, that for a sufficiently nice function Φ we have

$$(F(\Phi(A)f))(p) = \Phi(c^2p^2)\hat{f}(p). \tag{2.60}$$

Consequently, combining (2.60) and (2.58), we obtain

$$u(x,t) = \int_R w(x,p)\{(\cos pct)\hat{\varphi}(p) + \frac{1}{pc}(\sin ptc)\hat{\psi}(p)\}\,dp, \tag{2.61}$$

which is the same as (2.52) obtained by other means. We shall see later that in this particular method the "outrageous assumption" can be justified, thus making the approach mathematically respectable.

Finally, we notice that the Fourier kernel $w(x, p)$, given by (2.47), satisfies

$$(A + c^2 p^2) w(x, p) = 0. \tag{2.62}$$

Consequently, $w(x, p)$ appears to be, in some sense, an *eigenfunction* of A with *eigenvalue* $(-c^2 p^2)$. (See chapters 3 and 4 for more details concerning this aspect.)

2.6 REDUCTION TO A FIRST-ORDER SYSTEM

We have pointed out that an alternative method frequently used when discussing wave motion is to replace the IVP for a partial differential equation by an equivalent problem for a first-order system. This approach has a number of advantages. We shall see that, once sufficient mathematical structure is available, existence and uniqueness results for solutions can be readily obtained and, furthermore, that energy considerations can be included quite automatically. We shall illustrate this approach here in an entirely formal manner. Precise analytical details will be provided in later chapters.

The initial value problem (2.43), (2.44) can be written in the form

$$\begin{bmatrix} u \\ u_t \end{bmatrix}_t (x, t) + \begin{bmatrix} 0 & -I \\ A & 0 \end{bmatrix} \begin{bmatrix} u \\ u_t \end{bmatrix} (x, t) = \begin{bmatrix} 0 \\ 0 \end{bmatrix}, \tag{2.63}$$

$$\begin{bmatrix} u \\ u_t \end{bmatrix} (x, 0) = \begin{bmatrix} \varphi \\ \psi \end{bmatrix} (x), \tag{2.64}$$

where, as before, $A = -c^2 \partial_x^2$.

These equations can be conveniently written in the form

$$\{\partial_t - iM\} \Psi(x, t) = 0, \qquad \Psi(x, 0) = \Psi_0(x), \tag{2.65}$$

where

$$\Psi(x, t) = \begin{bmatrix} u \\ u_t \end{bmatrix} (x, t), \qquad \Psi_0(x) = \begin{bmatrix} \varphi \\ \psi \end{bmatrix} (x), \tag{2.66}$$

$$-iM = \begin{bmatrix} 0 & -I \\ A & 0 \end{bmatrix}. \tag{2.67}$$

If we again make the outrageous assumption that A is a constant, then it will follow that M is a constant matrix and hence (2.65) can be reformulated as an IVP for an ordinary differential equation of the form

$$\{d_t - iM\} \Psi(t) = 0, \qquad \Psi(0) = \Psi_0, \tag{2.68}$$

where we have used the notation

$$\Psi(x,t)=\Psi(.,t)(x)=:\Psi(t)(x). \tag{2.69}$$

The solution of (2.68) can be obtained, by using an integrating factor technique, in the form

$$\Psi(t)=\exp(itM)\Psi(0). \tag{2.70}$$

Writing the exponential term in a series form, we obtain

$$\begin{aligned}
e^{itM} &= \sum_{n=0}^{\infty}\frac{(itM)^n}{n!}=\left\{\sum_{n=\text{even}}+\sum_{n=\text{odd}}\right\}\frac{(itM)^n}{n!}\\
&=\left\{I-\frac{t^2M^2}{2!}+\frac{t^4M^4}{4!}-\cdots\right\}+i\left\{tM-\frac{t^3M^3}{3!}+\frac{t^5M^5}{5!}-\cdots\right\}.
\end{aligned}$$

Now, using

$$M=i\begin{bmatrix}0 & -I\\ A & 0\end{bmatrix},\qquad M^2=A\begin{bmatrix}I & 0\\ 0 & I\end{bmatrix} \tag{2.71}$$

and recalling the series expansions for $\sin x$ and $\cos x$, we obtain

$$\begin{aligned}
\exp\, itM &= \left\{I-\frac{t^2A}{2!}+\frac{t^4A^2}{4!}+\cdots\right\}\begin{bmatrix}I & 0\\ 0 & I\end{bmatrix}\\
&\quad -A^{-1/2}\left\{tA^{1/2}-\frac{t^3A^{3/2}}{3!}+\frac{t^5A^{5/2}}{5!}+\cdots\right\}\begin{bmatrix}0 & -I\\ A & 0\end{bmatrix}\\
&=\cos tA^{1/2}\begin{bmatrix}I & 0\\ 0 & I\end{bmatrix}-A^{-(1)/2}\sin tA^{1/2}\begin{bmatrix}0 & -I\\ A & 0\end{bmatrix}\\
&=\begin{bmatrix}\cos tA^{1/2} & A^{-(1)/2}\sin tA^{1/2}\\ -A^{1/2}\sin tA^{1/2} & \cos tA^{1/2}\end{bmatrix}.
\end{aligned} \tag{2.72}$$

If we bear in mind the notation (2.70), then it is clear that the first component of the solution (2.70) yields the same solution of the given IVP as obtained earlier.

This approach can be given a rigorous mathematical development, as we shall see. We shall make considerable use of it in this book since, on the one hand, it provides a relatively easy means of settling questions of existence and uniqueness and, on the other hand, it offers good prospects for developing constructive methods.

So far we have only been discussing IVPs for the one-dimensional wave equation, that is, the FP for waves on a string. In the next few sub-sections we turn our attention to some PPs, in both the AP and NAP cases associated with this FP, and indicate how the various methods discussed so far are either inadequate or need to be modified in certain ways.

2.7 SOME PERTURBED PROBLEMS FOR WAVES ON STRINGS

In our investigations of waves on strings so far, we have considered the string to have uniform density throughout. This generated what we came to call a free problem associated with the classical wave equation. Associated with this FP is a whole hierarchy of perturbed problems. Perhaps the most immediate PP arises when we investigate waves on a string that has piecewise uniform density.

2.7.1 Waves on a Nonuniform String

Consider two semi-infinite strings Ω_1 and Ω_2 of (linear) density ρ_1 and ρ_2, respectively, that are joined at the point $x = r$ and stretched at tension T with Ω_1 occupying the region $x < r$ and Ω_2 the region $x > r$. As the two strings have different (linear) densities, it follows that their associated wave speeds c_1, c_2 are also different.

We shall see that this problem is a one-dimensional version of the more general interface problems. In this latter problem a (given) incident wave travels in a homogeneous medium that terminates at an interface with another, different, homogeneous medium in which the wave can also travel. Examples of such problems are given, for instance, by electromagnetic waves travelling in air meeting the surface of a dielectric and by acoustic waves travelling in air meeting an obstacle. In this class of problems the interface "scatters" the given incident wave and gives rise to reflected and transmitted waves. When all these waves combine, the resulting wave fields are readily seen to be quite different from those occurring in the related FP. Such problems are examples of *target scattering* problems. We illustrate some of the features of such problems by considering the following one-dimensional problem.

The governing wave equation is

$$\{\partial_t^2 - c^2(x)\partial_x^2\}u(x,t) = 0, \qquad (x,t) \in \mathbf{R} \times \mathbf{R}, \tag{2.73}$$

$$\begin{aligned} c(x) &= c_1, \qquad x < r, \qquad x \in \mathbf{R} \\ &= c_2, \qquad x > r, \qquad x \in \mathbf{R}, \end{aligned} \tag{2.74}$$

$$u(x,0) = \varphi(x), \qquad u_t(x,0) = \psi(x). \tag{2.75}$$

At the interface we shall require continuity of the displacement and of the transverse forces. This leads to boundary conditions of the form

$$u(r^-,t) = u(r^+,t), \qquad u_x(r^-,t) = u_x(r^+,t). \tag{2.76}$$

We remark that we assume here, unless otherwise stated, that $r > 0$. This is simply because in later illustrations we shall find it a convenient means of keeping track of the target (i.e., interface).

Let $f(x - c_1 t)$ denote a given incident wave in Ω_1. We assume that Ω_2 is initially at rest, so that $u(x,0)=0$ and $u_t(x,0)=0$ for $x>r$.

The wave field $u(x,t)$, which must satisfy (2.73)–(2.75), has the general form

$$u(x,t)=\begin{cases} f(x-c_1t)+g(x+c_1t), & x<r, \\ h(x-c_2t)+H(x+c_2t), & x>r. \end{cases} \tag{2.77}$$

However, since Ω_2 is initially at rest, we must have

$$h'(\zeta)=0 \quad \text{and} \quad H'(\zeta)=0 \quad \text{for } \zeta>0. \tag{2.78}$$

Hence $H(x+c_2t)$ is a constant for $t>0$ and therefore may be discarded. The appropriate wave field is thus

$$u(x,t)=\begin{cases} f(x-c_1t)+g(x+c_1t), & x<r, \quad t>0, \\ h(x-c_2t), & x>r, \quad t>0. \end{cases} \tag{2.79}$$

We also notice that

$$\begin{aligned} f(\zeta)=h(\zeta)=0, &\qquad \zeta>r, \\ g(\zeta)=0, &\qquad \zeta<r. \end{aligned} \tag{2.80}$$

At the interface certain boundary conditions will always have to be satisfied by solutions of (2.73). The most immediate conditions are, as we have already mentioned, the following.

Continuity of displacement:

$$u(r^-,t)=u(r^+,t). \tag{2.81}$$

Continuity of transverse force:

$$u_x(r^-,t)=u_x(r^+,t), \tag{2.82}$$

where

$$u(r^-,t)=\lim_{\substack{x\to r\\ x<r}} u(x,t), \qquad u(r^+,t)=\lim_{\substack{x\to r\\ x>r}} u(x,t), \tag{2.83}$$

and similarly for the derivatives.

Substitute (2.79) into (2.81), (2.82) to obtain

$$f(r-c_1t)+g(r+c_1t)=h(r-c_2t), \tag{2.84}$$

$$f'(r-c_1t)+g'(r+c_1t)=h'(r-c_2t), \tag{2.85}$$

where the primes denote differentiation with respect to the argument.

Integrate (2.85) to obtain

$$\frac{f(r-c_1t)}{-c_1}+\frac{g(r+c_1t)}{c_1}=\frac{h(r-c_2t)}{-c_2}, \tag{2.86}$$

where the integration constant has been set to zero in order to ensure (2.81) is satisfied.

Solving (2.84), (2.86) for the unknowns g and h, we obtain

$$g(x+c_1t)=\left(\frac{c_2-c_1}{c_1+c_2}\right)f(2r-x-c_1t), \qquad x<r, \tag{2.87}$$

$$h(x-c_2t)=\left(\frac{2c_2}{c_1+c_2}\right)f\left(r\left(1-\frac{c_1}{c_2}\right)+\frac{c_1}{c_2}x-c_1t\right), \qquad x>r. \tag{2.88}$$

It is worth noticing a number of interesting features of the solutions represented by (2.87) and (2.88). For convenience of presentation and without any loss of generality at this stage, we shall assume that $r=0$.

1. When $c_2=0$, there is no transmitted wave; the reflected wave has the form

   $$g(x+c_1t)=-f(-x-c_1t),$$

   and the required solution is

   $$u(x,t)=f(x-c_1t)-f(-x-c_1t). \tag{2.89}$$

   Thus, as expected, the reflected wave travels in the opposite direction to the incident wave. However, although the incident and reflected waves have the same shape, they are seen to have opposite signs.

   Also, we notice from (2.89) that $u(0,t)=0$ for all t. Thus the solution (2.89) describes waves on a semi-infinite string with a fixed point at $x=0$.
2. When $c_1=c_2$, there is no reflected wave, again as would be expected.
3. When $c_2>c_1$, the incident and reflected waves are seen to travel in opposite directions with the same profile but no change in sign.
   Thus the reflected and incident waves are in phase at the junction (interface) provided $c_2>c_1$ and are otherwise totally out of phase.

We will be able to obtain more detailed information about the wave field once we have introduced the notions of eigenfunction expansions and Green's functions.

Finally in this subsection, we consider the case when the incident wave is a simple harmonic wave of angular frequency ω. In this case we will have an incident wave of the form

$$u_i(x,t)=f(x-ct)=\exp\{ik(x-ct)\}=\exp\{i(kx-\omega t)\} \tag{2.90}$$
$$=\exp\{ikx\}\exp\{-i\omega t\},$$

where $\omega=kc$.

We notice that the incident wave separates into the product of two components, one only dependent on x, the other only dependent on t. It is natural to expect that this will be the case for the complete wave field. Consequently, for the nonhomogeneous string problem we are considering, we can expect the complete wave field to be separable and to have the form

$$u(x,t)=\begin{cases} e^{-i\omega_1 t}v_1(x), & \omega_1=c_1k_1, \quad x<0,\\ e^{-i\omega_2 t}v_2(x), & \omega_2=c_2k_2, \quad x>0.\end{cases} \tag{2.91}$$

Therefore, bearing in mind the sign convention introduced just after (2.26), the space-dependent component of the wave field $v(x)$ can be written in the form

$$v(x)=\begin{cases} e^{ik_1x}+Re^{-ik_1x}, & x<0,\\ Te^{ik_2x}, & x>0.\end{cases} \tag{2.92}$$

If we now apply the boundary conditions (2.81), (2.82), which must hold for all $t>0$, then we readily find

$$R=\frac{k_2-k_1}{k_1+k_2}, \qquad T=\frac{2k_2}{k_1+k_2}\exp\{i(\omega_2-\omega_1)t\}. \tag{2.93}$$

Here R and T are known as the *reflection* and *transmission coefficients*, respectively. In the case when we are only interested in solutions that have the same frequency, these coefficients assume the simpler form

$$R=\frac{c_2-c_1}{c_1+c_2}, \qquad T=\frac{2c_2}{c_1+c_2}.$$

Combining (2.91)–(2.93), we see that we recover the solutions (2.87), (2.88) in the case when $r=0$.

2.7.2 Waves on a Semi-infinite String with a Fixed End

In the previous subsection we saw what could happen when a wave travelling on an infinite string meets a change in string density. This change in density presented an obstacle to the wave, and its influence on the final wave structure was determined by imposing suitable boundary conditions on solutions of the wave equation. The following subsections contain further illustrations of the influence of boundary conditions on the wave field. We will often use these as test cases in later chapters.

We consider the IVP

$$\{\partial_t^2-c^2\partial_x^2\}\,u(x,t)=0, \qquad x<0, \qquad t>0, \tag{2.94}$$

$$u(x,0)=\varphi(x), \qquad u_t(x,0)=\psi(x), \qquad x<0. \tag{2.95}$$

We shall denote by $v(x,t)$ the solution of this IVP when the initial data functions have been extended to the entire real line by defining

$$\varphi(x)=\psi(x)=0, \qquad x>0. \tag{2.96}$$

Recalling the results in section 2.4, we see that the solution of this extended problem can be written

$$v(x,t) = f(x-ct) + g(x+ct), \tag{2.97}$$

where, by (2.31), (2.32) and the definition of the data functions

$$f(\zeta) = \frac{1}{2}\varphi(\zeta) - \frac{1}{2c}\int_0^{\zeta} \psi(s)\,ds + A, \tag{2.98}$$

$$g(\zeta) = \frac{1}{2}\varphi(\zeta) + \frac{1}{2c}\int_0^{\zeta} \psi(s)\,ds - A. \tag{2.99}$$

We notice that since $\varphi(\zeta)$ and $\psi(\zeta)$ vanish, by definition, when their arguments are positive, $f(\zeta)$ and $g(\zeta)$ are no more than constants in these cases. As these constants do not appear in the final solution form (see (2.33)), we can, for convenience, often set them to zero. However, we should always remember that the arbitrary functions f and g are determined by the data functions φ and ψ only up to a constant as indicated in (2.98), (2.99).

We consider the IVP (2.94), (2.95) as a FP. We shall assume that this FP is perturbed by requiring that the string be fixed at $x=0$. Consequently, solutions of (2.94), (2.95) will also be required to satisfy the boundary condition

$$u(0,t) = 0. \tag{2.100}$$

Following the arguments used above, we find, via (2.97) and (2.100), that

$$f(-ct) + g(ct) = 0, \qquad t > 0, \tag{2.101}$$

which leads to the functional relation

$$g(\zeta) = -f(-\zeta), \qquad \zeta > 0. \tag{2.102}$$

Hence the reflected wave is

$$g(x+ct) = -f(-x-ct),$$

and the required solution (of the extended problem) is

$$v(x,t) = f(x-ct) - f(-x-ct). \tag{2.103}$$

From (2.103) we see that the incident and reflected waves have the same shape, that is, profile, characterised by f but have opposite signs. These two waves move in opposite directions and will coalesce in such a way as to ensure that $u(0,t)$ is always zero.

To illustrate these various features in a little more detail, let us assume that at time $t=0$ the incident wave, characterised by v_i, has a nonzero amplitude only in the region $x \in (\alpha, \beta)$.

At some later time $t \neq 0$ the incident wave will have moved (evolved) so that it now has nonzero amplitude only in the region $x \in (\alpha + ct, \beta + ct)$. This implies that

$$v_i(x, t) = f(x - ct) = 0 \quad \text{for } x > \beta + ct$$

or, equivalently,

$$v_i(x, t) = f(x - ct) = 0 \quad \text{for } x - ct > \beta. \tag{2.104}$$

Thus we take as our illustration the case characterised by the functional relation

$$f(\zeta) = 0 \quad \text{for } \zeta > \beta. \tag{2.105}$$

This will ensure that we are dealing with an incident wave that continues to vanish, as indicated above, for $t = 0$ as t increases from zero.

For the purpose of more detailed illustrations it is convenient to set

$$\alpha = -ct_1 \quad \text{and} \quad \beta = -ct_0.$$

We now notice that for the reflected wave, denoted v_r, we have, using (2.105),

$$v_r(x, t) = -f(-x - ct) = 0 \quad \text{for } -x - ct > -ct_0. \tag{2.106}$$

This relation implies

$$v_r(x, t) = 0 \quad \text{for } -x > -c(t_0 - t). \tag{2.107}$$

However, since $c > 0$, the final inequality in (2.107) is only meaningful provided $t_0 > t$. Hence we conclude that the solution for $t < t_0$ consists only of the incident wave $v_i(x, t) = f(x - ct)$ travelling towards the origin $x = 0$.

The reflected wave $v_r(x, t) = -f(-x - ct)$ appears at time $t = t_0$ and travels away from $x = 0$.

We also notice in this illustration that at time $t = 0$ the incident wave also vanishes when $x < \alpha$. Consequently, on repeating similar arguments to those used to obtain (2.105), we find that for this illustration we must also have

$$f(\zeta) = 0 \quad \text{for } \zeta < -ct_1. \tag{2.108}$$

This relation implies that the incident wave vanishes at $t = t_1$. Consequently, for $t > t_1$, only the reflected wave remains.

Therefore, in the time interval (t_0, t_1), the total waveform is a combination of v_i and v_r and is likely to have a complicated structure. However, for $t < t_0$ and $t > t_1$ we only have one wave to deal with and, clearly, they will each have a structure that depends on important features of the given problem.

We notice that if, in subsection 2.7.1, we set $c_2 = 0$ and $r = 0$, then, as might be expected, we recover the above results.

2.7.3 Waves on an Elastically Braced String

We shall assume in this subsection that a portion of a homogeneous string is subjected to an elastic restoring force $E(x)$ per unit length of the string. Newton's laws of motion then indicate that the equation governing wave motion on the string is

$$\{\partial_t^2 - c^2\partial_x^2\}u(x,t) - c^2\mu^2(x)u(x,t) = 0, \tag{2.109}$$

where $\mu^2(x) = E(x)/T_s$ and T_s denotes the string tension.

Equation (2.109) is of a form that is typical when investigating *potential scattering* problems. Here $c^2\mu^2(x)$ can be viewed as the potential term, and we shall only be interested in the case when $E(x)$ is localised; that is, $E(x)$ will be assumed either to vanish outside a finite region of the string or to decay exponentially away from some fixed reference point.

A wave incident on the elastic region (the "potential") will be partly reflected and partly transmitted. However, even if $\mu(x)$ has a constant value, the solution of (2.109) is not as easy to obtain as the solutions of (2.73). To see this, consider the case when $\mu(x)$ has the constant value μ_0 and we seek solutions of (2.109) that have one angular frequency ω. When this is the situation, we assume a solution of the form

$$u(x,t) = w(x,\omega)e^{-i\omega t}, \tag{2.110}$$

and by direct substitution into (2.109) we find that $w(x,\omega)$ must satisfy

$$\left\{d_x^2 + \left(\frac{\omega}{\nu}\right)^2\right\} w(x,\omega) = 0, \tag{2.111}$$

where

$$\frac{1}{\nu^2} = \frac{1}{c^2} + \left(\frac{\mu_0}{\omega}\right)^2. \tag{2.112}$$

We then obtain the string displacement $u(x,t)$, by (2.110), in the form

$$u(x,t) = a\exp\left\{-i\omega\left(t - \frac{x}{\nu}\right)\right\} + b\exp\left\{-i\omega\left(t + \frac{x}{\nu}\right)\right\}. \tag{2.113}$$

We see that in (2.113) ν is the phase velocity of the wave. Furthermore, we notice that the wave motion represented by (2.113) is dispersive since by (2.112) the phase velocity ν of the wave is frequency-dependent. It follows that distortionless propagation of the wave as described by f and g in the general solution of the classical wave equation is no longer possible.

We also notice that there is a "cutoff" frequency associated with the wave motions generated in this system. According to (2.111), (2.112), frequencies that are less than $\mu_0 c$ lead to an imaginary propagation constant. These low-frequency disturbances do not propagate as waves; they merely move the string up and down in phase. Thus it is possible that localised wave motion might be excited on a nonhomogeneous string.

To illustrate the scattering of an incident wave by the elastic region (potential), consider a string with a segment that has a constant elastic restoring force so that we have

$$\mu(x) = \begin{cases} 0, & |x| > R, \\ \mu_0, & |x| < r. \end{cases} \tag{2.114}$$

If a wave of frequency ω and unit magnitude is incident on this region, then the resulting spatial part of the wave field can be written, as in the previous subsection, in the form

$$v(x) = \begin{cases} e^{ikx} + Re^{-ikx}, & x < -r, \\ Te^{ik_2x}, & x > +r, \\ Ae^{i\alpha x} + Be^{-i\alpha x}, & |x| < r, \end{cases} \tag{2.115}$$

where $k = \omega/c$ and $\alpha = \sqrt{k^2 - \mu_0^2}$. It is clear that a wave will not propagate in the elastic region unless $k \geq \mu_0$.

The reflection and transmission coefficients, R and T, respectively, together with the constants A and B are determined by requiring continuity of displacement and slope at $x = \pm r$. It is a straightforward but rather lengthy matter to show that

$$R \exp(2ikr) = \frac{\mu_0}{D} \sin 2\alpha r, \tag{2.116}$$

$$T \exp(2ikr) = \frac{2i\alpha k}{D}, \tag{2.117}$$

where

$$D = (k^2 + \alpha^2) \sin 2\alpha r + 2ikr \cos 2\alpha r. \tag{2.118}$$

We see from (2.116) that perfect transmission (i.e., $R = 0$) occurs when $\sin 2\alpha r = 0$, that is, whenever an integral number of half-wavelengths of the wave on the elastic region fit into that region.

We also notice that R and T become unbounded at zeros of the denominator D. These will be identified as the *resonances* of the system.

The PPs we have discussed in the last three subsections are all APs. Since our main interest in this monograph is centred on NAPs involving time-dependent perturbations, the temptation now is to replace r in all the above APs by $r(t)$ and to claim that we have thus obtained the solution of the corresponding NAP. As a consequence, we would then be able, for example, to follow the "migration" of resonances or to determine how the reflection and transmission coefficients might be varying with time. However, this claim is false. It is false mainly because the expression we obtain in this manner as a "solution" of the governing equation does not in fact satisfy that equation. However, in many cases of practical interest, it turns out that this procedure can often provide good agreement between theory and practice. In a later subsection this aspect will be clarified under the heading *quasi-stationary approximation* (QSA) *methods*. An indication of why this is only an approximation procedure is illustrated below.

2.7.4 Waves on a String of Varying Length

In this subsection we assume that the extremities of the string are at $x=0$ and $x=r(t)$. The displacement of the string at a point x and at time t is denoted $u(x,t)$ and is assumed to satisfy the following IBVP.

$$\{\partial_t^2 - c^2\partial_x^2\}\, u(x,t)=0, \quad x\in(0,r(t)), \qquad t\in\mathbf{R}, \tag{2.119}$$

$$u(0,t)=F(t), \qquad u(r(t),t)=0, \qquad t_0<t\in\mathbf{R}, \tag{2.120}$$

$$u(x,t_0)=f_0(x), \qquad u_t(x,t_0)=g_0(x), \qquad x\in[0,r(t)]. \tag{2.121}$$

The d'Alembert solution of (2.119) in $[0,r(t)]$ is given by

$$u(x,t)=f\left(t-\frac{x}{c}\right)+g\left(t+\frac{x}{c}\right), \qquad t\in\mathbf{R}. \tag{2.122}$$

The precise forms of f and g are determined by ensuring that (2.120) and (2.121) are satisfied. To indicate a solution procedure, we consider a particular case by assuming the following.

(i) $F(t)\equiv 0$. Consequently, (2.120) and (2.122) imply

$$u(x,t)=f\left(t-\frac{x}{c}\right)-f\left(t+\frac{x}{c}\right). \tag{2.123}$$

(ii) $r(t)=vt$, where v is a constant. The boundary condition at $x=r(t)$ yields

$$f\left(\left[1-\frac{v}{c}\right]t\right)=f\left(\left[1+\frac{v}{c}\right]t\right), \qquad t\in\mathbf{R}. \tag{2.124}$$

This is a homogeneous difference equation of a geometric type that has to be solved to yield f. To this end, write (2.124) in the form

$$f(\gamma\tau)=f(\tau), \qquad \tau\in\mathbf{R}, \tag{2.125}$$

where

$$\gamma=\frac{1+v/c}{1-v/c}, \qquad \tau=\left(1-\frac{v}{c}\right)t.$$

Let

$$\tau=e^{\eta} \quad \text{and} \quad f(e^{\eta})=h(\eta). \tag{2.126}$$

Combine (2.125) and (2.126), and we obtain

$$f(\tau)=h(\eta)=f(\gamma\tau)=f(e^{\log\gamma}\tau)=f(e^{\log\gamma}e^{\eta})=h(\eta+\log\gamma).$$

Thus

$$h(\eta)=h(\eta+\log\gamma), \qquad \eta\in\mathbf{R}, \tag{2.127}$$

which indicates that h is a periodic function of η with period

$$\log \gamma =: \frac{2\pi}{\delta} \quad \text{(say)}. \tag{2.128}$$

Consequently, since

$$\begin{matrix}\sin\\ \cos\end{matrix}\left(n\delta\left[\eta+\frac{2\pi}{\delta}\right]\right)=\begin{matrix}\sin\\ \cos\end{matrix}\,(n\delta\eta+2n\pi])=\begin{matrix}\sin\\ \cos\end{matrix}\,(n\delta\eta) \tag{2.129}$$

and since h has period $2\pi/\delta$, we thus have the characterisation

$$f(\tau)=f(e^{\eta})=h(\eta)=\sum_{n=0}^{\infty} A_n \sin(n\delta\log\tau)+B_n\cos(n\delta\log\tau). \tag{2.130}$$

Using (2.130) in conjunction with (2.123), we obtain after some straightforward manipulations

$$u(x,t)=2\sum_{n=1}^{\infty}\sin\left(\frac{n\delta}{2}\log\left\{\frac{t-x/c}{t+x/c}\right\}\right)\left[A_n\cos\left(\frac{n\delta}{2}\log\left(t^2-x^2/c^2\right)\right)+\right.$$
$$\left.-B_n\sin\left(\frac{n\delta}{2}\log(t^2-x^2/c^2)\right)\right]. \tag{2.131}$$

We must now choose the coefficients A_n and B_n to ensure that the initial conditions (2.121) hold. To this end, recall that from the standard d'Alembert solution we have

$$f\left(t_0+\frac{x}{c}\right)=-\frac{1}{2}\left\{f_0\left(x+\frac{1}{c}\right)\int_0^x g_0(s)\,ds\right\}, \qquad -vt_0<x<vt_0. \tag{2.132}$$

We use this relation to determine the coefficients in (2.131) as follows.

1. Notice the orthogonality relation

$$\int_{z_1}^{z_2}\begin{matrix}\sin\\ \cos\end{matrix}\,(m\delta z)\,\begin{matrix}\sin\\ \cos\end{matrix}\,(n\delta z)\,dz=\begin{matrix}0, & m\neq n,\\ \pi/\delta, & m=n,\end{matrix} \tag{2.133}$$

where $z_2=z_1\pm 2\pi/\delta$.

2. Anticipate the use of (2.132) and (2.133) and obtain from (2.130) the result

$$f\left(t_0+\frac{x}{c}\right)=\sum_{n=0}^{\infty}A_n\sin\left(n\delta\log\left\{t_0+\frac{x}{c}\right\}\right)+B_n\cos\left(n\delta\log\left\{t_0+\frac{x}{c}\right\}\right). \tag{2.134}$$

3. The result (2.134) suggests that we set, in (2.132),

$$z=\log\left(t_0+\frac{x}{c}\right) \quad \text{which implies } dz=\frac{dx}{ct_0+x}.$$

4. Since (2.132) is defined for $-vt_0 < x < vt_0$, we take

$$\begin{aligned} z_1 &= \log\left(\left(1-\frac{v}{c}\right)t_0\right), \\ z_2 &= \log\left(\left(1+\frac{v}{c}\right)t_0\right) \\ &= \log\left(\left(1-\frac{v}{c}\right)t_0\right) - \log\gamma \\ &= z_1 - \frac{2\pi}{\delta}. \end{aligned}$$

From (2.134), (2.133) and (2.132) we then obtain

$$\begin{aligned} \left.\begin{matrix} A_n \\ B_n \end{matrix}\right\} &= \frac{\delta}{\pi}\int_{-vt_0}^{vt_0} f\left(t_0+\frac{x}{c}\right) \begin{matrix}\sin\\ \cos\end{matrix} \left(n\delta\log\left(t_0+\frac{x}{c}\right)\right)\frac{dx}{ct_0+x} \\ &= -\frac{\delta}{\pi}\int_{-vt_0}^{vt_0}\left[f_0(x)+\frac{1}{c}\int_0^x g_0(s)\,ds\right] \begin{matrix}\sin\\ \cos\end{matrix} \left(n\delta\log\left(t_0+\frac{x}{c}\right)\right)\frac{dx}{ct_0+x}. \end{aligned}$$

This has indicated aspects of the associated propagation problem. A related scattering problem will be considered in the next subsection.

2.7.5 A Scattering Problem on a Semi-infinite String

In the final subsection of this chapter we consider in detail a one-dimensional scattering problem. This will provide an illustration of some of the aspects that will be introduced when investigating more general problems.

We consider wave motions on an infinite string occupying the x-axis to the right of a bead that fixes the string at that point. The *bead is allowed to move*, and its position at time t relative to a fixed origin is denoted by $x = r(t)$. The string is assumed to vibrate only transversely to the x-axis and in one plane. Wave motions on the string can be modelled in terms of a boundary value problem of the form

$$\begin{aligned} \{\partial_t^2 - \partial_x^2\}u(x,t) &= 0, \qquad x > r(t) \quad \text{for all } t \in \mathbf{R}, \\ u(r(t),t) &= 0. \end{aligned} \tag{2.135}$$

Here we have taken c, the wave propagation speed on the string, to be such that $c = 1$. We further assume that there are constants ρ and α such that

$$|r(t)| \leq \rho \quad \text{and} \quad \left|\frac{dr(t)}{dt}\right| \leq \alpha \leq c = 1. \tag{2.136}$$

The conditions in (2.136) ensure that the motion of the bead remains in a bounded region and that the velocity of the bead is less than the wave propagation speed.

The solution of (2.135) has the general form

$$u(x,t) = g(t+x) + f(t-x) = u_-(x,t) + u_+(x,t), \tag{2.137}$$

where $g \equiv u_-$ characterises the "incident" wave travelling from right to left along the string and $f = u_+$ characterises the reflected wave travelling from left to right along the string.

Paralleling the analysis of waves on the string when the bead is fixed for all $t \in \mathbf{R}$ (see [11]), we have that in the present case the wave energy in the string, at time t, can be expressed in the form

$$E(t)u = \frac{1}{2}\int_{r(t)}^{\infty} \{u_t^2 + u_x^2\}(x,t)\,dx \tag{2.138}$$

Direct calculation yields

$$\{u_t^2 + u_x^2\}(x,t) = 2(g'(z))^2 + 2(f'(w))^2, \tag{2.139}$$

where $z = (t+x)$ and $w = (t-x)$. If now we assume that g' and f' are square-integrable, then $g'(z) \to 0$ as $|z| \to \infty$ and we conclude that

$$E(t)u \to E(t)u_+ \quad \text{as } t \to \infty.$$

Similarly, we find that

$$E(t)u \to E(t)u_- \quad \text{as } t \to -\infty.$$

The boundary condition in (2.135) requires

$$\begin{aligned} 0 = u(r(t),t) &= f(t - r(t)) + g(t + r(t)) \\ &= (f \circ h)(t) + (g \circ k)(t), \end{aligned} \tag{2.140}$$

where the usual function composition algebra implies, for instance, that

$$(f \circ h)(t) = f(h(t)). \tag{2.141}$$

In (2.140) we have taken

$$h(t) = t - r(t) \quad \text{and} \quad k(t) = t + r(t).$$

It then follows that the boundary condition is satisfied when

$$(f \circ h)(t) = -(g \circ k)(t), \tag{2.142}$$

which implies

$$f = -g \circ p, \qquad p = k \circ h^{-1}. \tag{2.143}$$

From (2.138), using the standard properties of integrals and the defining equation in (2.135), we obtain

$$\frac{dE(t)u}{dt} = -\frac{1}{2}r'(t)\{u_t^2 + u_x^2\}(r(t),t) + [u_t u_x]_{x=r(t)}^{\infty}$$

We now make use of the fact that we have assumed that f' and g' are square-integrable together with the chain rule for differentiation of composed functions and the boundary condition result (2.143) to obtain after a straightforward but rather lengthy calculation

$$\frac{dE(t)u}{dt} = 2r'(t)\{g'(t+r(t))\}^2 \left\{\frac{1+r'(t)}{1-r'(t)}\right\}. \tag{2.144}$$

Three cases are now immediate.

Case 1: $r'(t) > 0$. In this case all terms on the right-hand side of (2.144) are positive. Consequently, when the bead (scattering object) is moving to the right (i.e., $v = dx/dt = r'(t) > 0$), the energy is increasing.

Case 2: $r'(t) < 0$. Arguing as in case 1, we can infer that when the bead is moving to the left, the energy is decreasing.

Case 3: $r'(t) = 0$. In this case the bead is stationary and the energy is constant in time.

Now, for illustration, assume that the incident wave $u_i(x, t)$ has the specific form

$$u_i(x, t) = g(t+x) = \exp\{-i\omega(t+x)\}. \tag{2.145}$$

We shall investigate how this wave is scattered by the moving bead.

Let $u(x, t)$ denote the total wave field at the point (x, t). Then

$$u(x, t) = u_i(x, t) + \psi(x, t),$$

where $\psi(x, t)$ represents the scattered wave.

The total field $u(x, t)$ is required to satisfy

$$\{\partial_t^2 - \partial_x^2\}u(x, t) = 0, \qquad x > r(t) \quad \text{for all } t \in \mathbf{R}, \tag{2.146}$$

$$u(r(t), t) = 0. \tag{2.147}$$

Consequently,

$$\{\partial_t^2 - \partial_x^2\}\psi(x, t) = 0, \qquad x > r(t) \quad \text{for all } t \in \mathbf{R}, \tag{2.148}$$

$$\psi(r(t), t) = -u_i(r(t), t) = -\exp\{-i\omega(t+r(t))\}. \tag{2.149}$$

The reflected wave $\psi(x, t)$ is required to be an outgoing solution of (2.148), (2.149) in the sense that it will travel from left to right on the string. Consequently, we require solutions that have the general form

$$\psi(x, t) = F(t-x), \tag{2.150}$$

where F is to be determined from the imposed conditions (2.149). Therefore we must have

$$F(t-r(t)) = -\exp\{-i\omega(t+x)\}$$

or, equivalently,

$$F(y) = -\exp\{-i\omega(y + 2r(t))\}, \tag{2.151}$$

where we have defined

$$y := t - r(t) =: G(t). \tag{2.152}$$

To complete the determination of F, it remains to express $r(t)$ entirely in terms of y. To this end, notice that if G has an inverse H, then

$$(G \circ H)(t) = (H \circ G)(t) = t.$$

In the present case, on combining these various results, we obtain

$$(G \circ H)(t) = G(H(t)) = H(t) - r(H(t)) = t, \tag{2.153}$$

$$(H \circ G)(t) = H(G(t)) = H(y) = t. \tag{2.154}$$

Substitute (2.154) into (2.151), and we see that the function F in (2.151) is characterised by the relation

$$F(y) = -\exp\{-i\omega(y + 2r(H(y))\}. \tag{2.155}$$

Consequently, the scattered wave is given by

$$\begin{aligned}\psi(x,t) &= -\exp\{-i\omega(t - x + 2r(H(t-x))\} \\ &= -\exp\{-i\omega(t-x)\}\exp\{-2ikr(H(t-x))\},\end{aligned} \tag{2.156}$$

where $k = \omega/c = \omega$.

When $r(t) = \text{const} = a$, we recover from (2.156) the familiar result

$$\psi(x,t) = -\exp\{-i\omega(t-x)\}\exp\{-2ika\}. \tag{2.157}$$

However, the function H in (2.155) has not yet been determined. In general this is a difficult task. Consequently, a quasi-stationary approximation is often used ([61]). This amounts to replacing the constant a in (2.157) by $r(t)$ to obtain

$$v(x,t) = -\exp\{-i\omega(t-x)\}\exp\{2ikr(t)\} \tag{2.158}$$

(see the discussion following (2.108)). We refer to $v(x,t)$ in (2.158) as the *quasi-stationary field* for the problem (2.148), (2.149).

We notice that (2.155) can be rewritten in the form

$$\begin{aligned}\psi(x,t) &= -\exp\{-i\omega t\}\exp\{-ik[2r(H(t-x) - x]\} \\ &= \exp\{-i\omega t\}p(x,t),\end{aligned} \tag{2.159}$$

where

$$p:(x,t) \to p(x,t) = -\exp\{-ik[2r(H(t-x)) - x]\}.$$

The result (2.159) is a natural extension to NAPs of the familiar separation of variables technique used extensively in studies of APs. In the present case it remains to determine the inverse function H, and this is generally a hard task. Nevertheless, (2.159) as it stands can yield useful information concerning the wave field. For example, consider the case when the boundary (bead) motion is periodic. In this case we assume there exists a $T > 0$ such that

$$r(t) = r(t+T) \quad \text{for all } t \in \mathbf{R}. \tag{2.160}$$

We notice that this means that the function $y \to r(H(y))$ is also periodic. To see this, replace t in (2.153) by y and add T to both sides of the result to obtain

$$H(y) + T - r(H(y)) = y + T,$$

which on using (2.160) yields

$$H(y) + T - r(H(y) + T) = y + T. \tag{2.161}$$

Using (2.153) again, but replacing t by $(y+T)$, yields

$$H(y+T) - r(H(y+T)) = y + T. \tag{2.162}$$

Combining (2.161) and (2.162), we obtain

$$H(y) + T - r(H(y) + T) = H(y+T) - r(H(y+T)). \tag{2.163}$$

Since $G(t): t \to t - r(t)$ has an inverse H defined in (2.153), (2.154), on comparing the right-hand sides of (2.163) and (2.153), we see that the existence of the inverse function requires that for the validity of (2.163) we should have

$$H(y) + T = H(y+T).$$

When this is the case, (2.163) implies

$$r(H(y+T)) = r(H(y) + T) = r(H(y)), \tag{2.164}$$

the last equality following from the periodicity of r. The result (2.160) establishes the required periodicity. The relation (2.160) now shows that for each fixed x the mapping $t \to p(x,t)$ that appears has period T.

The exact solution ψ, given by (2.156), is constant along the rays defined in space-time by $t = x + \tau$ for each τ. We shall set, recalling (2.159),

$$\Psi(\tau) = \psi(t-\tau, \tau) = -\exp\{-i\omega\tau\}\exp\{-ik2r(H(\tau))\} \tag{2.165}$$

and refer to $\Psi(\tau)$ as the far field of the solution $\psi(x,t)$. It is useful to think of $\Psi(\tau)$ as the limiting value of $\psi(x,t)$, as x and/or t becomes large, along the rays

$x = t - \tau$. This ray is the path, in space-time, followed by a signal with speed $c = 1$ that was transmitted from the origin $x = 0$ at time $t = \tau$.

For the quasi-stationary field we recall (2.158) and define the quasi-stationary far field to be

$$\Phi(\tau) = -\exp\{-i\omega\tau\}\exp\{-ik2r(\tau)\}. \tag{2.166}$$

Since $\exp\{-ik2r(H(\tau))$ and $\exp\{-ik2r(\tau)\}$ are periodic quantities with period T, we can expand Ψ and Φ in a Fourier series to obtain

$$\Phi(\tau) = -\sum_{-\infty}^{\infty} a_n \exp\left\{i\left(-\omega + \frac{2\pi n}{T}\right)\tau\right\},$$

$$\Psi(\tau) = -\sum_{-\infty}^{\infty} b_n \exp\left\{i\left(-\omega + \frac{2\pi n}{T}\right)\tau\right\}.$$

Thus we see that the far fields of both the exact and the quasi-stationary solutions are composed of the frequency of the incident wave plus multiples of the frequency of the periodic boundary motion. The above Fourier series can form the basis for the development of constructive methods. It has been shown by Kleinman and Mack [61] that the quasi-stationary approximation agrees well with experimental data. In most applications it appears that the first-order approximation is adequate.

Chapter Three

Mathematical Preliminaries

3.1 INTRODUCTION

The mathematical analysis of scattering phenomena and the development of associated scattering theories are conducted in the framework of *linear spaces*. These linear spaces involve generalisations of familiar concepts introduced in elementary courses in vector algebra and in the analysis of functions of a real variable. In these first courses we invariably worked with the value of a function at some point in its domain of definition rather than with the abstract quantity of the function itself. This strategy, as we shall see, will no longer be adequate for our purposes.

Some of the concepts and associated results that we indicate in the next few sections may seem to be somewhat abstract and heavyweight for some of the physical problems we might have in mind. However, we shall see that it is only by properly recognising and using these ideas that we will be able to make significant progress in the mathematical analysis of more demanding problems. However, in order to work with such abstract quantities, we will need generalisations of such familiar concepts as addition, multiplication by a scalar, magnitude and distance. These required generalisations are obtained by extending the familiar notions of geometry and vector algebra used in $\mathbf{R}^n$, the n-dimensional real Euclidean space. We shall see that this leads quite naturally to the notions of a *metric*, a generalisation of the familiar Euclidean distance function, a *metric space* and a *normed linear space*. Furthermore, we shall find that it will no longer be sufficient to work in finite-dimensional spaces such as $\mathbf{R}^n$ and $\mathbf{C}^n$, the n-dimensional complex Euclidean space. Instead, we must be prepared to work in infinite-dimensional spaces. This we shall find that we can do very conveniently in the structure of certain special linear spaces known as *Hilbert spaces*.

In this chapter we indicate these various generalisations, and in doing so we shall prove as little as possible. Our aim here is to collect together and illustrate by examples the various concepts, basic definitions and key theorems with which the reader might not be immediately familiar but which we shall use frequently in later chapters. References to more detailed accounts will be given in the text and more fully in chapter 11.

3.2 NOTATION

We begin by introducing and explaining some standard mathematical notation frequently used throughout the book.

Let X and Y be any two sets. The *inclusion* $x \in X$ denotes that the quantity x is an element in the set X. The (*set*) *inclusion* $Y \subset X$ denotes that the set Y is contained in the set X, in which case the set Y is said to be a (*proper*) *subset* of the set X. The possibility that Y might actually be the same as Y exists, and in this case we write $Y \subseteq X$ and simply refer to Y as a *subset* of X. The set that consists of elements belonging to X or Y or both is called the *union* of X and Y and is denoted by $Y \cup X$. The set consisting of all elements belonging to X and Y simultaneously is called the *intersection* of X and Y and is denoted by $X \cap Y$. The set consisting of all the elements of X that do not belong to Y is called the *difference* of X and Y is denoted by $X \setminus Y$. In particular, if $Y \subset X$, then $X \setminus Y$ is called the *complement* of Y in X.

Finally, at this stage we write $X \times Y$ to denote the set of elements of the form (x, y) where $x \in X$ and $y \in Y$. We recall that a simple example of this notation is the Euclidean space $\mathbf{R}^2$, where we identify $X = Y = \mathbf{R}$. This notation is frequently used when dealing with real and complex functions of several variables.

With this understanding it is clear that $X \subset X \cup Y$, $Y \subset X \cup Y$, $X \cap Y \subset X$, $X \cap Y \subset Y$ and that if $Y \subset X$, then $X = Y \cup (X \setminus Y)$.

3.3 VECTOR SPACES

The concept of a vector space will provide a framework within which abstract quantities can be manipulated *algebraically* in a meaningful manner.

DEFINITION 3.1 *A vector space (linear space) over a set of scalars* $\mathbf{K}$ *is a nonempty set* X *of elements* $x, y, \ldots$ *called vectors, together with two algebraic operations called vector addition and multiplication by a scalar that satisfy the following.*

(i) $x + y + z = (x + y) + z = x + (y + z)$, $x, y, z \in X$.

(ii) There exists a zero element $\theta \in X$ *such that* $x + \theta = x$, $x \in X$.

(iii) If $x \in X$, *then there exists an element* $(-x) \in X$ *such that* $x + (-x) = \theta$, $x \in X$.

(iv) $x + y = y + x$, $x, y \in X$.

(v) $(\alpha + \beta)x = \alpha x + \beta y$, $\alpha, \beta \in \mathbf{K}$, $x \in X$.

(vi) $\alpha(x + y) = \alpha x + \alpha y$, $\alpha \in \mathbf{K}$, $x, y \in X$.

(vii) $\alpha(\beta x) = \alpha\beta x$, $\alpha, \beta \in \mathbf{K}$, $x \in X$.

(viii) There exists a unit element $I \in \mathbf{K}$ *such that* $Ix = x$, $x \in X$.

We remark that in the following chapters $\mathbf{K}$ will usually be either $\mathbf{R}$ or $\mathbf{C}$.

We will often need the notion of the distance between abstract quantities. This we can introduce by mimicking familiar processes in Euclidean geometry.

DEFINITION 3.2 *A metric space* M *is a set* X *and a real-valued function* d, *called a metric or distance function, defined on* $X \times X$ *such that for all* $x, y, z \in X$,

(i) $d(x, y) \geq 0$,

(ii) $d(x, y) = 0$ *if and only if* $x = y$,

(iii) $d(x, y) = d(y, x)$,

(iv) $d(x, z) \leq d(x, y) + d(y, z)$.

Axiom (iv) is known as the triangle inequality.

We emphasise that a set X can be made into a metric space in many different ways simply by employing different metric functions. Consequently, for the sake of clarity we shall sometimes denote a metric space in the form (X, d) in order to make explicit the metric employed.

Example 3.3 *Let $X = \mathbf{R}^2$ and d be the usual Euclidean distance function (metric). Then if $x = (x_1, x_2, x_3, \ldots, x_n)$ and $y = (y_1, y_2, y_3, \ldots, y_n)$ denote any two points in $\mathbf{R}^n$, the distance between these two points is given by*

$$d(x, y) = \left\{ \sum_{i=1}^{n} (x_i - y_i)^2 \right\}^{1/2}.$$

It is this particular distance function and its properties that we mimicked when formulating definition 3.2. Clearly, (X, d) in this example is a metric space.

Example 3.4 *Let $X = C[0, 1]$, the set of all real-valued continuous functions defined on the subset $[0, 1] \subset \mathbf{R}$. This set can be made into two metric spaces, M_1 and M_2, where*

$$M_1 := (X, d_1),$$

$$d_1(f, g) := \max_{x \in [0,1]} |f(x) - g(x)|, \qquad f, g \in X,$$

$$M_2 := (X, d_2),$$

$$d_2(f, g) := \int_0^1 |f(x) - g(x)|\, dx, \qquad f, g \in X.$$

In constructing M_1 and M_2 we must, of course, prove that d_1 and d_2 each satisfy the metric space axioms (i)–(iv) in definition 3.2. In these two examples axioms (i)–(iii) are obviously satisfied. Axiom (iv), usually the hardest property to establish, is seen to hold in these two cases by virtue of well-known properties of the modulus and of Riemann integrals.

We remark that the metric space $C[0, 1]$ is an example of an infinite-dimensional space and is one that we shall frequently have occasion to use.

Once we have introduced the notion of distance into the abstract spaces, we are well placed to give a precise meaning to what is meant by convergence in such spaces.

Definition 3.5 *A sequence of elements $\{x_n\}_{n=1}^{\infty}$ in a metric space $M := (X, d)$ is said to converge to an element $x \in X$ if*

$$d(x, x_n) \to 0 \quad \text{as } n \to \infty.$$

In this case we write either $x_n \to x$ as $n \to \infty$ or $\lim_{n\to\infty} x_n = x$, where it is understood that the limit is taken with respect to the distance function d.

We would emphasise that the use of different metrics can induce different convergence results. For instance, in example 3.4 we have

$$d_2(f, g) \le d_1(f, g), \qquad f, g \in X.$$

Therefore, given a sequence $\{f\}_{n=1}^{\infty} \subset X$ such that $f_n \to f$ with respect to d_1, it follows that we also have $f_n \to f$ with respect to d_2. However, if we are given that the sequence converges with respect to d_2, then it does not follow that the sequence also converges with respect to d_1.

DEFINITION 3.6 *A sequence $\{f_n\}_{n=1}^{\infty} \subset (X, d)$ is called a Cauchy sequence if for all $\varepsilon > 0$ there exists a number $N(\varepsilon)$ such that $n, m \geqslant N(\varepsilon)$ implies $d(f_n, f_m) < \varepsilon$.*

The following is a standard result.

THEOREM 3.7 *In any metric space $M := (X, d)$ every convergent sequence*
(i) has a unique limit,
(ii) is a Cauchy sequence.

It must be emphasised that the converse of (ii) is false. It is possible that there are Cauchy sequences in (X, d) that might not converge to a limit element $x \in X$. This difficulty can be avoided if attention is restricted to certain preferred classes of metric spaces.

DEFINITION 3.8 *A metric space in which all Cauchy sequences converge to an element of that space is called a complete metric space.*

It can be shown that for the metric spaces M_1, M_2 in example 3.4, the metric space M_1 is complete but M_2 is incomplete.

We see from the above discussion that a promising strategy when analysing problems is the following. If when analysing a given problem we decide that the analysis should be conducted using elements from some set X, then to make the mathematical analysis easier and still maintain control over the requirements of the given problem, we enlarge X by adding to it the limits of all Cauchy sequences in X. The original set X will then be contained in some larger set, say $\tilde{X}$. The set $\tilde{X}$ will then have "nicer" properties than those of X when taken alone yet still be able to "control" the requirements of the given problem. The relation between X and $\tilde{X}$ is described as follows.

DEFINITION 3.9 *Given a metric space $M := (X, d)$, a set $Z \subset X$ is said to be dense in X if every element $y \in X$ is the limit, with respect to d, of a sequence of elements in Z.*

Consequently, in most practical situations we try to work with nice elements belonging to a set Z that is dense in some larger set X whose elements can provide more general results.

As an illustration of this concept, assume we are required to deal with a second-order ordinary differential equation. Then, as we have seen in elementary courses on differential equations, it is quite natural to begin any analysis of the problem by working with elements that are continuous and twice continuously differentiable; let this be the set Z. However, in practice it is often quite difficult to work with simply the elements of Z. More detailed information concerning the existence of solutions, convergence properties and the development of approximation methods can be obtained if we work with elements that are square-integrable over the range of interest. Let this be the set X. It turns out that in many cases of practical interest $Z \subset X$ and, furthermore, Z is dense in X.

To give an indication that this enlargement, or *completion* as it is more properly called, can indeed be made, we first need to introduce the following notions.

DEFINITION 3.10 *(i) Let X_1, X_2 be sets and let $M \subseteq X_1$ be a subset. A mapping f from M into X_2 is a rule that assigns to an element $x \in M \subseteq X_1$ an element $f(x) =: y \in X_2$, and we write*

$$f : X_1 \supseteq M \rightarrow X_2.$$

The element $y = f(x) \in X_2$ is the image of x with respect to f. The set M is called the domain of definition of f, which is denoted $D(f)$. Consequently, we write

$$f : X_1 \supseteq D(f) \rightarrow X_2$$

and

$$x \rightarrow f(x) = y \in X_2 \quad \text{for all } x \in D(f) \subseteq X_1.$$

The set of all images with respect to f is the range of f, denoted $R(f)$, where

$$R(f) := \{y \in X_2 : y = f(x) \quad \text{for } x \in D(f) \subseteq X_1\}.$$

(ii) A mapping f is called injective or an injection or one-to-one (denoted 1-1) if for every $x_1, x_2 \in D(f) \subseteq X_1$ we have that

$$x_1 \neq x_2 \quad \text{implies } f(x_1) \neq f(x_2).$$

This means that different elements in $D(f)$ have different images in $R(f)$.

(iii) A mapping f is called surjective or a surjection or a mapping of $D(f)$ onto X_2 if $R(f) = X_2$.

(iv) For an injective mapping $T : X_1 \supseteq D(T) \rightarrow X_2$, the inverse mapping T^{-1} is defined to be the mapping $R(T) \rightarrow D(T)$ such that $y \in R(T)$ is mapped onto that $x \in D(T)$ for which $Tx = y$. Less generally, but more conveniently here, we define T^{-1} only if T is one-to-one and onto X_2.

(Note: In (iv) we have written Tx rather than $T(x)$. This anticipates the notation used in a later section.)

In the following chapters we will deal with a variety of different mappings. For example, a *function* is usually understood to be the mapping of one (real or complex) number into some other number. It is a rule that assigns to a number in one set a number in some other set. The term *operator* will be reserved for mappings between sets of abstract elements such as functions themselves rather than their numerical values.

Definition 3.11 *A mapping f from a metric space (X_1, d_1) to a metric space (X_2, d_2) is said to be continuous if for $\{x_n\}_{n=1}^{\infty} \subset X_1$ we have $f(x_n) \to f(x)$ with respect to the structure of (X_2, d_2) whenever $x_n \to x$ with respect to the structure of (X_1, d_1).*

Definition 3.12 *Let $M_j = (X_j, d_j)$, $j = 1, 2$, be metric spaces. A mapping f that satisfies*

(i) $f : X_1 \to X_2$ is one-to-one and onto (bijection),

(ii) preserves metrics in the sense

$$d_2(f(x), f(y)) = d_1(x, y), \qquad x, y \in X_1,$$

is called an isometry, and M_1, M_2 are said to be isomorphic.

It is clear that an isometry is a continuous mapping. Furthermore, isometric spaces are essentially identical as metric spaces in the sense that any result that holds for a metric space $M = (X, d)$ will also hold for all metric spaces that are isometric to M. It is for this reason that we always try to work in a framework of isometric spaces. This will be particularly important when we come to develop some of the finer points of scattering theory.

With this preparation we can now state a fundamental result that will indicate in what sense an incomplete metric space can be made complete.

Theorem 3.13 *If $M = (X, d)$ is an incomplete metric space, then it is possible to find a complete metric space $\tilde{M} = (\tilde{X}, \tilde{d})$ so that M is isometric to a dense subset of $\tilde{M}$.*

The familiar concepts of *open* and *closed sets* on the real line extend to arbitrary metric spaces according to the following definition.

Definition 3.14 *If $M = (X, d)$ is a metric space, then*

(i) the set

$$B(y, r) := \{x \in X : d(x, y) < r\}$$

is called the open ball in M, of radius $r > 0$ and centre y,

(ii) a set $G \subset X$ is said to be open with respect to d if for all $y \in G$ there exists $r > 0$ such that $B(y, r) \subset G$,

(iii) a set $N \subset G$ is called a neighbourhood of $y \in N$ if $B(y, r) \subset N$ for some $r > 0$,

(iv) a point x is called a limit point of a subset $Y \subset X$ if

$$B(x, r) \cap \{Y \setminus \{x\}\} \neq \emptyset \quad \text{for all } r > 0,$$

where $\emptyset$ denotes the empty set,

(v) a set $F \subset X$ is said to be closed if it contains all its limit points,

(vi) the union of F and all its limit points is called the closure of F and will be denoted $\bar{F}$,

(vii) a point $x \in Y \subset X$ is an interior point of Y if Y is a neighbourhood of x.

These various abstract notions can be quite simply illustrated by considering subsets of the real line.

Finally in this section, we introduce a particularly important class of metric spaces.

Definition 3.15 *A normed linear space is a vector space X, defined over $\mathbf{K}=\mathbf{R}$ or $\mathbf{C}$, together with a mapping $\|.\|: X\to \mathbf{R}$, known as a norm on X, satisfying the following.*

(i) $\|x\| \geqslant 0$ for all $x \in X$.

(ii) $\|x\| = 0$ if and only if $x = \theta$, the zero element in X.

(iii) $\|\lambda x\| = |\lambda|\, \|x\|$ for all $\lambda \in \mathbf{K}$ and $x \in X$.

(iv) $\|x+y\| \leq \|x\| + \|y\|$ (triangle inequality).

The pair $(X, \|.\|)$ is referred to as a real or complex normed linear (vector) space depending on whether the underlying field $\mathbf{K}$ is $\mathbf{R}$ or $\mathbf{C}$.

Example 3.16

(i) $X=\mathbf{R}^n$ is a real normed linear space with a norm defined by

$$\|x\| = \|(x_1, x_2, x_3, \ldots , x_n)\| := \left\{ \sum_{k=1}^{n} |x_k|^2 \right\}^{1/2} .$$

(ii) $X=C[0, 1]$, the set of all continuous functions defined on $[0, 1]$, is a real normed linear space with a norm defined by either

$$\|f\| := \sup_{x\in[0,1]} |f(x)| , \qquad f \in X,$$

or

$$\|f\| := \int_0^1 |f(y)|\, dy, \qquad f \in X.$$

We notice that any normed linear space $(X, \|.\|)$ is also a metric space with the metric (distance function) d defined by

$$d(x, y) = \|x - y\| .$$

This is the *induced metric* on X. With this understanding we see that such notions as convergence, continuity, completeness and open and closed sets introduced earlier for metric spaces carry over to normed linear spaces. Typically, we have the following.

Definition 3.17 *Let $(X, \|.\|)$ be a normed linear space. A sequence $\{x_n\}_{n=1}^{\infty} \subset X$ is said to converge to $x \in X$ if, given $\varepsilon > 0$, there exists $N(\varepsilon)$ such that*

$$\|x_n - x\| < \varepsilon \quad \text{whenever } n \geqslant N(\varepsilon),$$

in which case we write either $\|x_n - x\| \to 0$ as $n \to \infty$ or $x_n \to x$ as $n \to \infty$.

Definition 3.18

(i) The normed linear space $(X, \|.\|)$ is complete if it is complete as a metric space in the induced metric.

(ii) A complete normed linear space is called a Banach space.

We would emphasise that there are metric spaces that are not normed linear spaces. A comparison of definitions 3.2 and 3.15 clearly indicates this.

3.4 DISTRIBUTIONS

A distribution is a generalisation of the concept of a classical function. It is a powerful mathematical tool for at least the following three reasons. First, in terms of the theory of distributions, it is possible to give a precise mathematical description of such idealised physical quantities as, for example, point charges and instantaneous impulses. Second, distribution theory provides a means for interchanging limiting operations when such interchanges might not be valid for classical functions. For instance, in contrast to classical analysis, in distribution theory there are no problems arising from the existence of nondifferentiable functions. Indeed, we shall see that all distributions, or generalised functions as they are sometimes called, can be treated as being infinitely differentiable. Third, distribution theory enables us to use series that in classical analysis would be considered divergent.

Distribution theory arises as a result of the following observation. A continuous real- or complex-valued function f of the real variable $x = (x_1, x_2, x_3, \ldots, x_n) \in \mathbf{R}^n$ can be defined on $\mathbf{R}^n$ in one or the other of two distinct ways. First, we can prescribe its value $f(x)$ at each point $x \in \mathbf{R}^n$. Alternatively, we could prescribe the value of the integral

$$I_f(\varphi) := \int_{\mathbf{R}^n} f(x)\varphi(x)\,dx$$

for *each* continuous complex-valued function φ whose value $\varphi(x)$ is zero for sufficiently large $|x|$ (the latter to ensure that the integral exists).

These two definitions have distinct and quite different characterisations. In the first a function is considered a rule that assigns *numbers to numbers.* In the second, which leads to something we will eventually call a distribution, we have a rule that assigns *numbers to functions*.

When working with the second approach, instead of dealing with the pointwise values $f(x)$ of the function f, we consider the *functional* I_f and its "values" $I_f(\varphi)$ at each of the *test functions* φ.

These two descriptions of f are equivalent. To see this, assume that for two continuous functions f and g, the functionals I_f and I_g, defined as indicated above, are equal. That is, for any test function φ we have $I_f(\varphi) = I_g(\varphi)$. It then follows from elementary properties of the integral that $f(x) = g(x)$ for all $x \in \mathbf{R}^n$. Hence the required equivalence is established.

A partial rationale for introducing this second way of defining a function can be given as follows. A distributed physical quantity cannot be characterised by its value at a point but rather by its averaged value in a sufficiently close neighbourhood of that point. For instance, it is impossible to measure the density of a material at a point. In practice we can only measure the average density of the material in a small neighbourhood of the point and then call this the density at the point. Thus we can think of a generalised function as being defined by its "average values" in a neighbourhood of the point of interest. Consequently, from a physical standpoint it is more convenient to consider continuous functions as functionals of the form indicated above.

To clarify matters, we give some examples and introduce a little more notation.

Example 3.19 *Let $C(\Omega)$ denote the set of all complex-valued functions that are continuous on the region Ω. For $f \in C(\Omega)$, let*

$$\|f\|_\infty := \sup\{|f(x)| : x \in \Omega\}.$$

It is readily verified that $C(\Omega)$ is a complex vector space with respect to the usual pointwise operations on functions. Furthermore, $C(\Omega)$ is a complete normed linear space with respect to $\|.\|_\infty$.

DEFINITION 3.20 *Let a real- or complex-valued function f, defined on a domain $D \subset \mathbf{R}^n$, be nonzero only for points belonging to a subset $\Omega \subset D$. Then the closure of Ω, denoted $\bar{\Omega}$, is called the support of f and is denoted supp f.*

If Ω is a compact set, that is, a closed and bounded set, then the function f is said to have compact support in D.

In order to simplify the notation when working in more than one dimension, we shall frequently use *multi-index notation*. This can be easily introduced if we consider a partial differential expression of the form

$$L = \sum_{q_1+\cdots+q_N \le p} a_{q_1,\ldots,q_N}(x)\partial_1^{q_1}\partial_2^{q_2}\ldots\partial_N^{q_N}, \qquad p \ge 0,$$

where $\partial_j := \partial/\partial x_j$, $j = 1, 2, \ldots, N$, and q_j, $j = 1, 2, \ldots, N$, are non-negative integers with $a_{q_1,\ldots,q_N}$ denoting differentiable (to sufficient order) functions.

Any set $q := \{q_1, q_2, \ldots, q_N\}$ of non-negative integers is called a *multi-index*. The sum $|q| = \sum_{k=1}^N q_k$ is called the *order* of the multi-index. We denote $a_{q_1,\ldots,q_N}(x)$ by $a_q(x)$ and $\partial_1^{q_1}\partial_2^{q_2}\ldots\partial_N^{q_N}$ by D^q. Consequently, the above partial differential expression can be written as

$$L = \sum_{|q| \le p} a_q(x) D^q.$$

DEFINITION 3.21 *Let Ω be an open set in $\mathbf{R}^n$.*

(i) $C_0^\infty(\Omega) := \{\varphi \in C^\infty(\Omega) : \text{supp } f \subset \Omega\}$

Here $\varphi \in C^\infty(\Omega)$ indicates that φ and all its partial derivatives of all orders exist and are continuous. An element $\varphi \in C^\infty(\Omega)$ is referred to as a smooth element or as an infinitely differentiable element.

(ii) The set $C_0^\infty(\Omega)$ is called the set of test functions defined on Ω.

(iii) A sequence of test functions $\{\varphi_n\}_{n=1}^\infty$ is said to be convergent to a test function φ if

(a) φ_n and φ are defined on the same compact set,

(b)

$$\left\|\varphi_n - \varphi\right\|_\infty \to 0 \qquad \text{as } k \to \infty,$$
$$\left\|D^\alpha\varphi_n - D^\alpha\varphi\right\|_\infty \to 0 \qquad \text{as } k \to \infty,$$

for all multi-indices α. Here $\|\varphi\|_\infty := \sup_{x\in\Omega}|\varphi(x)|$.

(iv) The set $C_0^\infty(\mathbf{R}^n)$ *together with the topology (a concept of convergence) induced by the convergence defined in (iii) is called the space of test functions and is denoted by* $\mathfrak{D}(\mathbf{R}^n)$.

Unless certain subsets of $\mathbf{R}^n$ *have to be emphasised, we shall write* $\mathfrak{D} \equiv \mathfrak{D}(\mathbf{R}^n)$ *unless otherwise stated.*

(v) A linear functional f *on* $\mathfrak{D}$ *is a mapping* $f:\mathfrak{D} \to \mathbf{K} = \mathbf{R}$ *or* $\mathbf{C}$ *such that*

$$f(a\varphi + b\psi) = af(\varphi) + bf(\psi)$$

for all $a, b \in C$ *and* $\varphi, \psi \in \mathfrak{D}$.

A mapping $f:\mathfrak{D} \to \mathbf{K}$ is a *rule* that, given any $\varphi \in \mathfrak{D}$, produces a *number* $z \in \mathbf{K}$, and we write

$$z = f(\varphi) \equiv \langle f, \varphi \rangle.$$

This indicates the *action* of the functional f on φ.

We remark that if $\varphi \in C_0^\infty(\Omega)$, $\Omega \subset \mathbf{R}^n$, then φ vanishes on $\partial\Omega$, the "boundary" of the set Ω.

Example 3.22 *Let* φ *be the function defined on* $\Omega := \mathbf{R}$ *by*

$$\varphi(x) = \begin{cases} 0, & |x| \geqslant a, \\ \exp\{\frac{1}{x^2 - a^2}\}, & |x| < a. \end{cases}$$

It is readily shown that φ *is infinitely continuously differentiable on* $\mathbf{R}$ *and that supp* $\varphi = [-a, a]$. *Hence* $\varphi \in C_0^\infty(\mathbf{R})$.

DEFINITION 3.23

(i) A functional f *on* $\mathfrak{D}$ *is continuous if it maps every convergent sequence in* $\mathfrak{D}$ *into a convergent sequence in* $\mathbf{K} = \mathbf{R}$ *or* $\mathbf{C}$; *that is,*

$$f(\varphi_n) \to f(\varphi) \quad \textit{whenever } \varphi_n \to \varphi \textit{ in } \mathfrak{D}.$$

(ii) A continuous linear functional on $\mathfrak{D}$ *is called a distribution or a generalised function.*

(*Note:* For our immediate convenience at this stage we shall use a bold typeface to indicate a distribution and will write $\langle \mathbf{f}, \varphi \rangle$ to denote the action of the distribution $\mathbf{f}$ on the test function φ. We will be able to relax this notation shortly.)

DEFINITION 3.24 *(convergence of distributions) A sequence* $\{\mathbf{f}_n\}$ *of distributions is said to be convergent if the sequence of numbers* $\{\langle \mathbf{f}_n, \varphi \rangle\}$ *is convergent for all* $\varphi \in \mathfrak{D}$.

This definition implies that if $\{\mathbf{f}_n\}$ is a convergent sequence of distributions, then there is a distribution $\mathbf{f}$ such that for $n \to \infty$,

$$\langle \mathbf{f}_n, \varphi \rangle \to \langle \mathbf{f}, \varphi \rangle \quad \text{for all } \varphi \in \mathfrak{D}.$$

In this case $\mathbf{f}_n$ is said to *converge weakly* to $\mathbf{f}$ as $n \to \infty$.

DEFINITION 3.25 *(i) A function f that is integrable on every open bounded subset $\Omega \subset \mathbf{R}^n$ is said to be locally integrable on $\mathbf{R}^n$.*

(ii) For every locally integrable function f there is a distribution $\mathbf{f}$ defined by

$$\langle \mathbf{f}, \varphi \rangle = \int_{\mathbf{R}^n} f(s)\varphi(s)\, ds.$$

The distribution $\mathbf{f}$ is said to be generated by the function f.

(iii) A distribution generated by a locally integrable function is called a regular distribution. Distributions that are not regular distributions are called singular distributions.

We remark that definition 3.25(ii) is meaningful since $B := \operatorname{supp} \varphi \subset \mathbf{R}^n$. In this case we have

$$|\mathbf{f}(\varphi)| = \left| \int_B f(s)\varphi(s)\, ds \right| \leq \sup_{x \in B} |\varphi(x)| \int_B |f(s)|\, ds.$$

The right-hand side is bounded since f is locally integrable. Hence we can conclude that $\mathbf{f}(\varphi)$ is well defined.

Thus we see that the class (set) of all distributions will contain elements that correspond to ordinary (classical) functions as well as singular distributions that do not.

DEFINITION 3.26 *The set of all distributions on $\mathfrak{D}$ together with the topology indicated in definition 3.24 is called the dual of $\mathfrak{D}$ and is denoted $\mathfrak{D}'$.*

For the sake of our convenience in later sections we summarise here the main properties of a distribution that we have introduced so far.

(i) *Linearity:* For any test functions φ, ψ and complex numbers α, β,

$$\mathbf{f}(\alpha\varphi + \beta\psi) = \alpha\mathbf{f}(\varphi) + \beta\mathbf{f}(\varphi).$$

(ii) *Continuity:*

$$\mathbf{f}(\varphi_n) \to \mathbf{f}(\varphi) \quad \text{whenever } \varphi_n \to \varphi.$$

(iii) *Equality:* Two distributions $\mathbf{f}$ and $\mathbf{g}$ are equal provided $\mathbf{f}(\varphi) = \mathbf{g}(\varphi)$ for any test function φ. They are said to be different if there exists a test function φ such that $\mathbf{f}(\varphi) \neq \mathbf{g}(\varphi)$.

We remark that, strictly speaking, equality here means equality almost everywhere (ae); that is, if the set $\{x : \mathbf{f}(\varphi) \neq \mathbf{g}(\varphi)\}$ has measure zero, then $\mathbf{f}(\varphi) = \mathbf{g}(\varphi)$. The notion of measure is introduced in the next chapter. For the time being we just think of the situation in $\mathbf{R}^1$ when measure can be identified with the length of an interval and a set of measure zero is a point.

(iv) *Linear combinations:* The linear combination $(\alpha\mathbf{f} + \beta\mathbf{g})$ of two distributions $\mathbf{f}$ and $\mathbf{g}$ is defined as

$$\langle \alpha\mathbf{f} + \beta\mathbf{g}, \varphi \rangle = \alpha\langle \mathbf{f}, \varphi \rangle + \beta\langle \mathbf{g}, \varphi \rangle.$$

(v) *Product of a distribution:* The product of a distribution **f** and a smooth function h is defined in a natural manner as

$$\langle h\mathbf{f}, \varphi \rangle = \langle \mathbf{f}, h\varphi \rangle \quad \forall \varphi \in \mathfrak{D}.$$

If $\varphi \in \mathfrak{D}$ and h is a smooth function, then $h\varphi$ is also a test function. However, if h is not smooth, then $h\varphi \notin \mathfrak{D}$. Therefore we cannot define the product of a distribution with a function that is discontinuous or has discontinuous derivatives.

We notice that our definition is a generalisation of the familiar identity

$$\int_{\Omega} \{h(x) f(x)\}\varphi(x)\, dx = \int_{\Omega} h(x)\{f(x)\varphi(x)\}\, dx,$$

which always holds when f is locally integrable.

We have already pointed out that a generalised function does not have values *at a point.* However, it is possible to give meaning to the statement that a distribution becomes zero in a *region*. The distribution **f** becomes zero in the region Ω if $\langle \mathbf{f}, \varphi \rangle = 0$ for all $\varphi \in \mathfrak{D}(\Omega)$, and we write $\mathbf{f} = \boldsymbol{\theta}$ in Ω.

(vi) *Support of a distribution:* The set of all points such that in no neighbourhood of each point does $\mathbf{f} \neq \boldsymbol{\theta}$ is known as the *support* of the distribution **f**. We denote the support of **f** by supp **f**. If supp **f** is bounded, then the distribution **f** is said to have *compact support.*

We now turn to the differential calculus of distributions. Our plan here, as indeed it will be when developing most properties of distributions, is to start with regular distributions and then generalise the results, whenever possible, to all distributions. For ease of presentation at this stage we restrict attention to processes in $\mathbf{R}^1 = \mathbf{R}$.

If f is a differentiable function that generates a regular distribution **f** and if $d\mathbf{f}/dx$ denotes the distribution generated by f', then we obtain, by integration by parts,

$$\left\langle \frac{d\mathbf{f}/}{dx}, \varphi \right\rangle = \int_{\mathbf{R}} f'(s)\varphi(s)\, ds = -\int_{\mathbf{R}} f(s)\varphi'(s)\, ds.$$

The integrated terms in the above vanish since φ is a test function and as such vanishes at infinity. Consequently, for regular distributions, corresponding to differentiable functions, we have

$$\langle d\mathbf{f}/dx, \varphi \rangle = -\langle \mathbf{f}, \varphi' \rangle.$$

Example 3.27 *For any distribution* **f**, *the functional* $\varphi \to -\langle \mathbf{f}, \varphi' \rangle$ *with* $\varphi \in \mathfrak{D}$ *is a distribution.*

Proof. If $\varphi \in \mathfrak{D}$, then $\varphi' \in \mathfrak{D}$ and hence $-\langle \mathbf{f}, \varphi' \rangle$ defines a functional on $\mathfrak{D}$. It is clearly linear (see definition 3.21(v)). Further, if $\varphi_n \to \varphi$ in $\mathfrak{D}$, then by a standard result for the differentiation of uniformly convergent sequences, $\varphi_n' \to \varphi'$ in $\mathfrak{D}$. Therefore we have that $\mathbf{f}$ is continuous since

$$-\langle \mathbf{f}, \varphi_n' \rangle \to -\langle \mathbf{f}, \varphi' \rangle,$$

which establishes the required result. □

The following definition now follows naturally.

DEFINITION 3.28 *The derivative of a generalised function* $\mathbf{f}$ *is the generalised function* $\mathbf{f}'$ *defined by*

$$\langle \mathbf{f}', \varphi \rangle = -\langle \mathbf{f}, \varphi' \rangle, \qquad \forall \varphi \in \mathfrak{D}.$$

We see from the above that the distribution generated by the derivative f' of a differentiable function f is the same as the derivative of the distribution $\mathbf{f}$. These two ways of interpreting the symbol $\mathbf{f}'$ for a differentiable function f are consistent with classical calculus.

The advantage of distribution theory over classical calculus is that *every* generalised function is differentiable; this follows from example 3.27.

If a function f is locally integrable but not differentiable in the classical sense, the associated distribution $\mathbf{f}'$ is called the *generalised derivative* of f.

It is a straightforward matter to obtain corresponding results when the underlying space is $\mathbf{R}^n$. For instance, let $\Omega \subset \mathbf{R}^n$ be an open set and let $\mathfrak{D}(\Omega)$ denote the space of test functions defined on Ω. Let α denote a multi-index. The αth distributional derivative of a distribution $\mathbf{f}$ on $\mathfrak{D}(\Omega)$ is the distribution $D^\alpha \mathbf{f}$ defined by

$$D^\alpha \mathbf{f}(\varphi) = \langle D^\alpha \mathbf{f}, \varphi \rangle = (-1)^{|\alpha|} \langle \mathbf{f}, D^\alpha \varphi \rangle = (-1)^{|\alpha|} \mathbf{f}(D^\alpha \varphi).$$

This follows using integration by parts (i.e., Green's theorem).

We now give a number of examples to illustrate these various ideas.

Example 3.29 *The Dirac delta* δ, *defined in* $\boldsymbol{R}^1$ *by*

(i) $\delta(x) = 0$, $x \neq 0$,

(ii) $\int_{-\infty}^{\infty} \delta(x)\, dx = 1$,

(iii) $\int_{-\infty}^{\infty} \delta(x)\varphi(x)\, dx = \langle \boldsymbol{\delta}, \varphi \rangle = \varphi(0)$, $\varphi \in \mathfrak{D}$,

generates a continuous linear functional $\boldsymbol{\delta}$. *That it is a linear functional follows immediately from (iii). The continuity of* $\boldsymbol{\delta}$ *follows from*

$$|\boldsymbol{\delta}(\varphi)| = |\langle \boldsymbol{\delta}, \varphi \rangle| = |\varphi(0)| \leq \sup |\varphi(x)| = \|\varphi\|_\infty.$$

However, $\boldsymbol{\delta}$ *is a singular distribution. To see this, let* φ *be the test function defined by*

$$\varphi(x) = \begin{cases} 0, & b > |x| \geqslant a, \\ \exp\{\frac{1}{x^2 - a^2}\}, & |x| < a, \end{cases}$$

where $b > a > 0$.

If we assume that $\boldsymbol{\delta}$ *is a regular distribution, then we can easily obtain*

$$\exp\left\{\frac{-1}{a^2}\right\} = |\varphi(0)| = \left|\int_{-b}^{b} \delta(s)\varphi(s)\,ds\right| \leq \frac{1}{e}\int_{-a}^{a} \delta(s)\,ds,$$

and by taking the limit $a \to 0$, *we obtain a contradiction.*

The derivative of $\boldsymbol{\delta}$ *is, following definition 3.28, defined by*

$$\langle \boldsymbol{\delta}', \varphi\rangle = -\langle \boldsymbol{\delta}, \varphi'\rangle = -\varphi'(0).$$

Similarly, the n*th derivative of* $\boldsymbol{\delta}$ *is given by*

$$\langle \boldsymbol{\delta}^{(n)}, \varphi\rangle = (-1)^{(n)}\langle \boldsymbol{\delta}, \varphi^{(n)}\rangle = (-1)^{(n)}\varphi^{(n)}(0).$$

A particularly useful example is provided by the functional $\mathbf{f}\boldsymbol{\delta}$ *with* $f(x) = x \in (-1, 1) =: \Omega$. *In this case we have, for all* $\varphi \in C_0^\infty(\Omega)$,

$$(\mathbf{f}\boldsymbol{\delta})(\varphi) = (\mathbf{x}\boldsymbol{\delta})(\varphi) = \langle \mathbf{x}\boldsymbol{\delta}, \varphi\rangle = \langle \boldsymbol{\delta}, x\varphi\rangle = [x\varphi]_{x=0} = 0.$$

Thus $\mathbf{x}\boldsymbol{\delta} = \boldsymbol{\theta} \in (C_0^\infty(\Omega))'$ *is the zero distribution on* $\mathfrak{D}(\Omega)$.

Every continuous function is locally integrable and hence generates a distribution (see definition 3.25). However, there are many discontinuous functions that are also locally integrable.

Example 3.30 *Consider the function* f *defined by*

$$f(x) = |x|^{-1/2}, \qquad x \in [-1, 1].$$

This function has a singularity at the origin. However, it is locally integrable since

$$\int_a^b |f(s)|\,ds = \int_a^b |s|^{-1/2}\,ds$$

is bounded in each interval $(a, b) \subset [-1, 1]$. *Thus* f *generates a distribution* $\mathbf{f}$ *defined by*

$$\mathbf{f}(\varphi) = \langle \mathbf{f}, \varphi\rangle = \int_{-1}^{1} |s|^{-1/2}\varphi(s)\,ds, \qquad \varphi \in C_0^\infty[-1, 1].$$

Hence $\mathbf{f}$ *is a regular distribution.*

Example 3.31 *The function defined by* $|x|$ *is a locally integrable function that is differentiable for all* $x \neq 0$. *However, it is not differentiable at* $x = 0$. *Nevertheless, it has a generalised derivative calculated as follows. For any test function* φ, *we have*

$$\begin{aligned}\langle |\mathbf{x}|', \varphi\rangle &= -\langle |\mathbf{x}|, \varphi'\rangle \\ &= -\int_{-\infty}^{\infty} |s|\, \varphi'(s)\, ds \\ &= -\int_{-\infty}^{0} |s|\, \varphi'(s)\, ds - \int_{0}^{\infty} |s|\, \varphi'(s)\, ds \\ &= -\int_{-\infty}^{0} \varphi(s)\, ds + \int_{0}^{\infty} \varphi(s)\, ds,\end{aligned}$$

where we have integrated by parts and used the fact that the test function φ vanishes at infinity. We now introduce a function sgn defined by

$$\operatorname{sgn}(x) = \begin{cases} -1 & \text{for } x < 0, \\ 1 & \text{for } x > 0. \end{cases}$$

It is not necessary here for us to specify $\operatorname{sgn}(0)$ *since it can be shown as an easy exercise that the above function will generate the same distribution* **sgn** *for any choice of* $\operatorname{sgn}(0)$. *Consequently, we have from the above that for all $\varphi \in \mathfrak{D}$,*

$$\begin{aligned}\langle |\mathbf{x}|', \varphi\rangle &= \int_{-\infty}^{\infty} \operatorname{sgn}(s)\varphi(s)\, ds \\ &= \langle \mathbf{sgn}, \varphi\rangle.\end{aligned}$$

Therefore we say that $|\mathbf{x}|' = \mathbf{sgn}$ in the sense of distributions. We remark that since the generalised derivative is a distribution, it is meaningless to talk about its value at a point; it can only be given any meaning over an interval.

Example 3.32 *The Heaviside function H, defined on $[-a, a]$ by*

$$H(x) = \begin{cases} 0, & -a \le x < 0, \\ 1, & 0 \le x < a, \end{cases}$$

is locally integrable and generates the distribution **H** *according to*

$$\mathbf{H}(\varphi) = \langle \mathbf{H}, \varphi\rangle = \int_{-a}^{a} H(s)\varphi(s)\, ds = \int_{0}^{a} \varphi(s)\, ds, \qquad \varphi \in C_0^{\infty}(-a, a).$$

Hence **H** *is a regular distribution.*

The distributional derivative of **H** *is calculated, in the now familiar manner, as follows, bearing in mind that the test function φ vanishes at infinity.*

$$\langle \mathbf{H}', \varphi\rangle = -\int_{0}^{\infty} \varphi'(s)\, ds = \varphi(0) = \int_{-\infty}^{\infty} \delta(s)\varphi(s)\, ds = \langle \boldsymbol{\delta}, \varphi\rangle = \boldsymbol{\delta}(\varphi).$$

Therefore we see that $\mathbf{H}' = \boldsymbol{\delta}$ in the sense of distributions.

We would remark that although the previous discussion has been conducted with respect to ordinary derivatives, nevertheless, similar results can be obtained for partial derivatives using the multi-index notation introduced earlier.

When we come to deal with differential equations having the typical form

$$Lu = f,$$

where L is a given differential expression, then various types of solutions must be considered. If f is a regular distribution generated by a function that is locally integrable but not continuous, then the above equation cannot be expected to have any meaning in the classical sense. A similar observation holds if f is a singular distribution. In these cases we say that the equation holds in the *sense of distributions*. The solutions to this equation that might be obtained in these cases will be distributions and are known as *weak* or *generalised solutions* of the equation.

Example 3.33 *Consider the differential equation*

$$xu'(x) = 0, \qquad x \in (-1, 1).$$

This equation has a classical solution

$$u(x) = constant.$$

However, if we regard the equation as a distributional differential equation, then it has a weak or generalised solution of the form

$$\mathbf{u} = c_1\mathbf{H} + c_2,$$

where $\mathbf{H}$ *is the distribution generated by the Heaviside function* H, *and* c_1, c_2 *are constants. To see that this is indeed the case, we first notice that (see example 3.32)*

$$\mathbf{u}' = c_1\boldsymbol{\delta}.$$

This in turn indicates that for any test function φ,

$$\mathbf{xu}'(\varphi) = \langle \mathbf{xu}', \varphi\rangle = \langle \mathbf{u}', x\varphi\rangle = \langle c_1\boldsymbol{\delta}, x\varphi\rangle = c_1\{(x\varphi)(0)\} = 0,$$

and we conclude that $\mathbf{xu}' = \mathbf{0}$ *or, equivalently,* $xu'(x) = 0$ *in the sense of distributions.*

There are various kinds of solutions of an equation of the form $Lu = f$ when it is considered an equation for a generalised function f. These are classified as follows.

Definition 3.34

(i) A distribution $\mathbf{u}$ *satisfying the equation* $L\mathbf{u} = \mathbf{f}$ *is called a distributional or generalised solution of the equation.*

(ii) A function u *that is sufficiently continuously differentiable and thus generates a regular distribution* $\mathbf{u}$ *that satisfies* $L\mathbf{u} = \mathbf{f}$ *in the generalised sense is called a classical solution of the equation* $Lu = f$.

(iii) A function u that is not n times continuously differentiable, and therefore cannot be a classical solution of $Lu = f$, but which generates a regular distribution $\mathbf{u}$ *that is a generalised solution of* $\mathbf{Lu} = \mathbf{f}$ *is called a weak solution of* $Lu = f$.

(iv) A distributional solution of $\mathbf{Lu} = \mathbf{f}$ *is a solution* $\mathbf{u}$ *that is a singular distribution.*

Finally in this subsection, we briefly sketch the notion of the *convolution* of two distributions $\mathbf{f}$ and $\mathbf{g}$.

The classical formula for the convolution of two continuous functions f and g defined on $\mathbf{R}$ is given by

$$(f * g)(x) := \int_{\mathbf{R}} f(x-y)g(y)\,dy = \int_{\mathbf{R}} f(y)g(x-y)\,dy.$$

For the distributional case we adopt the following.

DEFINITION 3.35 *The convolution* $\mathbf{f} * \mathbf{g}$ *of two distributions* $\mathbf{f}$ *and* $\mathbf{g}$ *is defined to be*

$$\langle (\mathbf{f} * \mathbf{g})(x), \varphi(x) \rangle = \langle \mathbf{f}(x), \langle \mathbf{g}(y), \varphi(x+y) \rangle \rangle,$$

where $\varphi \in \mathfrak{D}$. *We have abused notation slightly in order to emphasise the "integration" variable. Furthermore, we have assumed in this definition that the distribution* $\mathbf{g}$ *has compact support. Consequently, the right-hand side of this expression is well defined since*

$$\psi(x) := \langle \mathbf{g}(y), \varphi(x+y) \rangle$$

is a test function.

We notice that when $\mathbf{f}$ and $\mathbf{g}$ are regular distributions, then $\mathbf{f} * \mathbf{g}$ is also a regular distribution and we recover the classical formula for the convolution of two functions.

Example 3.36 *The Dirac delta concentrated at* $x = a$ *is denoted* $\delta_a(x)$ *and defined to be*

$$\delta_a(x) = \delta(x-a).$$

This has the property (see definition 3.24)

$$\langle \boldsymbol{\delta}_a(x), \varphi(x) \rangle = \int_R \delta_a(x)\varphi(x)\,dx = \varphi(a).$$

Furthermore, if $\mathbf{f}$ *is a regular distribution, then*

$$\begin{aligned} \langle (\mathbf{f} * \boldsymbol{\delta}_a)(x), \varphi(x) \rangle &= \langle \mathbf{f}(x), \langle \boldsymbol{\delta}_a(y), \varphi(x+y) \rangle \rangle \\ &= \langle \mathbf{f}(x), \varphi(x+a) \rangle \\ &= \langle \mathbf{f}(x-a), \varphi(x) \rangle. \end{aligned}$$

Thus, remembering that the integration variables are only written in for convenience, we have

$$\mathbf{f}(x) * \boldsymbol{\delta}_a(x) = \mathbf{f}(x - a)$$

in a distributional sense.

Since it can be readily shown that different locally integrable functions define different distributions, it follows that the set of locally integrable functions can be embedded in the set of all distributions. An even more powerful result can be obtained, namely, that $\mathfrak{D}$ is dense in $\mathfrak{D}'$. Hence every distribution is a weak limit of test functions. For a more detailed discussion of these remarks, see chapter 11 and the references cited there.

With these various observations in mind we see that when a function f is considered a distribution, then it is identical to all functions that can be obtained by changing the values of $f(x)$ on isolated points, more precisely, on sets of measure zero. Hence a distribution is *not* associated with *a* function but with an *equivalence class* of functions that are equal almost everywhere (ae), that is, everywhere except on sets of measure zero.

The use of bold type to indicate a distribution will sometimes be suppressed from now onwards. Whether a quantity f is to be regarded as either a function or as a distribution will usually be clear from the text. However, the bold type will be restored if clarification is needed.

3.5 FOURIER TRANSFORMS AND DISTRIBUTIONS

Integral transforms play an important role in mathematical analysis and its applications. Perhaps one of their most impressive properties is that they can transform differentiation into the algebraic operation of multiplication by a scalar. Consequently, if we are faced with having to solve an ordinary differential equation for an unknown function u, then we can use a suitable integral transform to obtain an equivalent algebraic equation for $\hat{u}$, the transform of u. We then solve the algebraic equation for $\hat{u}$ and recover the required solution function u by means of an associated inverse integral transform. Similarly, a partial differential equation involving an unknown function v can be transformed into an equivalent ordinary differential equation for the transformed function $\hat{v}$. The required function v is recovered by solving for $\hat{v}$ and then using an associated inverse integral transform.

The prototype of such transforms is the Fourier transform. Here we give a *very* brief sketch of some of the more important aspects of the classical Fourier transform. Full details of the theory of Fourier transforms and their applications can be found in the references cited in chapter 11. In this connection we would particularly recommend [31], [129] and, for the mathematical theory, [144]. Once we have introduced the classical Fourier transform, we shall turn our attention to the notion of generalised Fourier transforms for distributions.

Definition 3.37 *A function $f:\mathbf{R}\to\mathbf{C}$ is said to be absolutely integrable if $\int_{-\infty}^{\infty} |f(x)|\,dx$ exists.*

Example 3.38

(i) Every test function is absolutely integrable.

(ii) Every continuous function that tends to zero faster than $|x|^{-(1+\alpha)}$, where $\alpha>0$, as $0<\alpha\le|x|\to\infty$ is absolutely integrable.

(iii) No polynomial other than the trivial polynomial that is everywhere zero is absolutely integrable.

The *classical Fourier transform* is defined as follows.

Definition 3.39 *For any absolutely integrable function f, the (classical) Fourier transform of f, denoted $\hat{f}$, is defined by*

$$\hat{f}(p)=\frac{1}{\sqrt{2\pi}}\int_{-\infty}^{\infty} f(x)e^{-ipx}\,dx=(\hat{F} f)(p).$$

The integral is convergent since f is absolutely integrable. Indeed, it can be shown to be uniformly convergent with respect to p.

Definition 3.40 *A function $f:\mathbf{R}\to\mathbf{C}$ is said to be piecewise-smooth if*

(i) all its derivatives exist and are continuous except possibly at a set of points $x_1, x_2, x_3, \ldots$ such that any finite interval contains only a finite number of the x_i and if

(ii) the function and all its derivatives have at most a finite number of jump discontinuities.

(iii) The function is said to be n times continuously differentiable if (i), (ii) are satisfied with "all" replaced by "its first n."

Associated with definition 3.39 is the *inverse transform*.

Definition 3.41 *If f is absolutely integrable, continuous and piecewise-smooth, then*

$$f(x)=\frac{1}{\sqrt{2\pi}}\int_{\mathbf{R}} \hat{f}(p)e^{+ipx}\,dp=(\overset{\wedge}{F^*} f)(x)\equiv\overset{\wedge}{F^*}(x)f,$$

where $\overset{\wedge}{F^*}$ is the transform inverse to $\hat{F}$.

Definitions 3.39 and 3.41 can be combined to provide what we will come to call an *inversion theorem* of the form

$$\hat{f}(p)=\frac{1}{\sqrt{2\pi}}\int_{\mathbf{R}} f(x)e^{-ipx}\,dx=(\hat{F} f)(p), \tag{3.1}$$

$$f(x)=\frac{1}{\sqrt{2\pi}}\int_{\mathbf{R}} \hat{F}(p)e^{+ipx}\,dp=(\overset{\wedge}{F^*} f)(x). \tag{3.2}$$

In practice we will always have to demonstrate that definitions 3.39 and 3.41 are available; that is, we will have to *prove* the inversion theorem for the problem being considered. The proof of such a theorem is not easy. Nevertheless, Fourier transforms play a central role in modern applied mathematical analysis. Their important property is the reciprocal nature of the transform as indicated by the proved inversion theorem. A glance at definitions 3.39 and 3.41 indicates that f is related to $\hat{f}$ in the same way that $\hat{f}$ is related to f apart from a minus sign. We shall return to this aspect in chapter 6 and subsequent chapters. In these later chapters we will see that it is possible to construct an integral transform that is particularly appropriate for the specific problem under consideration in that it has an associated inverse theorem.

The Fourier transform we have just introduced has the following properties.

Example 3.42 *Let f be an absolutely integrable and piecewise-smooth function.*

(i) $f(-x) = (\hat{\hat{f}})(x)$.

(ii) $(\hat{f'})(p) = ip\hat{f}(p)$.

(iii) $\frac{d}{dp}[(\hat{F}f)(p)] = -i[\hat{F}(xf)](p)$.

(iv) If $f_a(x) := f(x-a)$, *then* $(\hat{F}f_a) = \exp(-ipa)(\hat{F}f)(p)$.

An alternative definition of the Fourier transform that is frequently used is given by

$$\tilde{f}(p) = \int_{\mathbf{R}} f(x)e^{ipx}\,dx = (\tilde{F}f)(p) \equiv \tilde{F}(p)f,$$

$$f(x) = \frac{1}{2\pi}\int_{\mathbf{R}} \tilde{f}(p)e^{-ipx}\,dp = (\tilde{F}^*f)(x) \equiv \tilde{F}^*(x)f.$$

The two transforms are related in the following way

$$\hat{F}(p) = \frac{1}{\sqrt{2\pi}}\tilde{F}(-p),$$

$$\hat{F}^*(x) = \sqrt{2\pi}\,\tilde{F}^*(-x).$$

This follows directly from the definitions. Corresponding to the properties outlined in example 3.42, we have the following.

Example 3.43 *For f as in example 3.42, we have*

(i) $2\pi f(-x) = (\tilde{\tilde{f}})(x)$,

(ii) $(\tilde{f'})(p) = -ip\tilde{f}(p)$,

(iii) $\frac{d}{dp}[(\tilde{F}f)(p)] = i[\tilde{F}(xf)](p)$.

(iv) If $f_a(x) := f(x-a)$, *then* $(\tilde{F}f_a) = \exp(-ipa)(\tilde{F}f)(p)$.

In the following chapters we shall use the Fourier transform in the form $\hat{F}$. This transform can be interpreted by saying that any (sufficiently nice) function f can be regarded as a superposition, either as a sum or as an integral, of an infinite number of sinusoidal waves (characterised by $\exp(-ipx)$) with different frequencies p, where the wave of frequency p has amplitude $\hat{f}(p)/\sqrt{2\pi}$. For this reason the integral relation

$$f(x) = \frac{1}{\sqrt{2\pi}} \int_{\mathbf{R}} \hat{f}(p) e^{+ipx}\, dp = (\hat{F}^* f)(x)$$

is called the *spectral resolution* of the function f, and $\hat{f}$ is referred to as the *spectral density* of f.

The development of the theory of Fourier transforms based on definitions 3.39 and 3.40 is inadequate for our purposes. This is because the theory is restricted to absolutely integrable functions and these are not suitable for the analysis we have in mind. To see this, simply recall that we are mainly interested here in differential equations and that these have solutions that have the form of either polynomials or trigonometric functions or exponential functions and none of these are absolutely integrable. This means that the forms we are most interested in do *not* have Fourier transforms in the classical sense of definitions 3.40 and 3.41. Consequently, we might expect to get a better theory of Fourier transforms if we work in terms of generalised functions, and this indeed proves to be the case.

The intuitively natural way to define the Fourier transform of a generalised function $\mathbf{f}$ is to define $\hat{\mathbf{f}}$, the Fourier transform of $\mathbf{f}$, by

$$\langle \hat{\mathbf{f}}, \varphi \rangle = \langle \mathbf{f}, \hat{\varphi} \rangle \tag{3.3}$$

for all $\varphi \in \mathfrak{D}$. However, the right-hand side of (3.3) does not make any sense because, in general, $\hat{\varphi}$ is not a test function since the Fourier transform of a function of bounded support (see definition 3.21) is not usually also a function of bounded support. To overcome this difficulty and at the same time retain as much as possible of the methodology we have so far outlined, we could try to introduce a new space of test functions, denoted by $\mathbf{S}(\mathbf{R})$, with the property that Fourier transforms of functions in $\mathbf{S}(\mathbf{R})$ are also in $\mathbf{S}(\mathbf{R})$. We would then be able to define transforms of functionals on $\mathbf{S}(\mathbf{R})$ by the method in (3.3). This means that we would have to develop another version of distribution theory using exactly the same methodology as before but with certain technical modifications that would ensure that Fourier transforms do indeed fit into the theory. With this in mind we introduce the following.

Definition 3.44 *A smooth function $f : \mathbf{R} \to \mathbf{C}$ such that for all n, $r \geq 0$,*

$$x^n \varphi^{(r)}(x) \to 0 \quad as\ |x| \to \infty$$

is called a function of rapid decay.

The set of all functions of rapid decay is denoted by $\mathbf{S}(\mathbf{R})$.

The collection $\mathbf{S}(\mathbf{R})$ can readily be shown to have the following properties.

(i) Every test function is a function of rapid decay; that is, $\mathfrak{D} \subseteq \mathbf{S}(\mathbf{R})$.
(ii) If φ, $\psi \in \mathbf{S}(\mathbf{R})$, then $(a\varphi + b\psi) \in \mathbf{S}(\mathbf{R})$ for all constants a, b.
(iii) If $\varphi \in \mathbf{S}(\mathbf{R})$, then $x^n \varphi^{(r)}(x) \in \mathbf{S}(\mathbf{R})$ for all $n, r \geq 0$.

(iv) If $|x^n\varphi^{(r)}(x)|$ is bounded for each $n, r \geq 0$, then φ is a function of rapid decay.
(v) Every function of rapid decay is absolutely integrable.

We can now prove the result we want.

THEOREM 3.45 *If $\varphi \in \mathbf{S}(\mathbf{R})$, then $\overset{\wedge}{\varphi} \in \mathbf{S}(\mathbf{R})$.*

Proof.

- Property (v) $\Rightarrow \varphi$ has a Fourier transform.
- Properties (iii), (v) $\Rightarrow x^n\varphi(x)$ are absolutely integrable.
- Example 3.42(iii) applied n times shows that $\overset{\wedge}{\varphi}$ is differentiable n times for any n.
- Applying example 3.42(iii) r times, we obtain

$$\begin{aligned}\left|p^n \overset{\wedge}{\varphi}^{(r)}(p)\right| &= \left|p^n \int_{\mathbf{R}} (-ix)^r \varphi(x) e^{-ipx}\, dx\right| \\ &= \left|\int_{\mathbf{R}} x^r \varphi(x) \left(\frac{d}{dx}\right)^n e^{-ipx}\, dx\right| \\ &= \left|\int_{\mathbf{R}} e^{-ipx} \left(\frac{d}{dx}\right)^n (x^r\varphi(x))\, dx\right|,\end{aligned}$$

where we have integrated by parts n times. Hence

$$\left|p^n \overset{\wedge}{\varphi}^{(r)}(p)\right| \leq \int_{\mathbf{R}} \left|\left(\frac{d}{dx}\right)^n (x^r\varphi(x))\right| dx.$$

Properties (iii) and (v) guarantee the convergence of the integral on the right-hand side, and thus we obtain a bound on the left-hand side. It then follows by property (iv) that $\overset{\wedge}{\varphi}$ is a function of rapid decay. □

A notion of convergence in $\mathbf{S}(\mathbf{R})$ is introduced as follows.

DEFINITION 3.46 *If $\varphi, \varphi_1, \varphi_2,$ are functions of rapid decay, then we say that $\varphi_m \to \varphi$ in $\mathbf{S}(\mathbf{R})$ as $m \to \infty$ provided that for all integers r and n we have, uniformly in x,*

$$x^n \varphi_m^{(r)}(x) \to x^n \varphi^{(r)}(x) \quad \text{as } m \to \infty.$$

With this preparation we can now introduce the following.

DEFINITION 3.47 *A distribution of slow growth is a continuous linear functional on the space $\mathbf{S}(\mathbf{R})$. Alternatively, we say that it is a linear functional that maps every convergent sequence in $\mathbf{S}(\mathbf{R})$ into a convergent sequence in $\mathbf{C}$.*

We remark that every distribution of slow growth is a distribution in the sense introduced earlier. The converse, however, is not true.

It is generally true that functions of slow growth generate distributions of slow growth and that functions that do not grow slowly in the sense described above generate distributions that are *not* of slow growth.

Ordinary functions that grow too rapidly at infinity do not belong to the set of distributions of slow growth. This we can express more precisely in the following manner.

DEFINITION 3.48

(i) $f(x) = O(x^n)$ *as* $|x| \to \infty$ *means that there exist numbers* A *and* R *such that* $|f(x)| \leq A\,|x|^n$ *whenever* $|x| > R$.

(ii) A function $f : \mathbf{R} \to \mathbf{C}$ *that is locally integrable and such that* $f(x) = O(x^n)$ *for some* n *as* $|x| \to \infty$ *is called a function of slow growth.*

Example 3.49

(i) Every n*th degree polynomial is* $O(x^n)$.

(ii) e^{-x} *is not a function of slow growth since* $x^{-n}e^{-x} \to \infty$ *as* $|x| \to \infty$ *for any* n.

(iii) e^{iax} *is a function of slow growth if* x *and* a *are real.*

DEFINITION 3.50 *To each locally integrable function of slow growth* f *there corresponds a distribution of slow growth* **f**, *a regular distribution, defined by*

$$\langle \mathbf{f}, \varphi \rangle = \int_{\mathbf{R}} f(x)\varphi(x)\,dx, \qquad \varphi \in \mathbf{S}(\mathbf{R}).$$

We have introduced the above statement as a definition. However, it is frequently presented in the form of a theorem in which it is proved that the above functional has all the properties required to ensure that **f** is a distribution of slow growth.

With these several results and remarks in mind, a theory of distributions of slow growth can now be constructed in a similar manner to that used when dealing with ordinary distributions. We leave the details as an exercise.

The reason for introducing the space **S**(**R**) was to be able to define Fourier transforms of distributions in the same way as already discussed for ordinary functions.

We define the Fourier transform $\hat{\mathbf{f}}$ of a generalised function **f** by

$$\langle \hat{\mathbf{f}}, \varphi \rangle = \langle \mathbf{f}, \hat{\varphi} \rangle \quad \text{for } \varphi \in \mathbf{S}(\mathbf{R}).$$

In order that this definition make sense, we need the following technical result (see [7, 144]).

THEOREM 3.51 *If* **f** *is a distribution of slow growth, then the functional* $\hat{\mathbf{f}} : \varphi \to \langle \mathbf{f}, \hat{\varphi} \rangle$ *is a distribution of slow growth.*

With this result available we can introduce the following.

Definition 3.52

(i) If **f** *is a distribution of slow growth, then its Fourier transform is the distribution of slow growth* $\hat{\mathbf{f}}$ *defined by*

$$\langle \hat{\mathbf{f}}, \varphi \rangle = \langle \mathbf{f}, \hat{\varphi} \rangle \quad \text{for } \varphi \in \mathbf{S}(\mathbf{R}).$$

(ii) If f is a locally integrable function of slow growth, then the distribution $\hat{\mathbf{f}}$ *is called the generalised Fourier transform of f.*

We now have available symbolically the same structure as that used for ordinary functions. Indeed, examples 3.42 and 3.43 will hold with f replaced by **f** appropriately.

We conclude this section by giving examples of the Fourier transforms of some frequently occurring functions.

Example 3.53 *The simplest function is the constant function. Write* **1** *for the distribution generated by the constant function whose value everywhere is 1. Then*

$$\langle \hat{\mathbf{1}}, \varphi \rangle = \langle \mathbf{I}, \hat{\varphi} \rangle = \int_{\mathbf{R}} 1 . \hat{\varphi}(x)\, dx = \int_{\mathbf{R}} \hat{\varphi}(x)\, dx.$$

Now

$$(\hat{\hat{\varphi}})(q) = \frac{1}{\sqrt{2\pi}} \int_{\mathbf{R}} e^{-iqx} . \hat{\varphi}(x)\, dx,$$

which implies

$$(\hat{\hat{\varphi}})(0) = \frac{1}{\sqrt{2\pi}} \int_{\mathbf{R}} \hat{\varphi}(x)\, dx.$$

Using definition 3.43(i),

$$(\hat{\hat{\varphi}})(0) = \varphi(-0) = \varphi(0).$$

Combining these several results, we obtain

$$\langle \hat{\mathbf{I}}, \varphi \rangle = \frac{1}{\sqrt{2\pi}} \int_{\mathbf{R}} \hat{\varphi}(x)\, dx = \frac{1}{\sqrt{2\pi}} (\hat{\hat{\varphi}})(0) = \frac{1}{\sqrt{2\pi}} \varphi(0) = \frac{1}{\sqrt{2\pi}} \int_{\mathbf{R}} \hat{\varphi}(x) \delta(x)\, dx$$

and conclude that

$$\hat{\mathbf{I}} = \sqrt{2\pi}\, \delta.$$

Example 3.54

(i) Write **x** *for the distribution generated by the function f defined by $f(x) = x$. Then*

$$\hat{x} = -\sqrt{2\pi}\, \delta' i$$

(ii)

$$\hat{\boldsymbol{\delta}} = \frac{1}{\sqrt{2\pi}} \mathbf{1}$$

(iii)

$$\delta(x-a)^{\wedge} = \frac{1}{\sqrt{2\pi}} e^{-iax}$$

$$(e^{-iax})^{\wedge} = \sqrt{2\pi}\,\delta(x+a).$$

The details are left as an exercise.

3.6 HILBERT SPACES

In this section we take the first determined steps to generalise the familiar concepts of algebra and geometry that we have been using in Euclidean spaces. In the Euclidean space setting we have always dealt with numbers and the numerical values of functions. We now want to extend these ideas in such a way that we can discuss the functions themselves rather than their numerical values. Of course, in making such an extension we want to ensure that we retain as much as possible of the already familiar methodologies. It turns out that a particular type of normed linear space called a Hilbert space provides an ideal setting for this purpose.

We begin by defining an inner product that is an abstract version of the familiar scalar product of elementary vector algebra. This will allow us to define, just as in finite-dimensional Euclidean spaces, the notion of angles, particularly right angles, in abstract vector spaces. We then go on to introduce the idea of two elements of an (abstract) vector space being perpendicular. This is followed by a discussion of how sets of perpendicular elements of a vector space can form a basis for an infinite-dimensional space in a similar way that the x-, y- and z-axes form a basis for $\mathbf{R}^3$.

We end this section with a brief outline of the salient features of operators on Hilbert spaces. These we will use frequently in the following chapters.

Definition 3.55 *An inner product on a vector space X is a rule that assigns to elements $x, y \in X$ a real or complex number, denoted by (x, y) and called the inner product of $x, y \in X$, that has the properties*

(i) (x, x) is real and positive for all $x \neq \theta$ the zero element in X and $(\theta, \theta) = 0$.

(ii) $(x, y) = \overline{(y, x)}$ for all $x, y \in X$, where the overbar denotes a complex conjugate.

(iii) $(ax, y) = a(x, y)$ for all $x, y \in X$ and any scalar a.

(iv) $(x + y, z) = (x, z) + (y, z)$ for all $x, y, z \in X$.

A vector space X together with an inner product $(.,.)$ is called an inner-product space.

We notice a number of features of this inner product. First, for a real inner-product space the overbar is redundant since all the inner products are real. Next, we see that (ii) and (iv) imply that we also have $(x + y, z) = (x, z) + (y, z)$. The algebraic properties thus appear to be the same as for the scalar product in ordinary vector algebra. However, there is one important difference. In a complex space the inner product is *not* linear in both arguments. It is, in fact, *conjugate linear* in the sense that $(x, ay) = \bar{a}(x, y)$ for any scalar a and $x, y \in X$. This follows from definition 3.55(ii) and (iii)

$$(x, ay) = \overline{a(x, y)} = \bar{a}(x, y).$$

Example 3.56

(i) On $\mathbf{C}^n$*, the set of n-tuples of complex numbers* $x = (x_1, x_2, \ldots, x_n)$*,* $y = (y_1, y_2, \ldots, y_n)$*, where* x_k*,* $y_k \in \mathbf{C}$ *for* $k = 1, 2, \ldots, n$*, we can define an inner product*

$$(x, y) := \sum_{k=1}^{n} x_k \overline{y_k}.$$

For the corresponding real space $\mathbf{R}^n$*, we can define an inner product with the same symbolic form, but now all the quantities are real.*

(ii) On $C[a, b]$*, the set of complex-valued continuous functions defined on the interval* $[a, b] \in \mathbf{R}$*, we can define an inner product of the form*

$$(f, g) = \int_a^b f(x)\overline{g(x)}\, dx, \qquad f, g \in C[a, b].$$

Definition 3.57 *On any inner-product space X we can define a norm by*

$$\|x\|^2 = (x, x).$$

Exercise 3.58

(i) Verify that the inner products introduced in example 3.56 satisfy the inner-product axioms in definition 3.55.

(ii) Verify that the quantity $\|.\|$ *introduced in definition 3.57 is indeed a norm.*

The norm induced by the inner product introduced in example 3.56(ii) is referred to as the *square-integrable norm* since

$$\|f\|^2 = (f, f) = \int_a^b f(x)\overline{f(x)}\, dx = \int_a^b |f(x)|^2\, dx.$$

The set of all continuous functions defined on the interval $[a, b]$ endowed with the square-integrable norm is a normed space (verify). Convergence in this normed space is understood as follows.

Definition 3.59 *Let* $F, f_1, f_2, f_3, \ldots$ *be functions with action either* $\mathbf{R} \to \mathbf{R}$ *or* $\mathbf{R} \to \mathbf{C}$*. We say that* $f_n \to F$ *in the mean on* $[a, b]$ *as* $n \to \infty$ *if*

$$\int_a^b |f_n(x) - F(x)|^2\, dx \to 0.$$

Clearly, this definition only holds for functions that are sufficiently well behaved for the integral to exist.

Convergence in the mean is less demanding than uniform convergence. It is called convergence in the mean because it is the mean value of $(f_n - F)$ that tends to zero and not the value at particular points.

The idea of convergence in the mean is particularly important in mathematical physics. The equations in mathematical physics are often solved by series expansion methods, for example, Fourier series. These expansions do not always converge uniformly, but they generally can be shown to converge in the mean.

Some of the differences between the types of convergence mentioned above are given in the following example.

Example 3.60 *Define functions f_n, $n = 1, 2, \ldots,$ by $f_n(x) = \exp(-nx)$. It is clear that $f_n \in C[0, 1]$. Furthermore, for each $x \in [0, 1]$, we have*

$$f_n(x) \to F(x) \quad as\ n \to \infty, \tag{3.4}$$

where F is defined by

$$F(x) = 0 \quad for\ x \neq 0,$$

$$F(0) = 1.$$

We emphasise that the result (3.4) is a statement about the convergence of a sequence of *numbers* $f_n(x)$ for any $x \in [0, 1]$. It is *not* a statement about the sequence of functions $f_n, n = 1, 2, \ldots$. The statement (3.4) indicates that $f_n \to F$ *pointwise*. It is also the case that $f_n \to F$ in the sense of convergence in the mean. The integral involved is $\int_0^1 \exp(-2nx)\,dx$, which is easily evaluated and can be shown to tend to zero as $n \to \infty$.

In this example $F \notin C[0, 1]$ since it is discontinuous. To ease matters, we may be tempted to consider the function F_0 defined by

$$F_0(x) = 0 \quad \text{for all } x \in [0, 1].$$

Clearly, $F_0 \in C[0, 1]$. We can now ask if $f_n \to F_0$ in the sense of convergence in $C[0, 1]$. Different answers can be obtained depending on the norm used on $C[0, 1]$.

We have seen that the collection $C[0, 1]$ can be turned into a normed space in a number of ways. We consider two cases.

Case 1: $C[0, 1]$ is endowed with the *uniform norm* defined by

$$\|f\| := \sup\{|f(x)| : 0 \leq x \leq 1\}.$$

This is also referred to as the *sup norm*.

Case 2: $C[0, 1]$ is endowed with the *square-integrable norm* defined by

$$\|f\|^2 = \int_0^1 |f(x)|^2\,dx.$$

Using the uniform norm, we conclude that $f \nrightarrow F_0$. This follows because $|f_n(0) - F_0(0)| = 1$ for all n, and hence $\sup|f_n(x) - F_0(x)|$ cannot possibly tend to zero. However, if we use the norm in case 2, then a simple integration indicates that $f_n \to F_0$.

Limits in the mean are not unique. This is because the value of an integral is unaffected by changing the value of the integrand at a number of isolated points.

We see from the above that there are two important types of convergence in *function space*. The first is uniform convergence, which is convergence with respect to the *sup norm* introduced in case 1 above. The second is *convergence in the mean*, which is convergence with respect to the integral norm introduced in case 2.

We have seen above that a sequence of continuous functions can converge in the mean to a discontinuous function. However, such a sequence cannot converge uniformly to a discontinuous function. Yet we notice that any uniformly convergent sequence in $C[a, b]$ can be integrated term by term and as such converges in the mean. Therefore, we see that uniform convergence implies convergence in the mean but that the converse does not hold.

One of the powerful features of applied functional analysis is that it provides a means of introducing and working with a norm that is best suited to the problem at hand, and in much of the analysis in this monograph we shall be working in a Hilbert space rather than a Banach space structure. A particularly important and frequently occurring Hilbert space is $L_2[a, b]$.

Definition 3.61 *$L_2[a, b]$ is the completion of $C[a, b]$ with respect to the square-integrable norm. Thus $L_2[a, b]$ contains all functions that are the limits of continuous functions in the sense of mean convergence.*

Example 3.62 *We have seen that $L_2[a, b]$ contains discontinuous functions. Indeed, it can contain functions that have infinite discontinuities provided that the discontinuity is nice enough for it to be square-integrable. For instance, the function f defined by $f(x) = x^{-1/3}$ is an element of $L_2[0, 1]$. However, the function g defined by $g(x) = x^{-2/3}$ is not. We can conclude that every function f that is such that the integral $\int_a^b |f(x)|^2\, dx$ exists is an element of $L_2[a, b]$.*

There is an important aspect of $L_2[a, b]$ that must always be borne in mind, namely, the space $L_2[a, b]$ contains functions that are *not* integrable according to the theory of Riemann integration.

For example, the function F defined by $F(x) = 1$ for rational values of x and zero for all other values of x is a discontinuous function and is such that $\int_a^b |F(x)|^2\, dx$ does not exist in the Riemann sense. Now, Riemann integration is the "usual" method of integration that we use in practical problems, but this little example shows that Riemann integration is not always adequate. Fortunately, a more powerful integration theory is available, namely, Lebesgue integration theory [63, 114, 144], by means of which we can show that the above integral does exist and hence that $F \in L_2[a, b]$.

In this monograph there is no need to have a detailed knowledge of Lebesgue theory. It will be sufficient simply to know it is available. Consequently, $L_2[a, b]$ can be regarded as a Banach space that contains all ordinary, that is, Riemann, square-integrable, functions together with other functions that are highly discontinuous, just like the function F introduced above, which must be included to make the space complete.

This situation is eased by recalling that $L_2[a, b]$ is the completion of $C[a, b]$ with respect to the square-integrable norm. Thus every element of $L_2[a, b]$ can be approximated arbitrarily well by elements in $C[a, b]$. That is to say, $C[a, b]$ is dense in $L_2[a,b]$. (See definition 3.9.) For this reason we really need only work with elements of $C[a,b]$ initially and then, if necessary, use denseness arguments to obtain more general results.

Returning to the general notion of inner-product spaces, we see that the inner product is an abstract version of the familiar scalar or dot product of finite-dimensional vector algebra. In this connection simply think of the algebra used in n-dimensional Euclidean space $\mathbf{R}^n$.

We have mentioned several times that when analysing problems, particularly with practical applications in mind, it is often more profitable and easier to work with the actual functions involved rather than with the numerical values of such functions. Working in the structure of inner-product spaces provides a means of doing this. In order to complete this introduction of an analytical structure that will enable us to work meaningfully with functions rather than with their numerical values, whilst still retaining the familiar methods used in finite-dimensional Euclidean space, we need the additional ingredient of *orthogonality*. Once this concept has been introduced, it will lead naturally to the meaning of the angle between two functions and of a basis of an infinite-dimensional space. This will then complete the basic mathematical structure we need in order to work in an infinite- rather than a finite-dimensional space setting.

With these several remarks in mind we notice that, should we so wish, we could build up essentially all Euclidean geometry in the context of an inner-product space structure. In support of this statement we first recall the following familiar result [63].

THEOREM 3.63 *(Cauchy-Schwarz inequality) For any complex numbers x_k, y_k, $k=1,2,\ldots,n$,*

$$\left|\sum_{k=1}^{n} x_k \overline{y_k}\right|^2 \leq \left(\sum_{k=1}^{n} |x_k|^2\right)\left(\sum_{k=1}^{n} |y_k|^2\right),$$

where the overbar denotes complex conjugate.

COROLLARY 3.64 *(Cauchy-Schwarz inequality for integrals)*

$$\left|\int_a^b f(x)g(x)\,dx\right|^2 \leq \left(\int_a^b |f(x)|^2\right)\left(\int_a^b |g(x)|^2\right)$$

for any functions f and g that ensure the integrals exist.

An abstract version of these two results is available for general inner-product spaces. To indicate this, we need a little notational preparation.

DEFINITION 3.65 *Let V be a vector space and $S \subset V$ a subset of elements x_1, $x_2,\ldots,x_p$. Consider the relation*

$$\sum_{k=1}^{p} a_k x_k = 0, \tag{3.5}$$

where a_k, $k=1, \ldots, p$, are scalars. If (3.5) only holds for $a_k=0$, $k=1, \ldots, p$, then the elements x_k, $k=1, \ldots, p$, are said to be linearly independent and S is a linearly independent subset of V. The elements x_k, $k=1, \ldots, p$, are said to be linearly dependent if they are not linearly independent. That is, (3.5) will hold for some p-tuple of scalars not all of which are zero. Similarly, S is a linearly dependent subset of V if it is not linearly independent.

DEFINITION 3.66 *A vector space V is said to be finite-dimensional if there is a positive integer n such that V contains a linearly independent set of n elements but any set of $(n+1)$ or more elements of V is linearly dependent. The integer n is the dimension of V and is denoted* $\dim V = n$. *By definition, the vector space $V = \{\theta\}$, where θ is the zero element, is finite-dimensional and* $\dim V = 0$.

If V is not finite-dimensional, then it is infinite-dimensional.

Infinite-dimensional vector spaces are of greater interest than finite-dimensional ones. This is particularly true for practical problems defined in terms of partial differential equations. For instance, $C[a, b]$ is infinite-dimensional, whereas $\mathbf{R}^n$ and $\mathbf{C}^n$ are finite-dimensional.

With this preparation the abstract version of theorem 3.65, suitable for use in inner-product spaces, can be stated in the following form [63].

THEOREM 3.67 *(Schwarz inequality) Let V be an inner-product space.*

(i) $|(x, y)|^2 \le (x, x)(y, y)$ for any $x, y \in V$.

(ii) Equality in (i) only holds if x and y are linearly dependent.

We have already seen that in any inner-product space we can define a norm by

$$\|x\|^2 := (x, x).$$

Consequently, the Schwarz inequality can be written in the form

$$|(x, y)| \le \|x\| \, \|y\| . \tag{3.6}$$

This important result indicates that in a real inner-product space we have

$$-1 \le \frac{(x, y)}{\|x\| \, \|y\|} \le 1.$$

If we compare this result with the expression

$$\cos \varphi = \frac{x . y}{|x| \, |y|},$$

where φ denotes the angle between two vectors x and y in three-dimensional geometry, then we can define an angle between two elements of an inner-product space by

$$\varphi := \cos^{-1} \left\{ \frac{(x, y)}{\|x\| \, \|y\|} \right\}. \tag{3.7}$$

3.6.1 Orthogonality, Bases and Expansions

Previously we defined, in (3.7), the angle between two elements of an inner-product space. A particularly useful notion that comes out of this is that of *orthogonality*.

DEFINITION 3.68
(i) Two elements x, y of an inner-product space X are said to be orthogonal if $(x, y) = 0$.
(ii) A set of elements $\{x_k\}_{k=1}^{\infty} \subset X$ is called orthonormal if

$$(x_k, x_m) = \begin{cases} 1, & k = m, \\ 0, & k \neq m. \end{cases}$$

This definition applies to both real and complex inner-product spaces.

Example 3.69
(i) For $\mathbf{C}^2$ endowed with the inner product

$$(x, y) = \sum_{k=1}^{2} x_k \overline{y_k}, \qquad x, y \in \mathbf{C}^2,$$

the elements $x = [1, i]$ and $y = [1, -i]$ are orthogonal because

$$(x, y) = ([1, i], [1, -i]) = 1.1 + i\overline{(-i)} = 1 + i^2 = 0.$$

(ii) For $C[0, \pi]$ with inner product

$$(f, g) := \int_0^{\pi} f(x)\overline{g(x)}\, dx,$$

the functions defined by $f(x) = \sin mx$ and $g(x) = \sin nx$ are orthogonal for any positive integers m, n with $m \neq n$.

An abstract version of Pythagoras' theorem is as follows.

THEOREM 3.70 *If x and y are orthogonal elements of an inner-product space, then*

$$\|x + y\|^2 = \|x\|^2 + \|y\|^2.$$

Proof.

$$\begin{aligned} \|x + y\|^2 &= (x + y, x + y) = (x, x) + (y, y) + (x, y) + (y, x) \\ &= \|x\|^2 + \|y\|^2. \end{aligned}$$

□

It is perhaps interesting to recall the amount of labour involved in proving this result in elementary geometry classes!

We recall that all vectors in $\mathbf{R}^3$ can be expressed in terms of three unit vectors, one each in the direction of the x-, y- and z-axis, respectively. The three unit vectors are said to form a basis for $\mathbf{R}^3$, and the coefficients attached to each, when expressing an arbitrary vector v in terms of these basis elements, are referred as the *components* of v with respect to the basis. We want to parallel this situation when working in infinite-dimensional spaces. To this end, we introduce the following.

Definition 3.71 *A set of elements $\{x_k\}$ of an inner-product space is called an orthogonal set if*
(i) $(x_j, x_k) = 0$ whenever $j \neq k$,
(ii) for each k, we have $x_k \neq \theta =$ zero element.

Part (ii) of this definition excludes the zero element. The effect of this is indicated by the following.

Lemma 3.72 *A finite orthogonal set is linearly independent.*

Proof. Let $\{x_k\}_{k=1}^n$ be an orthogonal set. We want to show that $\sum_{k=1}^n c_k x_k = 0$ implies that the scalars c_k, $k = 1, 2, 3, \ldots, n$, are all zero.
If $\sum_{k=1}^n c_k x_k = 0$, then for any j,

$$0 = \left(\sum_{k=1}^{n} c_k x_k, x_j \right) = \sum_{k=1}^{n} c_k (x_k, x_j) = c_k \|x_k\|^2 ,$$

the last equality following by orthogonality. Therefore we conclude that $c_k = 0$, $k = 1, 2, 3, \ldots, n$, provided $\|x_k\| \neq 0$. This is ensured by part (ii), and the required result follows. □

This result indicates that any set containing the zero element is a linearly dependent set. We shall rely on this result later when we work with various types of expansions that are particularly useful in applications.

Definition 3.73 *An orthogonal basis for an inner-product space V is an orthogonal set $\{e_n\}$ that is such that for any $x \in V$ there are scalars c_n such that*

$$x = \sum_{k=1}^{\infty} c_k e_k .$$

It should be noted that not every (inner-product) space has a basis. It may be that a space is so large that infinite linear combinations such as those in definition 3.73 do not account for all elements of the space. However, those spaces, and especially inner-product spaces, that occur in applications usually do have a basis. Furthermore, in applications it turns out the spaces involved are usually complete. Consequently, for the remainder of this monograph we shall adopt the following definition.

Definition 3.74 *A complete inner-product space with a basis is a Hilbert space.*

When working in finite-dimensional spaces, a basis is a very useful commodity because instead of manipulating the vectors or elements of the finite-dimensional space, we can manipulate their components. These components are numbers, either real or complex, and as such are often more amenable to computation than the elements themselves. It turns out that much the same is true in infinite-dimensional spaces. Consequently, it is of prime importance to know how bases can be constructed for infinite-dimensional spaces. The following offers a systematic and constructive method for doing this.

THEOREM 3.75 *(Gram-Schmidt orthogonalisation process) Given a sequence $\{f_n\}$ in an inner-product space, there is an orthogonal sequence $\{g_n\}$ such that every finite linear combination of f_n is a finite linear combination of g_n.*

The proof of this theorem is straightforward and constructive. However, it is rather lengthy, and the details can be found in standard texts such as [63]. It is sufficient for us at the moment to know that this process exists.

We now turn our attention to representing elements of a Hilbert space as an infinite series of basis vectors. We shall see later that the ability to do this will enable us to develop a number of powerful, constructive methods for solving problems we encounter in applications. What we are about to do is very similar to developing a classical Fourier series, and it turns out that the required coefficients can be obtained in much the same way that classical Fourier coefficients are determined. To see this, we require the following property of inner products.

THEOREM 3.76 *Let V denote an inner product space with inner product $(.,.)$.*

(i) If $\{x_n\}$ is a sequence in V such that $x_n \to x \in V$ as $n \to \infty$, then $(x_n, y) \to (x, y)$ as $n \to \infty$ for all $y \in V$.

(ii) If $\{u_n\}$ is a sequence in V and if $S := \sum_{k=1}^{\infty} u_k$, then

$$\sum_{k=1}^{\infty} (u_k, y) = (S, y) \quad \text{for any } y \in V.$$

Proof. Using the Schwarz inequality, we obtain

$$|(x_n, y) - (x, y)| = |(x_n - x, y)| \leq \|x_n - x\| \, \|y\|,$$

and the right-hand side tends to zero by hypothesis.

The second part follows immediately by setting $x = S$ and $x_n = \sum_{k=1}^{n} u_k$ in the above. □

This result indicates that for fixed $y \in V$ the inner product is a continuous function of x, and vice versa.

THEOREM 3.77 *Let $\{e_n\}$ be an orthogonal basis for an inner-product space V. Any $x \in V$ can be expanded in the form $x = \sum_{k=1}^{\infty} c_k e_k$, where the coefficients c_k are determined by*

$$c_n = \frac{(x, e_n)}{\|e_n\|^2}. \tag{3.8}$$

Proof. Using the definition of basis, we see that there are scalars (numbers) c_n such that for any $x \in V$ we have a result of the form $x = \sum_{k=1}^{\infty} c_k e_k$. Consequently, using theorem 3.76, we have that for all n,

$$(x, e_n) = \sum_{k=1}^{\infty} c_k (e_k, e_n).$$

All terms on the right-hand side of this expression vanish except when $k = n$, and hence (3.8) follows immediately. □

The importance of this last result is that it offers a generalisation of the usual Fourier series expansion to inner-product spaces.

Definition 3.78 *Let $\{e_n\}$ be an orthogonal set in an inner-product space V. For any $x \in V$, the series $\sum_{k=1}^{\infty} c_k e_k$ is called a generalised Fourier series and the numbers c_n defined in (3.8) are called the generalised Fourier coefficients or expansion coefficients of x with respect to the set $\{e_n\}$.*

In the following chapters we will devote a considerable amount of effort towards obtaining these generalised Fourier expansions. Once established, these generalised Fourier expansions will enable us to decompose a given hard problem into a number of simpler problems. This aspect we shall deal with in detail in the chapter on spectral theory. Another benefit of having available a generalised Fourier expansion is that it provides a means of developing constructive approximation procedures for use in practical applications. To see this, notice first that the coefficients (3.8) have the property that $\sum_{k=1}^{N} c_k e_k$ can be made as close as we wish to x simply by taking N large enough. However, in practical numerical computations we do not have the luxury of having the ability to work with an infinite number of elements and consider what happens as $N \to \infty$. Instead, we are constrained to working with a finite number of elements. Consequently, we are faced with having to answer questions of the following form. Given a finite set of elements $e_1, e_2, \ldots, e_N$ of a Hilbert space, what linear combination of these elements is the best approximation to a given element x in the Hilbert space? The answer will be an expression of the form

$$\sum_{k=1}^{N} c_{kN} e_k. \tag{3.9}$$

In (3.9) the coefficient c_{kN} depends on N, hence the double subscripts. This means that if we use the coefficients c_{kN} when using the finite set $e_1, e_2, \ldots, e_N$, then we can expect to have to change *all* the coefficients to get the best approximation to x if we work with a finite set of the form $e_1, e_2, \ldots, e_N, e_{N+1}$. However, it turns out that when the elements $e_1, e_2, \ldots, e_N$ are pairwise-orthogonal, the coefficients in (3.9) are independent of N and are just the coefficients used in (3.8). For this reason we will always make determined efforts in applications to work with orthogonal sets and the associated orthogonal expansions.

This remarkable feature is encapsulated in the following theorem.

Theorem 3.79 *Let $\{e_k\}_{k=1}^{N}$ be an orthogonal set in an inner-product space V. For any $x \in V$, the coefficients c_k that minimise $\|x - \sum_{k=1}^{N} c_k e_k\|$ are given by (3.8).*

Proof. Set $c_k = (x, e_k)\ \|e_k\|^2 + d_k$ and expand $\|x - \sum_{k=1}^{N} c_k e_k\|$ as an inner-product. Direct calculation of the inner-product terms then clearly indicates that the required minimum is obtained when $d_k \equiv 0$ for all k. □

In this overview of the notion of a basis we have been using the intuitively obvious concept of the countability of a set. For the sake of completeness we make this more precise as follows.

DEFINITION 3.80

(i) A set is said to be countably infinite if it can be put in a one-to-one correspondence with the set of all positive integers.

(ii) A set is said to be countable if it is either finite or countably infinite. That is, the elements of the set can be put into one-to-one correspondence with either a finite set of integers or the set of all integers.

This definition simply means that we can label uniquely each element in the set with an integer. For more details see the texts listed in the bibliography, in particular [63, 114].

In many books dealing with Hilbert space theory, the following result is proved.

THEOREM 3.81 *An inner-product space has a basis if and only if it contains a countable dense set.*

We remark that in applications this result is often taken as axiomatic.

A natural question to ask about the coefficients in the generalised Fourier expansion of an element of an inner-product space concerns their behaviour as $k \to \infty$. It turns out that they tend to zero as indicated by the following theorem.

THEOREM 3.82 *(Bessel's inequality) Let $\{e_k\}$ be an orthonormal set in an inner-product space V. For any $x \in V$,*

$$\sum_{k=1}^{\infty} |c_k|,^2 \leq \|x\|,$$

where $c_k = (x, e_k)$, $k = 1, 2, \ldots$.

This result leads to the following criterion for deciding whether or not a given orthonormal sequence is a basis.

THEOREM 3.83 *(Parseval's relation) Let $\{e_k\}$ be an orthonormal sequence in an inner-product space V. This set is a basis for V if and only if for each $x \in V$,*

$$\sum_{k=1}^{\infty} |c_k|^2 = \|x\|,$$

where $c_k = (x, e_k)$, $k = 1, 2, \ldots$, are the expansion coefficients for $x \in V$ with respect to the set $\{e_k\}$.

The proofs of these last two results can be found in any of the texts on functional analysis listed in the bibliography.

An important result, which will lead the way to the spectral theorem and associated decomposition results in a Hilbert space setting, is the following.

THEOREM 3.84 *(Riesz-Fischer theorem) Let $\{e_k\}$ be an orthonormal basis for an infinite-dimensional Hilbert space H.*

If $\{c_k\}$ is a sequence of numbers with the property that the series $\sum_{k=1}^{\infty} |c_k|^2$ is convergent, then there exists an $x \in H$ such that $x = \sum_{k=1}^{\infty} c_k e_k$ with $c_k = (x, e_k)$.

REMARK 3.85 *In the above, H could be a real or a complex Hilbert space. Correspondingly, the numbers c_k could be either real or complex.*

Again the proof of this theorem is to be found in standard texts on functional analysis [63].

We now provide some additional geometric aspects of abstract Hilbert spaces. We begin with two frequently used notational features.

DEFINITION 3.86
(i) An inner-product space is called a pre-Hilbert space.
(ii) A complete inner-product space is called a Hilbert space.

DEFINITION 3.87
(i) A subset G of a Hilbert space H is called a linear manifold (or simply a manifold) if it is invariant with respect to linear operations; that is, if λ_1, λ_2 are scalars, then $\lambda_1 g_1 + \lambda_2 g_2 \in G$ whenever $g_1, g_2 \in G$.
(ii) A closed linear manifold is called as subspace of H.

(*Warning:* Always check the definition of "subspace" being used by an author; they do vary!)

If G_1, G_2 are two manifolds in a Hilbert space H, then the subset of H, denoted $G_1 + G_2$, consisting of all elements $g \in H$ of the form $g = g_1 + g_2$ with $g_1 \in G_1$, $g_2 \in G_2$ is called the *sum* of G_1 and G_2.

The subset $G_1 \cap G_2 \subset H$ consisting of elements of H that are simultaneously elements of both G_1 and G_2 is a manifold called the *intersection* of $G_1$and G_2.

DEFINITION 3.88 *Let G_1, G_2 be two manifolds in a Hilbert space H. The direct sum of G_1 and G_2, denoted $G_1 \oplus G_2$, is defined by*

$$G_1 \oplus G_2 = (G_1 + G_2 : G_1 \cap G_2 = \{\theta\}),$$

where θ denotes the (unique) zero element in H.

DEFINITION 3.89
(i) Let H_k, $k = 1, 2$, be Hilbert spaces with inner products $(.,.)_k$, $k = 1, 2$, respectively. The set of ordered pairs

$$\langle x, y \rangle, \qquad x \in H_1, \quad y \in H_2,$$

is a Hilbert space H with linear operations and inner product defined by

$$\lambda_1 \langle x_1, y_1 \rangle + \lambda_2 \langle x_2, y_2 \rangle = \langle \lambda_1 x_1 + \lambda_2 x_2, \lambda_1 y_1 + \lambda_2 y_2 \rangle$$

$$(\langle x_1, y_1 \rangle, \langle x_2, y_2 \rangle) = (x_1, x_2)_1 + (y_1, y_2)_2,$$

where $x_1, x_2 \in H_1$, $y_1, y_2 \in H_2$ and λ_1, λ_2 are scalars, either real or complex.

The Hilbert space H is called the direct product of H_1 and H_2 and is denoted $H := H_1 \times H_2$.

(ii) Let H_k, $k = 1, 2, \ldots$, be a sequence of Hilbert spaces with structure $(.,.)_k$, $\|.\|_k$, $k = 1, 2, \ldots$, respectively. Let H denote the set of sequences $\{x_k\}_{k=1}^{\infty}$, $x_k \in H_k$,

$k = 1, 2, \ldots$, *that satisfy* $\sum_{k=1}^{\infty} \|x_k\|_k^2 < \infty$. *The set* H *is a Hilbert space with inner product defined by*

$$(x, y) = \sum_{k=1}^{\infty} (x_k, y_k)_k$$

with $x, y \in H$ *and* $x_k, y_k \in H_k$, $k = 1, 2, \ldots$. *We write*

$$H = \mathop{\times}_{k=1}^{\infty} H_k.$$

One of the main reasons why Hilbert spaces are easier to handle than general Banach spaces is because they provide a very clear and easily workable extension of the familiar algebra and geometry of finite-dimensional spaces to infinite-dimensional spaces. In applications we very much want this extension. More truthfully perhaps, we want to know that such an extension is always available. This means that for much of the time in applications we can work as though we are making the analysis in finite-dimensional spaces. A limiting process would provide the more general results.

Let G be a subspace (or just a manifold) of a Hilbert space H. The elements of H that are orthogonal to G, that is, orthogonal to all elements of G, constitute a subspace of H called the *orthogonal complement* of G in H that is denoted $G^{\perp}$. The dimension of $G^{\perp}$ is called the *codimension* of G, and we write codimension $G = \dim G^{\perp}$.

One of the most powerful results of Hilbert space theory for use in applications is that a Hilbert space H has an orthogonal decomposition into the orthogonal direct sum of subspaces G and $G^{\perp}$. This feature is encapsulated in the following celebrated theorem.

THEOREM 3.90 *(projection theorem) Let*

(i) H *be a Hilbert space with structure* $(.,.)$, $\|.\|$,

(ii) M *be a subspace of* H,

(iii) $M^{\perp}$ *be the orthogonal complement of* M *in* H.

Then every element $x \in H$ *can be written uniquely in form*

$$x = y + z, \qquad y \in M, \quad z \in M^{\perp}.$$

This theorem can be written compactly in the form

$$H = M \oplus M^{\perp} = \{y + z : y \in M, z \in M^{\perp}\}.$$

Furthermore, direct calculation shows that

$$\|x\|^2 = \|y\|^2 + \|z\|^2.$$

The decomposition theorem 3.90 is the bedrock for much of the analysis in this monograph and indeed for the majority of practical applications. We shall see that it provides a means for decomposing not only Hilbert spaces but also various operations that are performed on them. This will often enable quite difficult problems to be broken down into simpler and more easily manageable components.

3.6.2 Linear Functionals and Operators on Hilbert Spaces

A *mapping* is a generalisation to vector spaces of the notion of a function (see definition 3.10).

Definition 3.91 *Let V_1, V_2 be vector spaces and let X_1, X_2 be subsets of V_1, V_2, respectively. A mapping T with action denoted by $T : X_1 \to X_2$ is a rule that, given any $x \in X_1$, associates with it an element of X_2 denoted Tx.*

We shall write $x \to Tx$ to denote that $x \in X_1$ is mapped into the element $Tx \in X_2$.

We now restrict our attention to Hilbert spaces. Let H be a Hilbert space with structure $(.,.)$, $\|.\|$.

A mapping f of a manifold $D(f) \subset H$ into a manifold $R(f) \subset \mathbf{K} = \mathbf{R}$ or $\mathbf{C}$ is called a *functional* on H, and we write

$$f : H \supset D(f) \to R(f) \subset \mathbf{K}.$$

This notation indicates that f can be either a real- or a complex-valued function on H. We shall make much use later of this type of observation.

This mapping will be a *linear functional* if

$$f(\lambda_1 h_1 + \lambda_2 h_2) = \lambda_1 f(h_1) + \lambda_2 f(h_2)$$

for any $h_1, h_2 \in D(f)$ and scalars λ_1, λ_2. The manifold $D(f)$ is called the *domain* (*of definition*) of the mapping f, whilst $R(f)$ is the *range* of the mapping f.

A more general type of mapping is an *operator.*

A mapping L of a manifold $D(L) \subset H$ onto a manifold $R(L)$ with action denoted by

$$L : H \supset D(L) \to R(L) \subset H$$

is called a *linear operator* in H if

$$L(\lambda_1 h_1 + \lambda_2 h_2) = \lambda_1 L(h_1) + \lambda_2 L(h_2)$$

for any $h_1, h_2 \in D(L)$ and scalars λ_1, λ_2. Similarly to the above, $D(L)$ and $R(L)$ are called, respectively, the *domain* and *range* of the *operator* L.

A linear operator $L : H \supset D(L) \to R(L) \subset H$ is said to be *densely defined* on H if $D(L)$ is dense in H, that is, if $\overline{D(L)} = H$.

Two of the simplest but quite important linear operators on H are

1. the *trivial operator* Θ, which is such that

$$\Theta h = \theta = \text{zero element in } H \quad \text{for all } h \in H,$$

2. the identity operator I, which is such that

$$Ih = h \in H \quad \text{for all } h \in H.$$

Given a linear operator L defined in a Hilbert space H, all solutions of the equation $Lh=\theta \in H$ form a manifold $N(L)\subset H$ known as the *null space* or *kernel* of the operator L.

When $N(L)=\{\theta\}$, the equation $Lh=\theta \in H$ has only the trivial solution $h=\theta$. In this case it is possible to define an *inverse operator* L^{-1}, which is such that [63, 114]

(i) $D(L^{-1})=R(L)$ and $R(L^{-1})=D(L)$,
(ii) $L^{-1}g=h$, where $g\in D(L^{-1})=R(L)$ and $h\in H$ is the unique solution of $Lh=g$,
(iii) the property (ii) implies

$$L^{-1}(Lh)=h\in D(L),$$
$$L(L^{-1}g)=g\in D(L^{-1}).$$

An important feature of operators on a Hilbert space is provided by the following definition, which offers a natural extension to infinite-dimensional spaces of familiar aspects of geometry in $\mathbf{R}^3$.

DEFINITION 3.92 *Let H be a Hilbert space and let $L: H\to H$ be a linear operator. The set of all elements $\langle h, Lh\rangle \in H\times H$ is denoted $\Gamma(L)$ and is called the graph of L.*

A linear manifold Γ in $H\times H$ is the graph of some linear operator L if and only if it does not contain elements of the form $\langle\theta, g\rangle$, $g\neq\theta$.

Two linear operators L_1, L_2 on a Hilbert space H are said to be equal if $D(L_1)$ $=D(L_2)=: D\subset H$ and $L_1h=L_2h$ for any $h\in D$. Equivalently, the operators are equal if $\Gamma(L_1)=\Gamma(L_2)$.

REMARK 3.93 *We would strongly emphasise that it should always be borne in mind that the operators L_1, L_2 with $D(L_1)\neq D(L_2)$ are different operators even if $L_1h=L_2h$ holds for any $h\in D(L_1)\cap D(L_2)$.*

An operator L' is said to be an *extension* of an operator L (alternatively, L is a *restriction* of L') if $D(L)\subset D(L')$ and $Lh=L'h$ for all $h\in D(L)$. Hence an extension L' of L is *any* operator that agrees with L when applied to elements of $D(L)$ but is arbitrary elsewhere. This arbitrariness is removed in practice by requiring that the extension have certain properties of the original operator, for instance, continuity. In the particular case when $D(L)$ is dense in H, it can be shown that the extension is unique.

Let L be a linear operator in a Hilbert space H and let $G_1\subset H$ be a subspace. Let $G_2=G_1^{\perp}$ be the orthogonal complement of G_1 in H. Further, set

$$D_1=G_1\cap D(L),$$
$$D_2=G_2\cap D(L)=G_1^{\perp}\cap D(L).$$

The subspace G_1 is called a *reducing subspace* of L if

$$D(L)=D_1\oplus D_2, \qquad LD_1\subset G_1, \qquad LD_2\subset G_2.$$

It is clear that $G_2=G_1^{\perp}$ is a reducing subspace (see theorem 3.90).

We shall denote the restriction of L to G_1 and G_2 by L_1 and L_2, respectively, and refer to L_1 and L_2 as the *parts* (*components*) of L. It is clear that $D(L)=D_1 \oplus D_2$ and that

$$Lh = L_1 g_1 + L_2 g_2, \qquad h \in D(L),$$

where g_k, $k=1,2$, is the projection of h onto the subspace G_k, $k=1,2$. Consequently, $R(L)=R(L_1)\oplus R(L_2)$. We shall refer to the operator L as the *direct sum* of its parts L_1, L_2, and we write $L=L_1 \oplus L_2$.

It is clear from the above remarks that a study of the operator L is equivalent to a study of its components L_1 and L_2.

The decomposition of an operator into more parts than two can be similarly defined.

The above remarks provide the first indications of how a given operator can be decomposed into several more manageable parts. We shall return to this aspect in detail when we deal with the topic of spectral theory.

Let

$$L_k : H \supset D(L_k) \to R(L_k) \subset H, \qquad k=1,2,$$

be linear operators. Bearing in mind that usually these operators are not defined on all of H, we define their sum L_1+L_2 and their product $L_1 L_2$ as follows.

$$\begin{aligned} &D(L_1+L_2) = D(L_1)\cap D(L_2),\\ &(L_1+L_2)h = L_1 h + L_2 h, \qquad h \in D(L_1+L_2), \end{aligned} \tag{3.10}$$

and

$$\begin{aligned} &D(L_1 L_2) = \{h \in D(L_2) : L_2 h \in D(L_1)\},\\ &(L_1 L_2)h = L_1(L_2 h), \qquad h \in D(L_1 L_2). \end{aligned} \tag{3.11}$$

We remark that $D(L_1+L_2) = D(L_1 L_2) = D(L_2)$ when $D(L_1)=H$. Also, in general, $L_1 L_2 \neq L_2 L_1$; that is, L_1 and L_2 do not commute.

A linear *functional* f on a Hilbert space H is said to be *bounded* on H if

$$\|f\| := \sup\left\{\frac{|f(h)|}{\|h\|} : h \in D(f)\subset H,\ h \neq \theta\right\} < \infty, \tag{3.12}$$

where $\|f\|$ is the norm of the functional f.

Similarly, an *operator* L on H is said to be *bounded* on H if

$$\|L\| := \sup\left\{\frac{\|Lh\|}{\|h\|} : h \in D(L)\subset H, h \neq \theta\right\} < \infty, \tag{3.13}$$

where $\|L\|$ is the norm or, more fully, the *operator norm* of the operator L.

We remark that the operator L has a bounded inverse L^{-1} if and only if

$$\inf\left\{\frac{\|Lh\|}{\|h\|} : h \in D(L)\subset H, h \neq \theta\right\} > 0 \tag{3.14}$$

and

$$\|L^{-1}\| = \left\{ \inf \left\{ \frac{\|Lh\|}{\|h\|} : h \in D(L) \subset H, h \neq \theta \right\} \right\}^{-1}. \tag{3.15}$$

It is possible to describe the general form of linear functionals on a Hilbert space. Before doing this, we need to introduce the following two results.

Theorem 3.94 *If a linear functional is continuous at any one point, then it is uniformly continuous.*

Proof. Let H be a Hilbert space. If $f : H \to \mathbf{K} = \mathbf{R}$ or $\mathbf{C}$ is continuous at $g \in H$, then for any $\varepsilon > 0$ there is a $\delta > 0$ such that $|f(g+h) - f(g)| < \varepsilon$, $h \in H$ whenever $\|h\| < \delta$. It then follows from the linearity of f that for any $x \in H$,

$$|f(x+h) - f(x)| = |f(h)| = |f(g+h) - f(g)| < \varepsilon$$

whenever $\|h\| < \delta$. Since δ is independent of x, the required uniformity follows. □

This result indicates that a linear functional is either discontinuous everywhere or uniformly continuous everywhere. Hence we shall simply call a uniformly continuous functional a continuous functional.

Continuity and boundedness are essentially the same property for linear functionals but not for functionals in general. Indeed, the following result can be obtained [63, 114].

Theorem 3.95 *A linear functional is continuous if and only if it is bounded.*

We now turn to a description of the general form of bounded linear functionals defined either on a whole Hilbert space H or on a dense manifold D of H.

Consider the function F_h defined by

$$F_h(g) = (g, h), \qquad g \in H, \tag{3.16}$$

where $h \in H$ is given. This functional is linear by virtue of the properties of an inner product and also continuous for the same reason. Alternatively, we can use the Schwarz inequality to show that it is bounded with

$$\|F_h\| = \sup \left\{ \frac{|F_h(g)|}{\|g\|} : g \neq \theta \right\} \leq \|h\|.$$

Substituting $g = h$ in this inequality, we can conclude that $\|F_h\| = \|h\|$.

A general result in this connection is the following [63, 114].

Theorem 3.96 *(Riesz representation theorem) If F is a bounded linear functional defined on either all of a Hilbert space H or a dense manifold $D(F) \subset H$, then there exists a unique vector $h \in H$ such that*

$$F = F_h = (., h) \qquad \textit{with } \|F\| = \|F_h\| = \|h\|.$$

(An alternative, slightly more transparent, statement of this is as follows. For every continuous linear functional F on a Hilbert space H, there is a unique $h \in H$ such that $F(g) = (g, h)$ for all $g \in H$.)

A proof of this important result can be found in the standard texts cited in the bibliography.

The set of linear bounded functionals on H is denoted by H^* and is referred to as the *dual* of H (see definition 3.26). We define linear operations and an inner product on this set in the following manner.

DEFINITION 3.97 *(structure of H^*)*
(i) $\lambda_1 F_{h_1} + \lambda_2 F_{h_2} = F_{h_3},\ h_3 := \overline{\lambda_1} h_1 + \overline{\lambda_2} h_2,$
(ii) $(F_{h_1}, F_{h_2}) = (h_1, h_2),$
where $\lambda_k \in \mathbf{K}$ *and* $h_k \in H,\ k = 1, 2.$

As a consequence of definition 3.97, we consider H^* a Hilbert space. To see that this makes sense, we keep in mind (3.10). Then recognising that an inner product is conjugate-linear we have

$$\begin{aligned}(\lambda_1 F_{h_1} + \lambda_2 F_{h_2})(g) &= \lambda_1 F_{h_1}(g) + \lambda_2 F_{h_2}(g)\\ &= \lambda_1 (g, h_1) + \lambda_2 (g, h_2)\\ &= (g, \overline{\lambda_1} h_1) + (g, \overline{\lambda_2} h_2)\\ &= F_{h_3}(g), \qquad h_3 = \overline{\lambda_1} h_1 + \overline{\lambda_2} h_2.\end{aligned}$$

Furthermore, the Riesz representation theorem indicates that there is a one-to-one mapping of H onto H^* such that $h \to F_h$. This mapping is isometric because $\|F_h\| = \|h\|$. Finally, it is conjugate-linear because, from the above,

$$\lambda_1 h_1 + \lambda_2 h_2 \to \overline{\lambda_1} F_{h_1} + \overline{\lambda_2} F_{h_2}.$$

These several remarks indicate that to regard H^* as a Hilbert space does indeed make sense.

We now turn our attention to the corresponding features of bounded linear operators defined on the whole Hilbert space H. These play a particularly important part in much of the analysis in later chapters.

We denote by $B(H)$ the set of all linear bounded operators on H that map H into itself. This set is invariant with respect to linear operations; that is, for scalars $\lambda_k \in \mathbf{K}$, $k = 1, 2$, we have $(\lambda_1 L_1 + \lambda_2 L_2) \in B(H)$ whenever $L_1, L_2 \in B(H)$. Consequently, $B(H)$ can be considered a linear space. We also notice that $B(H)$ is invariant with respect to products of operators; that is, $L_1 L_2 \in B(H)$ whenever $L_1, L_2 \in B(H)$.

Much of the analysis that follows will involve approximation processes. In particular, we could become involved with sequences of operators. Consequently, we need an understanding of the convergence processes that are available.

In Hilbert spaces there are two principal notions of convergence.

DEFINITION 3.98 *Let H be a Hilbert space with structure $(.,.)$, $\|.\|$.*
(i) A sequence $\{x_k\}_{k=1}^{\infty} \subset H$ is said to be strongly convergent to an element $x \in H$ if

$$\lim_{k \to \infty} \|x_k - x\| = 0,$$

in which case we write

$$s-\lim_{k\to\infty} x_k = x.$$

(ii) A sequence $\{x_k\}_{k=1}^{\infty} \subset H$ *is said to be weakly convergent to an element* $x \in H$ *if, for all* $y \in H$,

$$\lim_{k\to\infty} (x_k, y) = (x, y),$$

in which case we write

$$w-\lim_{k\to\infty} x_k = x.$$

Bounded linear operators acting in Hilbert spaces will be of particular interest to us in later chapters. Consequently, it will be convenient to introduce the following convergence concepts in $B(H)$.

Definition 3.99 *Let* H *be a Hilbert space and let* $\{T_k\}_{k=1}^{\infty} \subset B(H)$.

(i) The sequence $\{T_k\}_{k=1}^{\infty}$ *is strongly convergent to* T *if, for all* $h \in H$,

$$\lim_{k\to\infty} \|T_k h - Th\| = 0,$$

and we write

$$s-\lim_{k\to\infty} T_k = T.$$

(ii) The sequence $\{T_k\}_{k=1}^{\infty}$ *is weakly convergent to* T *if* $\{T_k h\}_{k=1}^{\infty}$ *is weakly convergent to* Th *for all* $h \in H$, *and we write*

$$w-\lim_{k\to\infty} T_k = T.$$

(iii) The sequence $\{T_k\}_{k=1}^{\infty}$ *is uniformly convergent to* T *if*

$$\lim_{k\to\infty} \|T_k - T\| = 0,$$

and we write

$$u-\lim_{k\to\infty} T_k = T.$$

We remark that uniform convergence is sometimes called convergence in norm.

The following results will be useful later.

Theorem 3.100 *Let* H *be a Hilbert space and let* $T_k, T, S_k, S \in B(H)$, $k = 1, 2, \ldots$.

(i) If $s-\lim_{k\to\infty} T_k = T$ *and* $s-\lim_{k\to\infty} S_k = S$, *then* $s-\lim_{k\to\infty} T_k S_k = TS$.

(ii) If $u-\lim_{k\to\infty} T_k = T$ *and* $u-\lim_{k\to\infty} S_k = S$, *then* $u-\lim_{k\to\infty} T_k S_k = TS$.

For a proof of this theorem see the references cited in chapter 11, in particular [63, 102, 114]. We remark that a similar result does *not* hold for weak convergence.

3.6.3 Some Frequently Occurring Operators

In this subsection we gather together the salient features of some linear operators that we will often meet in the following chapters.

Definition 3.101 *A bounded linear operator T on a Hilbert space H is invertible if there exists a bounded linear operator T^{-1} on H such that*

$$TT^{-1} = T^{-1}T = I,$$

where I is the identity operator on H.

We remark that in the above definition T is *on* H; that is, $D(T) = H$. When $D(T) \subset H$, that is, when T is *in* H, the definition will have to be modified. We will return to this aspect when it is needed later in some applications.

The following result can be established [63, 102, 114]

. Indeed, it is often used as the definition of an inverse operator.

Theorem 3.102 *An invertible operator $T \in B(H)$ is one-to-one and maps H onto H. The inverse of T is unique.*

The next result, which provides a generalisation of the geometric series $(1 - a)^{-1}$, is fundamental to much of the material in later chapters.

Theorem 3.103 *(Neumann series) Let H be a Hilbert space and let $T \in B(H)$ have the property $\|T\| < 1$. Then*

(i) $(I - T)$ is invertible,

(ii) $(I - T)^{-1} \in B(H)$,

(iii) $(I - T)^{-1} = \sum_{k=0}^{\infty} T_k$ (Neumann series) is uniformly convergent,

(iv) $\|(I - T)^{-1}\| \leq (1 - \|T\|)^{-1}$.

The proof of this result is entirely straightforward but rather lengthy [63, 102].

The existence of an inverse operator is of fundamental importance when solving operator equations. Such operators are not always easy either to determine or to work with. However, associated with a given operator on a Hilbert space is an operator that has some of the flavour of an inverse operator, namely, an *adjoint operator*.

Let H be a Hilbert space and let $T \in B(H)$. Then, for $h \in H$, given the functional (Tg, h), $g \in H$, is linear and bounded. Consequently, by the Riesz representation theorem, there is a unique element h^* such that

$$(Tg, h) = (g, h^*).$$

The mapping $h \to h^*$ is defined on all of H and is readily seen to be linear and bounded. Consequently, we write

$$T^*h = h^* \quad \text{for any } h \in H,$$

so that

$$(Tg, h) = (g, T^*h), \qquad g, h \in H.$$

These various observations can be written compactly in the following form (see chapter 11).

THEOREM 3.104 *Let H be a Hilbert space with structure $(.,.)$ and let $T \in B(H)$. There exists a unique linear bounded operator T^* on H, called the adjoint of T, defined by*

$$(Tg, h) = (g, T^*h) \quad \text{for all } g, h \in H,$$

which is such that

$$\|T^*\| = \|T\|.$$

Some elementary properties of T^* are contained in the following.

THEOREM 3.105 *Let H be a Hilbert space and let $T, S \in B(H)$.*

(i) $T^{**} := (T^*)^* = T$.

(ii) $(\lambda T)^* = \bar{\lambda} T^*$, $\lambda \in \mathbf{C}$.

(iii) $(T + S)^* = T^* + S^*$.

(iv) $(TS)^* = S^* T^*$.

(v) If T is invertible, then so also is T^*, and $(T^*)^{-1} = (T^{-1})^*$.

(vi) $\|TT^*\| = \|T^*T\| = \|T\|^2$.

We emphasise, as before, that these several results have to be modified when $D(T) \neq H$.

DEFINITION 3.106 *Let H be a Hilbert space and let $T \in B(H)$.*

(i) If $T = T^*$, then T is self-adjoint or Hermitian.

(ii) If $T^* = T^{-1}$, then T is unitary.

(iii) If $TT^* = T^*T$, then T is normal.

(iv) If T is unitary and $R, S \in B(H)$ are such that $R = TST^{-1}$, then R and S are unitarily equivalent with respect to T.

(v) T is self-adjoint if and only if $(Tf, g) = (f, Tg)$ for all $f, g \in H$.

(vi) T is unitary if and only if T is invertible and $(Tf, g) = (f, T^{-1}g)$ for all f, $g \in H$.

REMARK 3.107

(i) It is instructive to examine the compatibility of (v) and (vi) above with the various results mentioned above in this subsection.

(ii) If T is self-adjoint, then (Tf, f) is real.

THEOREM 3.108 *Let H be a Hilbert space with structure $(.,.)$. If $T \in B(H)$ is self-adjoint, then $\|T\|$, the norm of T, can be determined by*

$$\|T\| = \sup_{\|f\|=1} |(Tf, f)| = \max\{|m|, |M|\},$$

where

$$M := \sup\{(Tf, f) : |f| = 1\},$$

$$m := \inf\{(Tf, f) : |f| = 1\},$$

and $f \in H$.

The numbers M and m are, respectively, the upper and lower bounds of the operator T.

A means of comparing the "size" of various operators on H is afforded by the following.

Definition 3.109 *Let H be a Hilbert space with structure $(.,.)$ and let $T \in B(H)$. If the operator T is self-adjoint, then it is said to be non-negative, and we write $T \geqslant 0$ if and only if, for all $f \in H$, we have $(Tf, f) \geqslant 0$. When the inequality is strict, T is said to be positive.*

It now follows that

(i) T is non-negative if $m \geqslant 0$,

(ii) operators T_1, T_2 are such that $T_1 \leq T_2$ if and only if $(T_2 - T_1) \geqslant 0$.

An important class of bounded operators are the *finite-dimensional* operators T that map H onto $R(T)$, where $\dim R(T) < \infty$. A finite-dimensional operator, or an *operator of finite rank* as it is sometimes called, can be expressed in the form

$$Th = \sum_{k=1}^{N} a_k(h, f_k) g_k, \tag{3.17}$$

where $\dim R(T) = N$. The real positive numbers a_k are ordered in the form $a_1 \geqslant a_2 \geqslant \cdots \geqslant a_N > 0$, and the sets of vectors $\{f_k\}, \{g_k\} \subset H$ are orthonormal.

In the general theory of operators a significant role is played by *compact* operators whose properties are similar to those of finite-dimensional operators.

Definition 3.110 *An operator $T \in B(H)$ is compact if it is the limit of a uniformly convergent sequence $\{T_k\}$ of operators of finite rank, that is, if $\|T - T_k\| \to 0$ as $k \to \infty$.*

An equivalent definition is as follows. An operator $T \in B(H)$ is compact if for every bounded sequence $\{f_k\}_{k=1}^{\infty} \subset H$ the sequence $\{Tf_k\}_{k=1}^{\infty}$ has a strongly convergent subsequence.

Example 3.111

(i) If T_1 is a compact operator and T_2 is a bounded operator on the same Hilbert space H, then T_1T_2 and T_2T_1 are also compact operators.

(ii) If T_k, λ_k, $k = 1, 2, \ldots N$, are, respectively, compact linear operators and complex-valued coefficients, then

$$T = \sum_{k=1}^{N} \lambda_k T_k$$

is a compact operator.

(iii) Operators T, T^*, TT^*, T^*T are simultaneously compact or noncompact.

Compact operators have a relatively simple structure similar to that of operators of finite rank. Specifically, a compact operator T can be written in the form

$$Th = \sum_{k=1}^{\infty} a_k(h, f_k) g_k, \tag{3.18}$$

where $\{a_k\}$ is a nonincreasing sequence of real positive numbers that tends to zero; that is, $a_k \to 0$ as $k \to \infty$. The numbers a_k are called the singular numbers of the operator T in (3.18). The sets $\{f_k\}, \{g_k\}$ are orthonormal.

If in (3.18) we have $a_k = 0$ for sufficiently large $k \geqslant N$, then T is a finite-dimensional operator as in (3.17); that is, T has a decomposition of the form (3.17).

If T is self-adjoint, then the sets $\{f_k\}$ and $\{g_k\}$ can be chosen to satisfy, for all k, either $f_k = g_k$ or $f_k = -g_k$.

An operator $T \in B(H)$ can also be usefully decomposed in the form

$$T = T_R + iT_I, \tag{3.19}$$

where

$$T_R := \frac{1}{2}(T + T^*) = T_R^*,$$

$$T_I := \frac{1}{2i}(T - T^*) = T_I^*.$$

This decomposition is similar to the decomposition of complex numbers.

One of the simplest self-adjoint operators is the *projection operator*, which is defined in the following way.

Let H be a Hilbert space with structure $(.,.)$ and let M be a subspace. The projection theorem (theorem 3.90) states that every element $f \in H = M \oplus M^{\perp}$ can be expressed *uniquely* in the form

$$f = g + h, \qquad g \in M, \qquad h \in M^{\perp}. \tag{3.20}$$

The uniqueness of this representation allows us to introduce a linear operator $P : H \to H$ defined by

$$Pf = g \in M, \qquad f \in H. \tag{3.21}$$

Such an operator is called a *projection operator* or a *projector*, as it provides a projection of H onto M.

We notice that (3.20) can also be written in the form

$$f = g + h = Pf + (I - Pf), \tag{3.22}$$

which indicates that

$$(I - P) : H \to M^{\perp}$$

is a projection of H onto $M^{\perp}$.

(*Note*: When it is important to emphasise the subspace involved in a decomposition of the form (3.22), we will write $P = P(M)$.)

Projection operators play a very important part in much of the analysis that is to follow. For convenience we collect together here their main properties. The proof of these various properties is a standard part of courses on the general theory of linear operators on Hilbert spaces; chapter 11 indicates a number of sources for the interested reader, in particular [4, 53].

THEOREM 3.112 *Let H be a Hilbert space with structure $(.,.)$.*

(i) A bounded linear operator $P : H \to H$ is a projection on H if and only if

$$P = P^2 = P^*.$$

An operator P such that $P = P^2$ is called idempotent.

(ii) Let $M \subset H$ be a subspace (closed linear manifold). The projection $P : H \to M$ has the following properties.

(a) $(Pf, f) = \|Pf\|^2$.

(b) $P \geqslant 0$.

(c) $\|P\| \leq 1$, $\|P\| = 1$ if $R(P) \neq \theta$.

(d) $(I - P) =: P^{\perp} : H \to M^{\perp}$ is a projection.

THEOREM 3.113 *Let H be a Hilbert space with structure $(.,.)$ and let $M_k \subset H$, $k = 1, 2$, be subspaces in H. The projections*

$$P(M_k) : H \to M_k, \qquad k = 1, 2,$$

have the following properties.

(i) The operator $P := P(M_1)P(M_2)$ is a projection on H if and only if

$$P(M_1)P(M_2) = P(M_2)P(M_1).$$

When this is the case, then

$$P : H \to R(P) = M_1 \cap M_2.$$

(ii) M_1, M_2 are orthogonal if and only if

$$P(M_1)P(M_2) = 0,$$

in which case $P(M_1)$ and $P(M_2)$ are said to be mutually orthogonal projections.

(iii) The operator $P := P(M_1) + P(M_2)$ is a projection on H if and only if M_1 and M_2 are orthogonal. In this case

$$P : H \to R(P) = M_1 \oplus M_2;$$

that is, $P(M_1) + P(M_2) = P(M_1 \oplus M_2)$.

(iv) The projections $P(M_1)$, $P(M_2)$ are partially ordered in the sense that the following statements are equivalent.

(a) $P(M_1)P(M_2) = P(M_2)P(M_1) = P(M_1)$.

(b) $M_1 \subseteq M_2$.

(c) $N(P(M_2)) \subseteq N(P(M_1))$, where $N(P(M_k))$ denotes the null space of $P(M_k)$, $k = 1, 2$.

(d) $\|P(M_1)f\| \leq \|P(M_2)f\|$ for all $f \in H$.

(e) $P(M_1) \leq P(M_2)$.

(v) A projection $P(M_2)$ is part of a projection $P(M_1)$ if and only if $M_2 \subset M_1$.

(vi) The difference $P(M_1) - P(M_2)$ *of two projections* $P(M_1)$, $P(M_2)$ *is a projection if and only if* $P(M_2)$ *is part of* $P(M_1)$. *When this is the case,*

$$P := (P(M_1) - P(M_2)) : H \to R(P) = M_1 \cap M_2^{\perp};$$

that is, $(P(M_1) - P(M_2)) = P(M_1 \cap M_2^{\perp})$.

(vii) A series of mutually orthogonal projections $P(M_k)$, $k = 1, 2, \ldots,$ *on* H, *denoted by* $\sum_k P(M_k)$, *is strongly convergent to the projection* $P(M)$, *where* $M = \oplus_k M_k$.

(viii) A linear combination of projections $P(M_k)$, $k = 1, 2, \ldots, N$, *on* H, *denoted by* $P := \sum_{k=1}^{N} \lambda_k P(M_k)$, *where* λ_k *are real-valued coefficients, is self-adjoint on* H.

(ix) Let $\{P(M_k)\}_{k=1}^{\infty}$ *denote a monotonically increasing sequence of projections* $P(M_k)$, $k = 1, 2, \ldots,$ *on* H. *Then*

(a) $\{P(M_k)\}_{k=1}^{\infty}$ *is strongly convergent to a projection* P *on* H; *that is,* $P(M_k) f \to Pf$ *as* $k \to \infty$ *for all* $f \in H$,

(b) $P : H \to R(P) = \overline{\cup_{k=1}^{\infty} R(P(M_k))}$,

(c) $N(P) = \cap_{k=1}^{\infty} N(P(M_k))$.

We remark that there exists an intimate connection between general self-adjoint operators and projections. This we will discuss later under the heading of spectral theory. As a consequence, we will be able to establish the various decomposition results and expansion theorem we have already mentioned.

In subsequent chapters we will quite often encounter the following particular type of compact operator.

Definition 3.114 *Let* H *be a Hilbert space with structure* $(.,.)$. *An operator* $T \in B(H)$ *is called a Hilbert-Schmidt operator if*

$$\|T\|^2 = \sum_{k=1}^{\infty} \|Te_k\|^2 < \infty,$$

where $\{e_k\}_{k=1}^{\infty}$ *is an orthonormal basis for* H.

If T *is a compact operator, then it is a compact Hilbert-Schmidt operator if its singular numbers* $a_k(T)$ *tend to zero sufficiently rapidly so that the series* $\sum_{k=1}^{\infty} a_k^2(T)$ *is convergent; that is,* $\sum_{k=1}^{\infty} a_k^2(T) < \infty$.

It is clear from this definition that any finite-dimensional operator is a compact Hilbert-Schmidt operator.

Some of the best known classes of Hilbert-Schmidt operators are integral operators. It will be convenient for later use to introduce these operators as operators on weighted L_2-spaces.

Definition 3.115 *The Hilbert space of functions square-integrable on the whole space* $\mathbf{R}^N$ *with respect to some weight function* ρ *is denoted* $L_{2,\rho}(\mathbf{R}^N)$. *Here* ρ *is a real-valued positive function in each open bounded region* $\Omega \subset \mathbf{R}^N$. *The structure*

$(.,.)_\rho$, $\|.\|_\rho$ is defined by

$$(f_1, f_2)_\rho := \int_{\mathbf{R}^N} f_1(x)\overline{f_2(x)}\rho(x)\,dx, \qquad f_1, f_2 \in C_0^\infty(\mathbf{R}^N),$$

$$\|f\|_\rho^2 := \int_{\mathbf{R}^N} |f(x)|^2\,\rho(x)\,dx, \qquad f \in C_0^\infty(\mathbf{R}^N).$$

The space $L_{2,\rho}(\mathbf{R}^N)$ is defined as the completion of $C_0^\infty(\mathbf{R}^N)$ with respect to this structure.

The space $L_{2,\rho}(\mathbf{\Omega})$, $\Omega \subset \mathbf{R}^N$ is defined similarly.

We define an *integral operator* T on $L_{2,\rho}(\mathbf{\Omega})$ according to

$$(Tf)(y) := \int_\Omega K(x, y) f(x)\rho(x)\,dx, \tag{3.23}$$

where K is called the *kernel* of T. The kernel is square-integrable on $\Omega \times \Omega$ in the sense

$$\int_\Omega \int_\Omega |K(x, y)|^2\,\rho(x)\rho(y)\,dx\,dy = k_0^2 < \infty. \tag{3.24}$$

The operator T defined in (3.23) is clearly linear and bounded. Further, its norm can be estimated according to $\|T\| \le k_0$. Moreover, it can be shown that T in (3.23) is a compact Hilbert-Schmidt operator.

The operator T^*, which is adjoint to T in (3.23), is defined by

$$(T^* f)(x) := \int_\Omega \overline{K(x, y)} f(y)\rho(y)\,dy. \tag{3.25}$$

The integral operator T in (3.23) is self-adjoint if and only if the kernel is symmetric in the sense that $K(x, y) = \overline{K(y, x)}$.

An integral operator is finite-dimensional if and only if its kernel $K(x, y)$ can be expressed in the form

$$K(x, y) = \sum_{k=1}^N g_k(x) f_k(y),$$

where f_k, g_k, $k = 1, 2, \dots, N$, are square-integrable functions. Kernels of this type are called *degenerate*. For an integral operator T with a degenerate kernel K, it can be shown that $\dim(R(T)) \le N$.

3.6.4 Unbounded Linear Operators on Hilbert Spaces

Not all linear operators are bounded. For instance, the differentiation operator defined on the space of continuous functions is not bounded; simply consider its action on x^n. However, it turns out that practically all the operators we encounter in applications are *closed operators*, and they retain much of the flavour of the continuity property displayed by bounded linear operators.

DEFINITION 3.116 *Let X_1, X_2 be normed linear spaces and let $T : X_1 \to X_2$ be a linear operator with domain $D(T) \subset X_1$. If*
(i) $x_k \in D(T)$ for all k,
(ii) $x_k \to x$ in X_1,
(iii) $Tx_k \to y$ in X_2,
when taken together imply $x \in D(T)$ and $Tx = y$, then T is said to be a closed operator.

We would emphasise that in definition 3.116 we require the *simultaneous* convergence of the sequences $\{x_k\}_{k=1}^{\infty}$ and $\{Tx_k\}_{k=1}^{\infty}$.

An alternative definition of a closed operator can be profitably given in terms of the graph of an operator.

DEFINITION 3.117 *Let X_1, X_2 be normed linear spaces with norms $\|.\|_1$, $\|.\|_2$, respectively. A linear operator*

$$T : X_1 \supseteq D(T) \to X_2$$

is a closed linear operator if and only if its graph $G(T)$,

$$G(T) := \{(x_1, x_2) \in X_1 \times X_2 : x_1 \in D(T),\ x_2 = Tx_1\},$$

is a closed subset of $X_1 \times X_2$. We recall that $X_1 \times X_2$ is a normed linear space where the structure is defined in the usual componentwise manner. For example, the norm is

$$\|(x_1, x_2)\|_{X_1 \times X_2} := \|x_1\|_1 + \|x_2\|_2 .$$

In applications it frequently turns out that a closed operator is in fact a bounded operator. To establish whether or not this is the case, we need some if not all of the following standard results (see [63, 102, 114] and chapter 11).

THEOREM 3.118 *(uniform boundedness theorem) Let I be an index set and let*
(i) X, Y be Banach spaces,
(ii) $\{T_\alpha\}_{\alpha \in I}$ be such that,

$$\sup_{\alpha \in I} \{\|T_\alpha x\|_Y\} < \infty \quad \textit{for all } x \in X.$$

Then $\sup_{\alpha \in I}\{\|T_\alpha\|\} < \infty$. That is, if $\{T_\alpha x\}_{\alpha \in I}$ is a bounded set in Y for all $x \in X$, then $\{T_\alpha\}$ is bounded in $B(X, Y)$. We recall that $B(X, Y)$ denotes the class of all bounded linear operators with action $X \to Y$.

THEOREM 3.119 *(bounded inverse theorem) Let X, Y be Banach spaces and let $T \in B(X, Y)$ be one-to-one and onto; then T^{-1} exists as a bounded operator.*

THEOREM 3.120 *(closed graph theorem) Let X, Y be Banach spaces and let $T : X \supset D(T) \to Y$ be a closed linear operator. If $D(T)$ is a closed set, then T is bounded.*

Example 3.121 *We shall show that the process of differentiation can be realised as a closed linear operator on the space of continuous functions. This will be in a form rather more general than we will require below, as we are principally concerned with working in Hilbert spaces. Nevertheless, this example will serve as a prototype.*

Let $X = Y = C[a, b]$, $-\infty < a < b < \infty$, be endowed with the usual supremum norm that for convenience and brevity we sometimes denote by $\|.\|_\infty$. Define, for example,

$$T : X \supset D(T) \to Y$$

$$(Tg)(x) = \frac{dg(x)}{dx} \equiv g'(x), \qquad g \in D(T), \qquad x \in [a, b],$$

$$D(T) := \{g \in X : g' \in X,\ g(a) = 0\}.$$

Let $\{g_k\} \subset D(T)$ be a sequence with the properties that as $k \to \infty$ we have $g_k \to g$ and $Tg_k \to h$ with respect to $\|.\|_\infty$. We can establish that T is closed if we can show that $g \in D(T)$ and that $Tg = h$. To this end, consider

$$f(x) = \int_a^x h(t)\,dt, \qquad x \in [a, b]. \tag{3.26}$$

The convergence $Tg_k \to h$ with respect to $\|.\|_\infty$ implies that the convergence is uniform with respect to x. Since $Tg_k \in C[a, b]$ for all k, it follows that $h \in C[a, b]$. Consequently, by the fundamental theorem of calculus, we deduce that f in (3.26) is continuous and differentiable with $f'(x) = h(x)$ for all $x \in [a, b]$. Furthermore, the properties of the Riemann integral in (3.26) indicate that $f(a) = 0$. Collecting these results, we can conclude that $f \in D(T)$.

Since $g_k \in D(T)$, the fundamental theorem of calculus implies

$$g_k(x) = \int_a^x g_k'(t)\,dt,$$

and we have

$$\begin{aligned} |g_k(x) - f(x)| &= \left| \int_a^x (g_k'(t) - h(t))\,dt \right| \\ &\le \|g_k' - h\|_\infty (b - a). \end{aligned}$$

Now the right-hand side of this expression tends to zero by virtue of the convergence $Tg_k \to h$, and it follows that $g_k \to f \in D(T)$. Furthermore, we will also have that $g' = f' = h$. Hence T, as defined, is a closed operator.

If an operator T is not closed, then it is sometimes possible to associate with T an operator that is closed. This parallels the process of associating with a set M in a metric space a closed set $\overline{M}$ called the closure of M.

DEFINITION 3.122 *A linear operator T is closable if whenever*
(i) $f_n \in D(T)$,
(ii) $f_n \to \theta$ as $n \to \infty$,
(iii) Tf_n tends to a limit as $n \to \infty$,
then $Tf_n \to \theta$ as $n \to \infty$.

If T is a closable operator defined on a normed linear space X, then we can define an extension of T (see the comments following definition 3.92), denoted $\overline{T}$ and called the *closure* of T, in the following manner.

1. Define

$$D(\overline{T}) := \{f \in X : \exists\{f_k\} \subset D(T) \text{ with } f_k \to f \text{ and } \{Tf_k\} \text{ a Cauchy sequence}\}.$$

Equivalently, we define $D(\overline{T})$ to be the closure of $D(T)$ with respect to the graph norm

$$\|f\|_G^2 = \|f\|_X^2 + \|Tf\|_X^2.$$

2. For $f \in D(\overline{T})$, set

$$\overline{T}f = \lim_{k\to\infty} Tf_k,$$

where $\{f_k\}$ is defined as in (i).

There are many connections between closed, bounded and inverse operators. The following theorem draws together a number of results that will be used frequently in later sections. Proofs of these results can be found in the texts cited in chapter 11; we would particularly mention [63, 102, 105, 114].

THEOREM 3.123 *Let X be a Banach space. An operator $T \subset B(X)$ is closed if and only if $D(T)$ is closed.*

THEOREM 3.124 *Let*
(i) X_1, X_2 be normed linear spaces,
(ii) $T := X_1 \supset D(T) \to X_2$ be a linear one-to-one operator.
Then T is closed if and only if the operator

$$T^{-1} := X_2 \supset R(T) \to X_1$$

is closed.

THEOREM 3.125 *Let*
(i) X_1, X_2 be normed linear spaces,
(ii) $T := X_1 \supset D(T) \to X_2$ be a linear operator.
(a) If there exists a constant $m \geqslant 0$ such that

$$\|Tf\|_2 \geqslant m\,\|f\|_1, \qquad f \in D(T),$$

then T is a one-to-one operator. Furthermore, T is closed if and only if $R(T)$ is closed in X_2.

(b) Let T be one-to-one and closed. Then $T \in B(X_1, X_2)$ if and only if $R(T)$ is dense in X_2 and there exists a constant $m > 0$ such that

$$\|Tf\|_2 \geqslant m \|f\|_1, \qquad f \in D(T).$$

(c) If T is closed, then

$$N(T) := \{f \in D(T) : Tf = \theta\}$$

is a closed subset of X_1.

3.6.5 Some Remarks Concerning Unbounded Operators

Many of the operators we encounter in applications are unbounded. Consequently, in this subsection we highlight this aspect and indicate a number of important results that we will have to recognise when dealing with such operators.

We begin by recalling definition 3.92 and emphasising its implications with some notation. We have seen that two operators T_1, T_2 are said to be equal, denoted $T_1 = T_2$, if $D(T_1) = D(T_2)$ and $T_1 f = T_2 f$ for all $f \in D(T_1) = D(T_2)$.

The *restriction* of an operator $T : X \supset D(T) \to Y$ to a subset $M \subset D(T)$ is denoted by $T|_M$ and is the operator defined by

$$T|_M : M \to Y, \qquad T|_M f = Tf \quad \text{for all } f \in M. \tag{3.27}$$

An *extension* of T to a set $S \supset D(T)$ is an operator $\tilde{T} : S \to Y$ such that

$$\tilde{T}\Big|_{D(T)} = T. \tag{3.28}$$

This last definition indicates that $\tilde{T} f = Tf$ for all $f \in D(T)$. Hence T is a restriction of $\tilde{T}$ to $D(T)$.

In this monograph we say that T is an operator *on* a Hilbert space H if its domain $D(T)$ is all of H. We say that T is an operator *in* H if $D(T)$ is not necessarily all of H.

Furthermore, if T_1 and T_2 are two linear operators, then we shall use the notation

$$T_1 \subset T_2 \tag{3.29}$$

to denote that $T_1 (T_2)$ is a restriction (extension) of $T_2 (T_1)$.

A particularly useful feature of a bounded linear operator T on a Hilbert space H is that it is associated with a bounded linear operator T^*, the adjoint of T, defined according to the equation (see theorem 3.104)

$$(Tf, g) = (f, T^* g) \quad \text{for all } f, g \in H,$$

where $(.,.)$ denotes the inner product on H. The proof of the existence of such an adjoint operator makes use of the Riesz representation theorem (theorem 3.96), and the proof breaks down when T is either unbounded or is not defined on all of H.

However, even when the proof fails, it may happen that for some element $g \in H$ there is an element $g^* \in H$ such that

$$(Tf, g) = (f, g^*) \quad \text{for all } f \in D(T). \tag{3.30}$$

If for some fixed $g \in H$ there is only one $g^* \in H$ such that (3.30) holds, then we can write

$$g^* = T^* g \tag{3.31}$$

and consider the operator T^* as being well defined for at least this $g \in H$. However, it remains to determine the conditions under which (3.30) yields a unique $g^* \in H$. Results in this direction are as follows.

Theorem 3.126 *Let*
(i) T be a linear operator on a Hilbert space H,
(ii) there exist elements $g, g^* \in H$ such that

$$(Tf, g) = (f, g^*) \quad \textit{for all } f \in D(T).$$

Then g^* is uniquely determined by g and (3.30) if and only if $D(T)$ is dense in H.

Theorem 3.127 *Let*
(i) T be a linear operator in a Hilbert space H with $\overline{D(T)} = H$,
(ii) $D(T^*) = \{g \in H : \text{there exists } g^* \in H \text{ satisfying (3.30)}\}$.
Then $D(T^*)$ is a subspace of H, and the operator T^* with domain $D(T^*)$ and defined by

$$T^* g = g^* \quad \textit{for all } g \in D(T)$$

is a linear operator.

These two results lead naturally to the following.

Definition 3.128 *Let T be a linear operator in a Hilbert space with $\overline{D(T)} = H$. The operator T^* with domain $D(T^*)$ defined as in theorem 3.127 is called the adjoint of T.*

When T is a bounded operator, this definition of an adjoint operator coincides with that given in theorem 3.104. However, we emphasise that in the general case when the operator may be unbounded, more attention must be given to the role and the importance of domains of operators. The following results illustrate this aspect.

Theorem 3.129 *Let T, S be linear operators in a Hilbert space H.*
(i) If $T \subset S$ and $\overline{D(T)} = H$ (which, incidentally, implies that $\overline{D(S)} = H$), then $T^* \supset S^*$.
(ii) If $\overline{D(T)} = \overline{D(T^*)} = H$, then $T \subset T^{**}$.
(iii) If T is one-to-one and such that $\overline{D(T)} = \overline{D(T^{-1})} = H$, then T^* is one-to-one and $(T^*)^{-1} = (T^{-1})^*$.

The following types of operators are important in applications.

Definition 3.130 *Let T be a linear operator in a Hilbert space H and assume $\overline{D(T)} = H$.*
(i) If $T = T^*$, then T is self-adjoint (in H).
(ii) If $T \subset T^*$, then T is symmetric (in H).

Some properties of these operators are indicated in the following.

Theorem 3.131 *Let T be a linear operator in a Hilbert space H.*
(i) If T is one-to-one and self-adjoint, then $\overline{D(T^{-1})} = H$ and T^{-1} is self-adjoint.
(ii) If T is defined everywhere on H, then T^* is bounded.
(iii) If T is self-adjoint and defined everywhere on H, then T is bounded.
(iv) If $\overline{D(T)} = H$, then T^* is closed.
(v) Every self-adjoint operator is closed.
(vi) If T is closable, then $(\overline{T})^* = T^*$ and $\overline{T} = T^{**}$.
(vii) If T is symmetric, then it is closable.

For proofs of these various results see the references cited in chapter 11 and in particular [105, 114].

Self-adjoint operators play a particularly important role in the following chapters. Here we give an indication how to decide whether or not a given operator has this property. First, we need the following notion.

Definition 3.132 *Let T be a linear operator in a Hilbert space H with $\overline{D(T)} = H$. The operator T is said to be essentially self-adjoint if $\overline{T} = T^*$.*

We notice that if T is essentially self-adjoint, then necessarily it must be closed since T^* is a closed operator (theorem 3.131(iv)). Furthermore, from theorem 3.31(vi), we conclude that $T^* = (\overline{T})^*$. Therefore T is essentially self-adjoint if and only if $\overline{T} = (\overline{T}\ )^*$, that is, if and only if $\overline{T}$ is self-adjoint.

From definition 3.130 we see that if T is symmetric, then it is a restriction of its adjoint T^*. However, we would emphasise that a symmetric operator need not be self-adjoint. It may well be that a symmetric operator *might* have a self-adjoint extension. Nevertheless, we must always remember that the extension is *not* unique; the extension process is usually followed in order to preserve, or even to provide, such properties as, for example, linearity, boundedness, self-adjointness and so on. We shall be particularly interested in the case when a symmetric operator has exactly one self-adjoint extension. In this connection we have the following results [105].

Theorem 3.133 *Let T be a linear symmetric operator in a Hilbert space H. Then*
(i) T is closable,
(ii) $R(\overline{T} \pm I) = \overline{R(T \pm I)}$, where I is the identity on H.

Theorem 3.134 *A symmetric operator $T : H \to H$ is self-adjoint if and only if*

$$R(T + iI) = H = R(T - iI).$$

Finally in this subsection, we give two examples of operators that are unbounded. These are the operators of multiplication by the independent variable and the differentiation operators. These operators frequently occur in mathematical physics, and perhaps the quickest way to see this is to recall that when integral transform methods are used, differentiation can be replaced by multiplication by the independent variable (see example 3.42).

Example 3.135 *Let $H := L_2(-\infty, \infty) = L_2(\mathbf{R})$ and let M denote the "rule" of multiplying by the independent variable, that is, $(Mf)(x) = xf(x)$. Consider the operator T defined by*

$$\begin{aligned} &T : H \supset D(T) \to H,\\ &Tf = Mf, \qquad f \in D(T),\\ &D(T) := \{f \in H : Mf \in H\}. \end{aligned}$$

Let

$$f_n(x) = \begin{cases} 1, & n \le x < n+1,\\ 0, & \text{elsewhere.} \end{cases}$$

Then clearly, $\|f_n\| = 1$, where $\|.\|$ denotes the usual norm in $L_2(\mathbf{R})$. Furthermore we have

$$\|Tf_n\|^2 = \int_{\mathbf{R}} |xf(x)|^2\, dx = \int_n^{n+1} x^2\, dx > n^2.$$

The two results imply that

$$\frac{\|Tf_n\|}{\|f_n\|} > n.$$

Consequently, since we can choose n as large as we please, it follows that T is unbounded on H.

Example 3.136 *Let $H = L_2(\mathbf{R})$ and consider the operator T defined by*

$$\begin{aligned} &T : H \supset D(T) \to H,\\ &Tf = i\frac{df}{dx} = if', \qquad f \in D(T),\\ &D(T) := \{f \in H : if' \in H\}. \end{aligned}$$

Further, let T be an extension of the operator T_0 defined to be

$$T_0 = T|_S,$$

where $S = D(T) \cap L_2[0,1]$ and $L_2[0,1]$ is a subspace of $L_2(\mathbf{R})$. Consequently, if T_0 is unbounded, then so is T. To show that T_0 is indeed unbounded, consider the sequence $\{f_n\}$ defined by

$$f_n(x) = \begin{cases} 1 - nx, & 0 \le x \le \frac{1}{n},\\ 0, & \frac{1}{n} < x \le 1. \end{cases}$$

The derivative of this function is

$$f_n'(x) = \begin{cases} -n, & 0 < x < \frac{1}{n}, \\ 0, & \frac{1}{n} < x < 1. \end{cases}$$

Straightforward calculation yields

$$\| f_n \|^2 = \int_0^1 |f_n(x)|^2 \, dx = \frac{1}{3n}$$

and

$$\| T_0 f_n \|^2 = \int_0^1 \left| f_n'(x) \right|^2 \, dx = n.$$

Hence

$$\frac{\| T_0 f_n \|}{\| f_n \|} = n\sqrt{3} > n.$$

It follows that T_0, and hence T, is unbounded on H.

We shall return to these examples and derive further properties of these two operators in later chapters.

Chapter Four

Spectral Theory and Spectral Decompositions

4.1 INTRODUCTION

Spectral theory provides mechanisms for decomposing quite complicated problems into a number of simpler problems with properties that are more manageable. We shall illustrate how this can be done in the following sections. The account is motivated by considering a typical abstract problem firstly when the underlying space is finite-dimensional and then when the space is infinite-dimensional. In both cases we will work mainly in a Hilbert space setting and assume that the operator that characterises the problem is self-adjoint. When working in an infinite-dimensional setting, we shall also require that the operator be compact on the Hilbert space. The discussion of these two cases leads quite naturally to a statement of the celebrated spectral theorem. We shall prove this theorem for bounded self-adjoint operators. The proof for more general operators is discussed in chapter 11.

In the final section of this chapter we indicate how spectral theory contributes to the definition and understanding of scattering theories.

4.2 BASIC CONCEPTS

In this chapter we will be concerned with abstract equations having the typical form

$$(A - \lambda I)u = f. \tag{4.1}$$

To be more precise, let X be a complex normed linear space and let

$$A : X \supseteq D(A) \rightarrow X$$

be a linear operator. In (4.1) I is the identity operator on X and $\lambda \in \mathbf{C}$. We assume that $f \in X$ is a given data element. The aim is to solve (4.1) for the unknown quantity $u \in X$.

Solutions of (4.1) can be written in the form

$$u = (A - \lambda I)^{-1} f =: R_\lambda f, \tag{4.2}$$

where $R_\lambda \equiv R_\lambda(A) = (A - \lambda I)^{-1}$ is known as the *resolvent* (operator) of A. Quite how useful the representation (4.2) may be depends crucially on the nature of the resolvent R_λ. This observation leads naturally to the following notions.

DEFINITION 4.1 *A regular value of A is a complex number λ such that*
(i) $R_\lambda(A)$ exists,
(ii) $R_\lambda(A)$ is bounded,
(iii) $R_\lambda(A)$ is defined on a dense subset of X.
The set of all regular values of A, denoted $\rho(A)$, is called the resolvent set of A.

DEFINITION 4.2 *The spectrum of A, denoted $\sigma(A)$, is the complement in the complex plane of the resolvent set $\rho(A)$; that is,*

$$\sigma(A) = \mathbf{C} \backslash \rho(A).$$

The spectrum of A is partitioned by the following disjoint sets.

(i) The point spectrum of A, denoted $\sigma_p(A)$, consists of all those $\lambda \in \mathbf{C}$ such that $R_\lambda(A)$ does not exist.

(ii) The continuous spectrum of A, denoted $\sigma_c(A)$, consists of all those $\lambda \in \mathbf{C}$ such that $R_\lambda(A)$ exists as an unbounded operator and is defined on a dense subset of X.

(iii) The residual spectrum of A, denoted $\sigma_r(A)$, consists of all those $\lambda \in \mathbf{C}$ such that $R_\lambda(A)$ exists as either a bounded or an unbounded operator but in either case is not defined on a dense subset of X.

The spectrum of A is the union of these three disjoint sets,

$$\sigma(A) = \sigma_p(A) \cup \sigma_c(A) \cup \sigma_R(A),$$

and any $\lambda \in \sigma(A)$ is referred to as a spectral value of A.

Before continuing, we recall the following properties of linear operators on Banach spaces. The proofs of these various results can be found in the standard texts cited in chapter 11.

THEOREM 4.3 *Let X, Y be Banach spaces and let $A : X \supseteq D(A) \to R(A) \subseteq Y$ denote a linear operator. Then*

(i) The inverse operator $A^{-1} : R(A) \to D(A)$ exists if and only if $Au = \theta_Y$ implies $u = \theta_X$, where θ_X, θ_Y are the zero elements in X and Y, respectively.

(ii) If A^{-1} exists, then it is a linear operator.

(iii) If $\dim D(A) = n < \infty$ and A^{-1} exists, then $\dim R(A) = \dim D(A)$.

THEOREM 4.4 *If a Banach space X is finite-dimensional, then every linear operator on X is bounded.*

These last two results combine to indicate that in the finite-dimensional case

$$\sigma_c(A) = \sigma_r(A) = \emptyset.$$

We thus see that the spectrum of a linear operator on a finite-dimensional space consists only of the point spectrum. In this case the operator is said to have a *pure point spectrum.*

The next few results are particularly useful in applications.

THEOREM 4.5 *The resolvent set $\rho(A)$ of a bounded linear operator on a Banach space X is an open set. Hence the spectrum $\sigma_c(A)$ is a closed set.*

THEOREM 4.6 *Let X be a Banach space and let A be a bounded linear operator on X. For all $\lambda_0 \in \rho(A)$ the resolvent operator $R_\lambda(A)$ has the representation*

$$R_\lambda(A) = \sum_{k=0}^{\infty} (\lambda - \lambda_0)^k R_{\lambda_0}^{k+1}(A). \tag{4.3}$$

The series is absolutely convergent for every λ in the open disc given by

$$|\lambda - \lambda_0| < \| R_{\lambda_0}(A) \|^{-1}$$

in the complex plane. This disc is a subset of $\rho(A)$.

THEOREM 4.7 *The spectrum $\sigma(A)$ of a bounded linear operator $A : X \to X$ on a Banach space X is compact and lies in the disc given by $|\lambda| \leq \|A\|$. Hence the resolvent set of A, $\rho(A)$, is not empty.*

DEFINITION 4.8 *Let X be a Banach space and let $A : X \supseteq D(A) \to X$ be a linear operator. If $(A - \lambda I)u = \theta$ for some nontrivial $u \in D(A)$, then u is an eigenvector of A with associated eigenvalue λ.*

For the remainder of this chapter we confine our attention to linear operators on a complex separable Hilbert space H.

The set M_λ consisting of the zero element in H and all eigenvectors of A corresponding to the eigenvalue λ is called the *eigenspace* of A corresponding to the eigenvalue λ.

The eigenspace M_λ is in fact a subspace, that is, a closed linear manifold of H. That it is a linear manifold is clear since for any $u_1, u_2 \in M_\lambda$ and $\alpha, \beta \in \mathbf{C}$ we have, by the linearity of A,

$$A(\alpha u_1 + \beta u_2) = \lambda(\alpha u_1 + \beta u_2).$$

To show that M_λ is a *closed* linear manifold, let $\{u_k\} \subset M_\lambda$ be a sequence such that $u_k \to u$ as $k \to \infty$. Now consider two cases.

Case 1: A is a bounded operator. A is bounded implies that A is continuous. Hence

$$Au = A \lim u_k = \lim Au_k = \lim \lambda u_k = \lambda u$$

and thus $u \in M_\lambda$, and we conclude that M_λ is closed.

Case 2: A is an unbounded closed operator. In this case, since $u_k \to u$ as $k \to \infty$, there will exist a w such that

$$w := \lim Au_k = \lim \lambda u_k = \lambda u.$$

However, A is a closed operator, which implies that $u \in D(A)$ and $Au = w$. Hence $u \in M_\lambda$, and we can conclude that M_λ is closed. Hence M_λ is a subspace of H.

In this monograph we will be largely concerned with linear operators on a Hilbert space that are either self-adjoint or unitary. Some of the more important properties of such operators are contained in the following.

Theorem 4.9 *The eigenvalues of a self-adjoint operator are real.*

Proof. Let $A : H \to H$ be a bounded self-adjoint operator and let $\lambda \in \sigma_p(A)$ with associated eigenvector u. Then

$$\lambda(u, u) = (\lambda u, u) = (Au, u) = (u, Au) = \overline{\lambda}(u, u),$$

which, because u is nontrivial, implies $\lambda = \overline{\lambda}$, and hence $\lambda \in \mathbf{R}$, and so $\sigma_p(A) \subset \mathbf{R}$. □

Theorem 4.10 *The eigenvalues of a unitary operator are complex numbers of modulus 1.*

Proof. Let $U : H \to H$ be a bounded unitary operator and let $\mu \in \sigma_p(U)$ with associated eigenvector w. Then

$$(w, w) = (Uw, Uw) = (\mu w, \mu w) = \mu\overline{\mu}(w, w),$$

which, since w is nontrivial, implies $|\mu|^2 = 1$. □

Theorem 4.11 *The eigenvectors of either a self-adjoint or a unitary operator corresponding to different eigenvalues are orthogonal.*

Proof. Let $A : H \to H$ be a self-adjoint operator and let $\lambda_1, \lambda_2 \in \sigma_p(A)$, $\lambda_1 \neq \lambda_2$, have associated eigenvectors u_1, u_2, respectively. Then

$$\begin{aligned}(\lambda_1 - \lambda_2)(u_1, u_2) &= (\lambda_1 u_1, u_2) - (u_1, \lambda_2 u_2)\\ &= (Au_1, u_2) - (u_1, Au_2)\\ &= (Au_1, u_2) - (Au_1, u_2) = 0,\end{aligned}$$

which implies $(u_1, u_2) = 0$ because $\lambda_1 \neq \lambda_2$.

Let $U : H \to H$ be a unitary operator and let $\mu_1, \mu_2 \in \sigma_p(A)$, $\mu_1 \neq \mu_2$, have associated eigenvectors w_1, w_2, respectively. Then

$$\mu_1 \mu_2 (w_1, w_2) = (\mu_1 w_1, \mu_2 w_2) = (Uw_1, Uw_2) = (w_1, w_2),$$

the last equality following from the defining property of a unitary operator (see definition 3.106). Hence we can conclude that $(w_1, w_2) = 0$ since $\mu_1 \overline{\mu_2} \neq 1$. □

In later chapters we will have occasion to make use of a result of the following form.

Theorem 4.12 *Let H be a complex separable Hilbert space and let $A, B : H \to H$ be linear operators. If B is bounded and B^{-1} exists, then A and BAB^{-1} have the same eigenvalues.*

Proof. If λ is an eigenvalue of A, then there exists a *nontrivial* $\varphi \in H$ such that $A\varphi = \lambda\varphi$.

If B^{-1} exists, then B must be a one-to-one onto operator (theorem 3.102). Consequently, $B\varphi$ cannot be zero for all nontrivial φ. Hence

$$BAB^{-1}B\varphi = BA\varphi = B\lambda\varphi = \lambda B\varphi,$$

which implies that λ is an eigenvalue of the operator BAB^{-1} with associated eigenvector $B\varphi$.

Conversely, let μ be an eigenvalue of BAB^{-1}. Then there exists a nontrivial $\psi \in H$ such that $BAB^{-1}\psi = \mu\psi$. Consequently,

$$B^{-1}BAB^{-1}\psi = \mu B^{-1}\psi,$$

which implies that μ is an eigenvalue of BAB^{-1} with associated eigenvector $B^{-1}\psi$. □

For self-adjoint and unitary operators on a complex separable Hilbert space H, eigenvectors corresponding to different eigenvalues are orthogonal (theorem 4.11). Consequently, the eigenspaces corresponding to different eigenvalues are orthogonal subspaces of H. This in turn implies that the operators act on the direct sum of the eigenspaces like a diagonal matrix. A hint of why this should be is provided by recalling that if λ is an eigenvalue of the linear operator $A : H \to H$, then the associated eigenspace M_λ is defined as

$$M_\lambda = \{\theta \neq \psi \in D(A) : A\psi = \lambda\psi\}.$$

More fully, let $A : H \to H$ be either a self-adjoint or a unitary operator and let $\lambda_1, \lambda_2, \ldots, \lambda_k, \ldots$ denote its different eigenvalues. For each λ we will write M_k to denote the eigenspace corresponding to λ_k. An orthonormal basis for M_k will be denoted by $\{\varphi_s^k\}_s$.

We remark that whilst the number of eigenvalues may be either finite or infinite, nevertheless, they are always countable. If this were not so, then there would be an uncountable number of different eigenvalues with an associated uncountable number of orthonormal basis vectors. This is impossible in a separable Hilbert space.

We also remark that the dimension of M_k, that is, the number of basis elements φ_s^k for M_k, may be either finite or infinite and, furthermore, may be different for different values of k.

Since the eigenvectors for different eigenvalues of A are orthogonal, the set of all eigenvectors φ_s^k for different k is orthonormal; that is,

$$(\varphi_r^k, \varphi_s^m) = \delta_{rs}\delta_{km},$$

where $(.,.)$ denotes the inner product in H and

$$\delta_{rs} = \begin{cases} 1, & r = s, \\ 0, & r \neq s, \end{cases}$$

is the Kronecker delta.

To proceed, we need the following concept.

DEFINITION 4.13 *Let E be a subset of a Hilbert space H and let D denote the set of all finite linear combinations of elements of E. The closure of D (in the topology of H) generates a subspace $G \subset H$. The subspace G is said to be spanned by E.*

Consider the subspace

$$H_p := \bigoplus_k M_k, \tag{4.4}$$

which consists of all linear combinations of the form $\sum_{k,s} a_s^k \varphi_s^k$. We shall refer to H_p as the *point subspace* of H. It is the subspace spanned by all the eigenvectors of A. Evidently, the set of vectors φ_s^k, $k, s = 1, 2, \ldots$, is an orthonormal basis for H_p. Thus we have

$$A\varphi_s^k = \lambda_k \varphi_s^k$$

and

$$(\varphi_r^m, A\varphi_s^k) = \lambda_k \delta_{mk} \delta_{rs}.$$

Thus on H_p the operator A acts like a diagonal matrix. The off-diagonal terms are all zero, whilst the diagonal terms are eigenvalues of A.

Summarising the above, we see that for any $\psi \in H_p$ there are scalars a_s^k such that

$$\psi = \sum_{k,s} a_s^k \varphi_s^k, \tag{4.5}$$

$$A\psi = \sum_{k,s} \lambda_k a_s^k \varphi_s^k. \tag{4.6}$$

The results (4.5), (4.6) provide an example of a *spectral representation* (*decomposition*) of the operator A.

We notice two things.

1. The spectral representation (4.5), (4.6) is only valid on H_p. For those operators that have a spectrum with more components than just eigenvalues (that is, $\sigma_c(A)$ and $\sigma_R(A)$ are not necessarily empty), (4.5), (4.6) are inadequate; more terms are required.
2. On finite-dimensional Hilbert spaces the spectrum of a linear operator is a pure point spectrum (theorem 4.4), in which case (4.5), (4.6) provide a perfectly adequate spectral representation.

4.3 CONCERNING SPECTRAL DECOMPOSITIONS

In the introduction to this chapter we said that one of the main reasons for introducing and using spectral theory is that it can provide mechanisms for decomposing quite complicated operators into simpler, more manageable components. In practice a full demonstration of this will involve working through the following stages.

- Determine a characterisation of a given physical problem in terms of an operator $A : H \to H$, where H denotes a complex separable Hilbert space.
- Determine $\sigma(A)$, the spectrum of A, as a subset of the complex plane **C**.
- Provide a decomposition of **C** into components intimately connected with the nature of $\sigma(A)$.
- Provide a decomposition of H into components, the *spectral components* of H, that are intimately connected with the nature of A and $\sigma(A)$.
- Provide an interpretation of A when it acts on the various spectral components of H. This will introduce the (*spectral*) *parts* of A, which are often more manageable than A itself.
- Show how results obtained when dealing with just the parts of A can be combined to provide meaningful and practical results for problems centred on A itself.

In this section we will confine our attention to decomposition aspects when H is finite-dimensional and when it is infinite-dimensional. In each case we will consider only self-adjoint and unitary operators.

4.3.1 Spectral Decompositions on Finite-Dimensional Spaces

The only spectral values of operators acting on a finite-dimensional space are eigenvalues.

Theorem 4.14 *On a finite-dimensional complex Hilbert space the eigenvectors of either a self-adjoint or a unitary operator span the space.*

Proof. Let H be a finite-dimensional complex Hilbert space and let $A : H \to H$ be a linear operator that is either self-adjoint or unitary. Let M denote the subspace spanned by the eigenvectors of A and let $P : H \to M$ be the projection operator onto M.

Since linear operators on an n-dimensional space ($n < \infty$) can always be represented in the form of an $n \times n$ matrix, the results of matrix algebra indicate that a linear operator on a finite-dimensional complex Hilbert space has at least one eigenvalue.

Suppose $M \neq H$ and consider the operator $A(I - P)$ on $M^{\perp}$ (which is clearly finite-dimensional). Then there must exist a scalar λ and a nontrivial element $v \in M^{\perp}$ such that

$$A(I - P)v = \lambda v.$$

Consequently, since $P : H \to M$ implies $Pv = \theta$, $v \in M^{\perp}$ (recall the projection theorem), we have

$$Av = APv + A(I - P)v = \lambda v,$$

which implies that v is an eigenvector of A, and this contradicts the assumption that M is not the whole space. □

This last theorem means that every self-adjoint or unitary operator A on an n-dimensional space ($n < \infty$) provides a basis for the space consisting entirely of orthonormal eigenvectors of A. Consequently, let

(i) $\lambda_1, \lambda_2, \ldots, \lambda_m$ denote the different eigenvalues of A,

(ii) M_k denote the eigenspace corresponding to the eigenvalue λ_k, $k = 1, 2, \ldots, m$,

(iii) $\{\varphi_s^k\}_s$ be an orthonormal basis for M_k, $k = 1, 2 \ldots, m$, where $s = 1, 2, \ldots, s(k)$ and $s(k)$ is a positive integer depending on k.

Then

$$A\varphi_s^k = \lambda_k \varphi_s^k, \qquad s = 1, 2, \ldots, s(k), \qquad k = 1, 2, \ldots, m,$$

and we can conclude that $\{\varphi_s^k\}_s$ is an orthonormal basis for the whole space. The total number of eigenvectors φ_s^k is n, and $m \leq n$. The number of orthonormal eigenvectors φ_s^k associated with the eigenvalue λ_k, namely, $s(k)$, indicates the dimension of M_k, denoted dim M_k, and, equivalently, is referred to as the *multiplicity* of λ_k. We remark that dim M_k may be different for different values of k.

A closer look at (4.5) suggests that we define the operator

$$P_k : H \to M_k, \qquad k = 1, 2, \ldots, m, \tag{4.7}$$

$$P_k : \psi \to P_k\psi = \sum_{s=1}^{s(k)} a_s^k \varphi_s^k, \qquad \psi \in H. \tag{4.8}$$

The representation (4.5) can now be written

$$\psi = \sum_{k=1}^{m} P_k \psi, \tag{4.9}$$

which in turn implies the completeness property

$$\sum_{k=1}^{m} P_k = I. \tag{4.10}$$

The operator P_k is a projection onto the eigenspace M_k. Since eigenspaces corresponding to different eigenvalues of a self-adjoint or a unitary operator are orthogonal, it follows that the projections P_k, $k = 1, 2, \ldots, m$, are orthogonal in the sense that

$$P_k P_m = \delta_{km} P_k. \tag{4.11}$$

Furthermore, (4.6) can now be written

$$A\psi = \sum_{k,s} \lambda_k a_s^k \varphi_s^k = \sum_{k=1}^{m} \lambda_k P_k \psi,$$

which implies

$$A = \sum_{k=1}^{m} \lambda_k P_k. \tag{4.12}$$

This is a representation of the operator A in terms of projection operators. It illustrates how the spectrum of A can be used to provide a representation of A in terms of simpler operators. This use of projections seems quite a natural way to obtain the required spectral decompositions.

Unfortunately, this particular approach does not generalise to an infinite-dimensional space setting. We now describe a slightly different way of obtaining a spectral decomposition of A that does generalise to an infinite-dimensional space setting where the spectrum of a linear operator can be very much more complicated than just a collection of eigenvalues of the type we have so far been considering.

We consider a self-adjoint operator on a finite-dimensional Hilbert space and order its distinct eigenvalues in the form

$$\lambda_1 < \lambda_2 < \cdots < \lambda_{m-1} < \lambda_m.$$

For each $\lambda \in \mathbf{R}$ we define an operator-valued function of λ by

$$E_\lambda = \begin{cases} 0, & \lambda < \lambda_1, \\ \sum_{k=1}^{r} P_k, & \lambda_r \leq \lambda < \lambda_{r+1}, \\ I, & \lambda \geq \lambda_m, \end{cases} \tag{4.13}$$

which we write more compactly in the form

$$E_\lambda = \sum_{\lambda_k \leq \lambda} P_k. \tag{4.14}$$

Clearly, E_λ is a projection operator onto the subspace of H spanned by all the eigenvectors associated with eigenvalues $\lambda_k \leq \lambda$.

It follows from (4.13) that

$$E_\mu E_\lambda = E_\lambda E_\mu = E_\mu, \qquad \mu \leq \lambda. \tag{4.15}$$

When (4.15) holds, we write

$$E_\mu \leq E_\lambda, \qquad \mu \leq \lambda. \tag{4.16}$$

These various properties indicate that E_λ changes from the zero operator in H to the identity operator on H as λ runs through the spectrum, that is, eigenvalues, of A. Furthermore, we notice that E_λ changes by P_k when λ reaches λ_k. With this in mind we define

$$dE_\lambda := E_\lambda - E_{\lambda-\varepsilon}, \qquad \varepsilon > 0. \tag{4.17}$$

It now follows that if $\varepsilon > 0$ is small enough to ensure that there is no λ_k such that

$$\lambda - \varepsilon < \lambda_k < \lambda,$$

then $dE_\lambda = 0$. Furthermore, if $\lambda = \lambda_k$, then

$$dE_\lambda = dE_{\lambda_k} = P_k. \tag{4.18}$$

We are now in the position to indicate a particularly important useful representation of self-adjoint operators.

First, we recall the definition of the Riemann-Stieltjes integral of a function g with respect to a function f, namely,

$$\int_a^b g(x)\, df(x) = \lim_{n\to\infty} \sum_{j=1}^{n} g(x_j)|f(x_j) - f(x_{j-1})|, \tag{4.19}$$

where $a = x_0 < x_1 < \cdots < x_n = b$ is a partition of the range of integration.

With (4.18) in mind we see that we have

$$\int_{-\infty}^{\infty} dE_\lambda = \lim_{n\to\infty} \sum_{j=1}^{n} 1(E_{x_j} - E_{x_{j-1}}),$$

where $\lambda_1 \leq x_0 < x_1 < \cdots < x_n = \lambda_m$. Consequently, bearing in mind (4.10) and (4.18), we obtain

$$\int_{-\infty}^{\infty} dE_\lambda = I. \tag{4.20}$$

Furthermore, with (4.12) in mind and arguing as above, we have

$$A = \sum_{k=1}^{m} \lambda_k P_k = \lim_{n\to\infty} \sum_{j=1}^{n} \lambda_k (E_{x_j} - E_{x_{j-1}}),$$

which implies

$$A = \int_{-\infty}^{\infty} \lambda\, dE_\lambda. \tag{4.21}$$

The expression (4.21) is the spectral representation of the self-adjoint operator A, which has eigenvalues $\lambda_1 < \lambda_2 < \cdots < \lambda_m$ on an n-dimensional complex Hilbert space H.

For arbitrary $\varphi, \psi \in H$ in the n-dimensional space H, the above results lead to

$$(\varphi, \psi) = \int_{-\infty}^{\infty} d(E_\lambda \varphi, \psi) = \int_{-\infty}^{\infty} dw(\lambda), \tag{4.22}$$

$$(A\varphi, \psi) = \int_{-\infty}^{\infty} \lambda\, d(E_\lambda \varphi, \psi) = \int_{-\infty}^{\infty} \lambda\, dw(\lambda), \tag{4.23}$$

where $w(\lambda) := (E_\lambda \varphi, \psi)$ defines a *complex-valued function of* λ that changes by $(P_k \varphi, \psi)$ at $\lambda = \lambda_k$.

For a unitary operator $U : H \to H$ that has eigenvalues $\lambda_k = \exp(i\theta_k)$ ordered in the form

$$0 < \theta_1 < \theta_2 < \cdots < \theta_m \leq 2\pi,$$

using similar arguments to those used above, we obtain the spectral representation

$$U = \int_0^{2\pi} \exp(i\lambda)\, dE_\lambda, \tag{4.24}$$

which leads to the expression

$$(U\varphi, \psi) = \int_0^{2\pi} \exp(i\lambda)\, d(E_\lambda \varphi, \psi). \tag{4.25}$$

Similar calculations are possible for compact linear operators on an infinite-dimensional space. (See definition 3.110 and example 3.111.) For the sake of illustration consider here only a positive compact operator. All its eigenvalues λ_k are non-negative, and we denote them in the form

$$\lambda_1 > \lambda_2 > \cdots > 0$$

with possibly the inclusion of $\lambda_0 = 0$. We can then write

$$A = \sum_{k=1}^{\infty} \lambda_k P_k, \tag{4.26}$$

where $\lambda_k \to 0$ as $k \to \infty$.

Denoting by P_0 the projection onto the null space M_0, we then have (compare (4.10))

$$\sum_{k=1}^{\infty} P_k + P_0 = I. \tag{4.27}$$

Then, as before (compare (4.13), (4.14)),

$$E_\lambda = \begin{cases} 0, & \lambda < 0, \\ P_0 + \sum_{j=k}^{\infty} P_j, & \lambda_{k-1} \leq \lambda < \lambda_k, \\ I, & \lambda \geq \lambda_1. \end{cases} \tag{4.28}$$

It is a straightforward matter to show, in a similar manner to that used above, that E_λ is a projection operator–valued function of λ. We can also conclude that, just as in the finite-dimensional case when we obtained (4.22), (4.23), we again have results of the form

$$(\varphi, \psi) = \int_{-\infty}^{\infty} d(E_\lambda \varphi, \psi) = \int_{-\infty}^{\infty} dw(\lambda), \tag{4.29}$$

$$(A\varphi, \psi) = \int_{-\infty}^{\infty} \lambda\, d(E_\lambda \varphi, \psi) = \int_{-\infty}^{\infty} \lambda\, dw(\lambda), \tag{4.30}$$

which implies

$$A = \int_{-\infty}^{\infty} \lambda \, dE_\lambda. \tag{4.31}$$

Consequently, we see that a self-adjoint operator on either a finite- or an infinite-dimensional space has, in the two special cases considered, an integral representation given by (4.21) and (4.31). In these cases the integral representations are really a means of re-expressing the diagonalisability property. However, for self-adjoint operators that do not belong to the two classes mentioned above, this notion of diagonalisability is no longer meaningful. Nevertheless, it might be possible to express any self-adjoint operator in the integral form (4.31) provided the spectral family $\{E_\lambda\}$, $\lambda \in \sigma(A)$, is appropriately defined. This is the content of the celebrated spectral theorem which we will discuss later.

4.3.2 Reducing Subspaces

We introduce this concept in terms of bounded operators. For unbounded operators the following definition and two theorems require a more careful statement that properly takes into account the domains involved. We return to this aspect at the end of this subsection.

Definition 4.15 *A subspace (closed linear manifold) $M \subseteq H$ reduces a bounded linear operator $A : H \to H$ if*
(i) $A\psi \in M$ for every $\psi \in M$,
(ii) $A\varphi \in M^\perp$ for every $\varphi \in M^\perp$.

The following two theorems indicate the main properties of reducing subspaces.

Theorem 4.16 *Let*
(i) H be a complex separable Hilbert space and let $M \subseteq H$ be a subspace,
(ii) $A : H \supseteq D(A) \to H$ be a linear operator,
(iii) $P : H \to M$ be a projection operator onto M.

The following statements are equivalent.
(a) M reduces A.
(b) $PA = AP$.
(c) $(I - P)A = A(I - P)$.

Proof. It is obvious that (b) and (c) are equivalent. We shall show that (a) $\Rightarrow$ (b) and that ((b),(c)) $\Rightarrow$ (a).

For any element $\psi \in H$ we have, by the projection theorem,

$$\psi = u + v, \qquad u \in M, \qquad v \in M^\perp.$$

Hence $A\psi = Au + Av$, and we conclude that (a) $\Rightarrow$ (b) as follows.
(a) $\Rightarrow$ (b): M reduces $A \Rightarrow Au \in M$ and $Av \in M^\perp$. Therefore

$$PA\psi = Au = AP\psi,$$

and hence (a) $\Rightarrow$ (b).

((b),(c)) $\Rightarrow$ (a): If $u \in M$ and $PA = AP$, then $Pu = u$ and

$$Au = APu = PAu.$$

Hence $Au \in M$ for all $u \in M$.

If $v \in M^{\perp}$ and $(I - P)A = A(I - P)$, then

$$(I - P)v = v.$$

Hence

$$Av = A(I - P)v = (I - P)Av,$$

and we conclude that $Av \in M^{\perp}$ for all $v \in M^{\perp}$.

Hence ((b), (c)) $\Rightarrow$ (a). □

This theorem indicates the important practical result

$$A = AP + A(I - P), \tag{4.32}$$

which implies that A is the sum of two parts, namely, AP as an operator on M and $A(I - P)$ as an operator on $M^{\perp}$.

Theorem 4.17 *Let A be a bounded linear operator that is either self-adjoint or unitary on a separable complex Hilbert space H. Let H_p be the subspace spanned by the eigenvectors of A. Then H_p reduces A.*

The operators induced by A in H_p and $H_p^{\perp}$ are again self-adjoint or unitary.

Proof. (i) Assume A is self-adjoint.

Let $u \in H_p$; then since $Au = \lambda u$ for some $\lambda \in \mathbf{R}$, we can conclude that $Au \in H_p$.

If $v \in H_p^{\perp}$, then

$$(u, Av) = (Au, v) = 0 \quad \text{for any } u \in H_p.$$

Hence $Av \in H_p^{\perp}$, and we conclude that H_p reduces A.

Let $P : H \to H$ be a projection. Then there exist operators A_1 in H_p and A_2 in $H_p^{\perp}$ with domains

$$D(A_1) = PH = H_p,$$
$$D(A_2) = (I - P)H = P^{\perp}H = H_p^{\perp},$$

such that (see (4.32))

$$AF = A_1(Pf) + A_2(P^{\perp}f), \qquad f \in H.$$

The operators A_1 and A_2 are the *operators induced* by A in H_p and $H_p^{\perp}$, respectively.

For $f, g \in H_p$ we obtain, recognising this last relation, theorem 4.16, and the defining properties of the operators

$$(A_1 f, g) = (APf, g) = (PAf, g) = (f, APg) = (f, A_1 g).$$

Hence A_1 is self-adjoint on H_p. Similarly, A_2 is self-adjoint on $H_p^{\perp}$.

(ii) Assume that A is unitary. As before, $u \in H_p$ implies $Au \in H_p$.

Let $v \in H_p^\perp$ and let w be an eigenvector of A with an associated eigenvalue λ. Then $Aw = \lambda w$ and

$$\lambda(w, Av) = (Aw, Av) = (w, v) = 0,$$

the third equality following from the unitarity of A. (See definition 3.106 (ii).) This last result implies $(w, Av) = 0$ because $|\lambda| = 1$ (theorem 4.10). Thus Av is orthogonal to every element in the orthonormal basis (of eigenvectors of A) for H_p. Therefore $Av \in H_p^\perp$, and hence H_p reduces A.

With A_1, A_2 defined as in part (i) we have for A unitary,

$$\|f\| = \|Af\| = \|A_1 f\| \quad \text{for all } f \in H_p.$$

Therefore A_1 is an isometric linear operator on H_p (definition 3.12). Similarly, A_2 is an isometric linear operator on $H_p^\perp$. To show that A_1 is unitary, it remains to show that A_1 maps H_p *onto* H_p. To this end, let $g \in H_p$ be given. Since A is unitary, then A maps H onto H and there exists an element $f \in H$ such that $Af = g$.

Since $Af = g \in H_p$ and $A_1 Pf \in H_p$ and $A_2 P^\perp f \in H_p^\perp$, we conclude, from the decomposition of A given in part (i), written in the form $g = A_1(Pf) + A_2(P^\perp f)$ and using theorem 3.112, that $A_2(P^\perp f) = \theta$ and $\|P^\perp f\| = \|A_2 P^\perp f\| = 0$. Consequently, $f \in H_p$ and

$$g = Af = A_1 f \in H_p.$$

□

We shall refer to H_p as the *point subspace* of A. Further, we shall denote $H_p^\perp$, the orthogonal complement of H_p, by H_c and refer to it as the *subspace of continuity* of A.

We see from (4.32) that a self-adjoint or unitary operator splits into two (smaller in some sense) parts. One part of the operator acts on the subspace spanned by eigenvectors and, as such, can be represented by a diagonal matrix of eigenvalues with respect to an orthonormal basis of eigenvectors of the given operator. The other part of the given operator acts on the orthogonal complement of the subspace spanned by the eigenvectors. The simplest form of this decomposition occurs when the eigenvectors of the given operators span the whole space. In this case the associated H_c will be empty.

Example 4.18 *Let H be a separable complex Hilbert space and let $N \subseteq H$ be a subspace. Let $P : H \to N$ be a projection of H onto N. Assume that there exists a nontrivial $\psi \in H$ such that*

$$P\psi = \lambda\psi, \qquad \lambda \in \mathbf{C}.$$

Then

$$\lambda^2\psi = \lambda P\psi = P^2\psi = P\psi = \lambda\psi,$$

which implies that $\lambda = 1$ or 0. We therefore conclude that a projection operator can only have one of two distinct eigenvalues, namely, $\lambda = 1$ or 0.

If $\psi \in N$, then $P\psi = \psi$ by virtue of the definition of the projection operator P. Also, if $\varphi \in N^{\perp}$, then $P\varphi = \theta$. Therefore, recalling that a projection operator is self-adjoint and that eigenspaces of self-adjoint operators corresponding to different eigenvalues are orthogonal, we can infer that

$N =$ eigenspace of P corresponding to the eigenvalue $\lambda = 1$.
$N^{\perp} =$ eigenspace of P corresponding to the eigenvalue $\lambda = 0$.

A basis for the whole space is obtained by combining the basis for N and the basis for $N^{\perp}$.

The projection operator P can be represented as a diagonal matrix with respect to this basis. The diagonal elements are one for the basis vectors of N and zero for the basis elements of $N^{\perp}$.

In this example we have completely described a projection operator in terms of its eigenvalues and eigenvectors.

We would emphasise that throughout this subsection we have assumed that all the linear operators are bounded.

For more general operators we can begin the discussion by, as before, assuming a decomposition of a separable complex Hilbert space H in the form

$$H = M + M^{\perp},$$

where M is a subspace of H. We shall denote by P the projection of H onto M.

Definition 4.19 *A possibly unbounded operator A, is said to be decomposed according to $H = M + M^{\perp}$ if*

$$PD(A) \subseteq D(A), \qquad APD(A) \subseteq M, \qquad A(I-P)D(A) \in M^{\perp}. \tag{4.33}$$

The results (4.33) imply that for any $f \in D(A)$ we have $Pf \in D(A)$ and $APf \in M$ and $A(I-P) \in M^{\perp}$. Hence

$$(I-P)APf = APf - PAPf = APf - APf = \theta$$

$$PA(I-P)f = \theta,$$

and we conclude $(I-P)APf = PA(I-P)f = \theta$. This leads to the conclusion that $APf = PAf$ for $f \in D(A)$. Thus we see that condition (4.33) is equivalent to the condition that A commutes with the projection P; that is,

$$PA \subseteq AP. \tag{4.34}$$

If one of the two equivalent conditions (4.33) or (4.34) is satisfied, then the restriction of A to $M \cap D(A)$ can be considered an operator in the Hilbert space M. This operator is called the *part* of A in H and is frequently denoted A/M.

Definition 4.20 *If the operator A is symmetric, then the operator A is said to be reduced by M if $PD(A) \subseteq D(A)$ and $APD(A) \subseteq M$.*

A result involving the reduction of possibly unbounded operators is the following.

THEOREM 4.21 *Let H be a separable complex Hilbert space and let $M \subseteq H$ a subspace. Let $A : H \supseteq D(A) \to H$ and let $P : H \to M$ be a projection.*

(i) If A is a symmetric operator, then A is reduced by M if and only if A and P commute.

(ii) If A is self-adjoint, then A/M is also self-adjoint.

The proof of this theorem is straightforward and can be found in the texts cited in chapter 11 [53, 95].

4.3.3 Spectral Decompositions on Infinite-Dimensional Spaces

Spectral studies on infinite-dimensional spaces are more complicated than similar studies on finite-dimensional spaces. For example, there are self-adjoint and unitary operators that have no eigenvalues, yet the spectrum of such operators consists of more than the point spectrum (see definition 4.2). Nevertheless, it is still possible to obtain spectral decompositions in terms of projection operators that have an integral form similar to that already obtained in subsection 4.3.1.

We shall assume, just as for the finite-dimensional case, that there exists a nondecreasing family of subspaces $\{M_\lambda\}$ of a complex separable Hilbert space H. These subspaces depend on a real parameter $\lambda \in (-\infty, \infty)$ such that

1. The intersection of all the M_λ is θ, the zero element in H.
2. The union of all the M_λ is a dense subset of H.

We now introduce a family of projection operators $\{E_\lambda\}$ associated with $\{M_\lambda\}$. First, we recall that the family is said to be *nondecreasing* if $M_\mu \subseteq M_\lambda$ for $\mu < \lambda$. Now, bearing in mind (4.13) and the discussion that followed, the following definition is natural.

DEFINITION 4.22 *A family of projection operators $\{E_\lambda\}$ depending on the parameter λ is said to be a spectral family or a resolution of the identity if it has the following properties.*

(i) $\{E_\lambda\}$ is nondecreasing in the sense that

$$E_\mu \leq E_\lambda \quad \text{for } \mu < \lambda.$$

Equivalently, we have

$$E_\lambda E_\mu = E_\mu E_\lambda = E_{\min(\lambda,\mu)} = E_\mu.$$

(ii) If $\varepsilon > 0$, then for any element $\varphi \in H$ and scalar λ

$$E_{\lambda+\varepsilon}\varphi \to E_\lambda \varphi \quad \text{as } \varepsilon \to 0.$$

(iii) $E_\lambda \psi \to \theta$ as $\lambda \to -\infty$, $E_\lambda \psi \to \psi$ as $\lambda \to +\infty$.
Equivalently, we write

$$s - \lim_{\lambda \to -\infty} E_\lambda = \Theta, \qquad s - \lim_{\lambda \to +\infty} E_\lambda = I,$$

where Θ denotes the zero operator on H.

With every spectral family $\{E_\lambda\}$ we can associate a self-adjoint or unitary operator. This is a statement of the celebrated spectral theorem, which can be quoted in the following form.

THEOREM 4.23 *(spectral theorem) Let H be a complex separable Hilbert space.*

(i) For each bounded self-adjoint operator A on H, there exists a unique spectral family $\{E_\lambda\}$ such that

$$(A\psi, \varphi) = \int_{-\infty}^{\infty} \lambda \, d(E_\lambda \psi, \varphi) \quad \text{for all } \varphi, \psi \in H.$$

Equivalently, we write

$$A = \int_{-\infty}^{\infty} \lambda \, dE_\lambda. \tag{4.35}$$

(ii) For each unitary operator U on H, there exists a unique spectral family $\{F_\lambda\}$ such that $F_\lambda = \Theta$ for $\lambda \leq 0$ and $F_\lambda = I$ for $\lambda \geq 2\pi$ such that

$$(U\psi, \varphi) = \int_{0}^{2\pi} e^{i\lambda} d(F_\lambda \psi, \varphi) \quad \text{for all } \varphi, \psi \in D(U) \subset H.$$

Equivalently, we write

$$U = \int_{0}^{2\pi} e^{i\lambda} \, dF_\lambda. \tag{4.36}$$

As before, we refer to (4.35) and (4.36) as the spectral decompositions of A and U, respectively.

We simply state this theorem here, as the proof of the theorem is quite technical and lengthy. Full details can be found in the texts cited in chapter 11; in this connection we would particularly mention ([4, 95, 102, 105]). In defence of this action we recall the sentiments expressed in the introductory chapter. This monograph is not a book on functional analysis or operator theory or spectral theory. However, that being said, we need many results from these three fields. Consequently, we only include proofs when they are needed, either for clarification in the development of material or because of potential practical usefulness in applications.

Example 4.24 *Let $H := L_2(0, 1)$ and define $A : H \to H$ by*

$$(A\psi)(x) = x\psi(x), \qquad \psi \in D(A) = H.$$

It is an easy exercise to show that A is linear and self-adjoint on H.

Let $\{E_x\}$ denote a family of projections defined by

$$(E_x\psi)(x) = \begin{cases} \psi(z), & z \leq x, \\ 0, & z > x. \end{cases}$$

The following are immediate.

(i) $E_x E_y = E_y E_x = E_x$, $x \le y$. *Equivalently,* $E_x \le E_y$, $x \le y$.

(ii)

$$\|E_{x+\varepsilon}\psi - E_x\psi\|^2 = \int_x^{x+\varepsilon} |\psi(y)|^2 \, dy \to 0 \quad \textit{as } \varepsilon \to 0^+.$$

Hence $E_{x+\varepsilon} \to E_x$ *as* $\varepsilon \to 0^+$.

We assume further that $\psi(x)$ *is zero for* x *outside the interval* $[0, 1]$. *In this case we have*

$$E_x = \Theta, \quad x < 0, \qquad E_x = I, \quad x > 1.$$

Consequently, $\{E_x\}$ *is a spectral family.*

We obtain a spectral decomposition by noticing

$$\begin{aligned}\int_{-\infty}^{\infty} x d(E_x\psi, \varphi) &= \int_{-\infty}^{\infty} xd\left\{\int_0^1 (E_x\psi)(y)\overline{\varphi(y)}\, dy\right\}\\ &= \int_0^1 xd\left\{\int_0^x \psi(y)\overline{\varphi(y)}\, dy\right\}\\ &= \int_0^1 x\psi(x)\overline{\varphi(x)}\, dx = (A\psi, \varphi).\end{aligned}$$

Thus A has the spectral representation

$$A = \int_{-\infty}^{\infty} x \, dE_x. \tag{4.37}$$

We also notice that

$$(E_x\psi, \varphi) - (E_{x-\varepsilon}\psi, \varphi) = \int_{x-\varepsilon}^{x} \psi(y)\overline{\varphi(y)}\, dy,$$

and the right-hand side tends to zero as $\varepsilon \to 0^+$. This implies that in this case $(E_x\psi, \varphi)$, as a function of x, is also continuous from the left. Hence the relation

$$w(x) = (E_x\psi, \varphi), \qquad \varphi, \psi \in H$$

defines a continuous function of x.

Example 4.25 *Let* $A : H \supseteq D(A) \to H = L_2(\mathbf{R})$ *be defined by*

$$(A\psi)(x) = x\psi(x), \qquad \psi \in D(A) = H.$$

Define the spectral family as in example 4.24. It is then readily shown that in this case A *also has a spectral decomposition. However, the spectral family* $\{E_x\}$ *in this case has the properties that* E_x *increases over the whole range* $-\infty < x < \infty$, *with* $E_x \to \Theta$ *as* $x \to -\infty$ *and* $E_x \to I$ *as* $x \to \infty$.

We see from these two examples that E_λ as a function of λ can increase continuously rather than by a series of jumps (steps), as was the case when working in a finite-dimensional setting. This is because A, as defined in each example, has no discrete eigenvalues.

4.4 SOME PROPERTIES OF SPECTRAL FAMILIES

The spectral family associated with a self-adjoint operator on a Hilbert space can provide information about the spectrum of the operator in a relatively simple manner. Indeed, we have seen that in a finite-dimensional setting the spectral family is discontinuous at the eigenvalues of the associated operator. This property carries over to an infinite-dimensional space setting. In addition, information about the parts of the spectrum, other than the eigenvalue spectrum which can exist in an infinite- dimensional space setting, can also be obtained. We collect here, simply as statements, a number of fundamental properties of spectral families. In doing this we assume that the spectral family has in fact been determined. The actual determination of a spectral family is a nontrivial exercise. We return to this point later.

Theorem 4.26 *Let H be a complex separable Hilbert space and let $A : H \to H$ be a bounded linear self-adjoint operator with an associated spectral family $\{E_\lambda\}$ and spectral decomposition $A = \int_{-\infty}^{\infty} \lambda \, dE_\lambda$. Then E_λ has a discontinuity at $\lambda = \mu$ if and only if μ is an eigenvalue of A.*

Let $P_\mu : H \to M_\mu$ denote the projection operator onto M_μ, the subspace spanned by the eigenvectors of A associated with the eigenvalue μ. Then

(i) $E_\lambda P_\mu = \begin{cases} P_\mu, & \lambda \geqslant \mu, \\ 0, & \lambda < \mu, \end{cases}$

(ii) for $\varepsilon > 0$,

$$E_\mu \psi - E_{\mu-\varepsilon} \psi \to P_\mu \psi$$

as $\varepsilon \to 0$ and for any $\psi \in H$.

For unitary operators the corresponding result is as follows.

Theorem 4.27 *Let H be a complex separable Hilbert space and let $U : H \to H$ be a linear unitary operator with an associated spectral family $\{F_\lambda\}$ and spectral decomposition $U = \int_0^{2\pi} e^{i\lambda} \, dF_\lambda$. Then F_λ has a discontinuity at $\lambda = \mu$ if and only if $e^{i\mu}$ is an eigenvalue of U.*

Let $P_\mu : H \to M_\mu$ denote the projection operator onto M_μ, the subspace spanned by the eigenvectors of U associated with the eigenvalue $e^{i\mu}$. Then

(i) $F_\lambda P_\mu = \begin{cases} P_\mu, & \lambda \geqslant \mu, \\ 0, & \lambda < \mu, \end{cases}$

(ii) for $\varepsilon > 0$,

$$F_\mu \psi - F_{\mu-\varepsilon} \psi \to P_\mu \psi$$

as $\varepsilon \to 0$ and for any $\psi \in H$.

These two theorems indicate that the jumps in the values of E_λ and F_λ are the same as in the finite-dimensional case. However, in the infinite-dimensional space setting a continuous increase in E_λ and F_λ is possible (see also examples 4.24 and 4.25).

The resolvent set of a self-adjoint operator can also be characterised in terms of the associated spectral family. Specifically, the following result can be obtained.

THEOREM 4.28 *Let H, A and $\{E_\lambda\}$ be as in theorem 4.26. A real number μ belongs to $\rho(A)$, the resolvent set of A, if and only if there exists a constant $c > 0$ such that $\{E_\lambda\}$ is constant on the interval $[\mu - c, \mu + c]$.*

The importance of this theorem is that it indicates that $\mu \in \sigma(A)$ if and only if the spectral family $\{E_\lambda\}$ is *not* constant in any neighbourhood of $\mu \in \mathbf{R}$.

We can say more about the spectrum of a self-adjoint operator. First, we need the following important property of self-adjoint operators.

THEOREM 4.29 *Let H be a complex separable Hilbert space and let $A : H \to H$ be a linear self-adjoint operator. The residual spectrum of A, denoted $\sigma_r(A)$, is empty.*

Proof. Assume $\sigma_r(A)$ is nonempty. By definition, if $\lambda \in \sigma_r(A)$, then the resolvent operator $R_\lambda(A) := (A - \lambda I)^{-1}$ exists as either a bounded or an unbounded operator on $\overline{D(R_\lambda(A))} \neq H$. This implies, by the projection theorem, that there exists a nontrivial element $\varphi \in H$ that is orthogonal to $D((A - \lambda I)^{-1}) = D(R_\lambda(A))$. However, $D(R_\lambda(A))$ is the range of $(A - \lambda I)$. Hence there exists a nontrivial element φ in H such that, for all $\psi \in D((A - \lambda I)) = D(A)$, we have

$$0 = ((A - \lambda I)\psi, \varphi) = (A\psi, \varphi) - \lambda(\psi, \varphi) = (\psi, A^*\varphi) - (\psi, \overline{\lambda}\varphi),$$

which implies that $A^*\varphi = \overline{\lambda}\varphi$, that is, $\overline{\lambda} \in \sigma_p(A^*)$. Consequently, since A is self-adjoint, we have that

$$\lambda \in \sigma r(A) \Rightarrow \overline{\lambda} \in \sigma_p(A^*) \Rightarrow \overline{\lambda} \in \sigma_p(A) \Rightarrow \lambda \in \sigma_p(A).$$

This is a contradiction, and we can conclude that $\sigma_r(A) = \emptyset$. □

This theorem indicates that for a self-adjoint operator A the spectrum of A, denoted $\sigma(A)$, decomposes in the form

$$\sigma(A) = \sigma_p(A) \cup \sigma_c(A).$$

Since points in $\sigma_p(A)$ correspond to discontinuities in $\{E_\lambda\}$, the spectral family of A, the following result follows immediately.

THEOREM 4.30 *Let A and $\{E_\lambda\}$ be defined as in theorem 4.26. A real number μ belongs to $\sigma_c(A)$, the continuous spectrum of A, if and only if $\{E_\lambda\}$ is continuous at μ and is not constant in any neighbourhood of μ.*

4.5 CONCERNING THE DETERMINATION OF SPECTRAL FAMILIES

The spectral theorem (theorem 4.23) indicates that a spectral family $\{E_\lambda\}$ determines a self-adjoint operator A according to the relation

$$A = \int_{-\infty}^{\infty} \lambda \, dE_\lambda.$$

Clearly, different spectral families lead to different self-adjoint operators.

In practical applications we are particularly interested in determining the spectral family $\{E_\lambda\}$ associated with a given self-adjoint operator. This can be achieved by means of the celebrated Stone's formula, which relates the spectral family of A and the resolvent of A.

THEOREM 4.31 *(Stone's formula) Let H be a complex separable Hilbert space and let $A : H \to H$ be a self-adjoint operator. The spectral family $\{E_\lambda\}$ associated with A and $(A - \lambda I)^{-1}$, the resolvent of A, are related as follows.*

For all $f, g \in H$ and for all $a, b \in \mathbf{R}$,

$$([E_b - E_a]f, g) = \lim_{\delta \downarrow 0} \lim_{\varepsilon \downarrow 0} \frac{1}{2\pi i} \int_{a+\delta}^{b+\delta} ([R(t + i\varepsilon) - R(t - i\varepsilon)]f, g)\, dt, \tag{4.38}$$

where $R(t \pm i\varepsilon) = [A - (t + i\varepsilon)I]^{-1}$.

The way in which this formula is used to provide the required spectral representations is indicated in the following example.

Example 4.32 *For the purpose of illustration we consider an operator that occurs frequently in scattering problems. Specifically, let $A : H \supset D(A) \to H = L_2(\mathbf{R})$ be defined by*

$$Au = u_{xx}, \qquad u \in D(A),$$
$$D(A) = \{u \in H : u_{xx} \in H \text{ and } u(0) = 0\}.$$

To use Stone's formula, we must compute the resolvents $R(t \pm i\varepsilon)$ of A. To this end, recalling the definition of A, we consider the boundary value problem

$$(A - \lambda I)v(x) = f(x), \qquad x \in (0, \infty), \qquad v(0) = 0. \tag{4.39}$$

This is an ordinary differential equation that has a solution given by

$$v(x) = (A - \lambda I)^{-1} f(x) = \int_0^\infty G(x, y) f(y)\, dy, \tag{4.40}$$

where $G(x, y)$, which is the Green's function for the problem (4.39), is readily found to have the form [104]

$$G(x, y) = \begin{cases} \dfrac{(\exp i\sqrt{x}) \sin \sqrt{\lambda} y}{\sqrt{\lambda}}, & 0 \le y \le x, \\[2ex] \dfrac{(\exp i\sqrt{y}) \sin \sqrt{\lambda} x}{\sqrt{\lambda}}, & 0 \le x \le y. \end{cases}$$

We now define

$$\lambda_+ := t + i\varepsilon = Re^{i\theta}, \qquad R = \sqrt{t^2 + \varepsilon^2}, \quad \theta = \tan^{-1}(\varepsilon/t),$$
$$\lambda_- := t - i\varepsilon = Re^{-i\theta},$$

and choose

$$\sqrt{\lambda_+} = R^{1/2}e^{i\theta/2}, \qquad \sqrt{\lambda_-} = R^{1/2}e^{i(\pi-\theta/2)}.$$

We then see that as $\varepsilon \downarrow 0$,

$$\sqrt{\lambda_+} \to +\sqrt{t}, \qquad \sqrt{\lambda_-} \to -\sqrt{t}.$$

Therefore

$$R(t+i\varepsilon) - R(t-i\varepsilon) = (A - \lambda_+ I)^{-1} - (A - \lambda_- I)^{-1},$$

and on first writing (4.40) out in full and collecting terms, we obtain

$$\lim_{\varepsilon\downarrow 0}\{R(t+i\varepsilon) - R(t-i\varepsilon)\}f(x) = \frac{2i\sin\sqrt{t}x}{\sqrt{t}}\int_0^\infty f(y)\sin\sqrt{t}y\,dy.$$

For convenience at this stage, assume $f, g \in C_0[a,b]$, $a < b < \infty$. *Stone's formula now reads*

$$([E_b - E_a]f, g) = \frac{1}{2\pi i}\int_a^b \int_0^\infty \frac{2i\sin\sqrt{t}x}{\sqrt{t}}\int_0^\infty f(y)\sin\sqrt{t}y\,dy\,\overline{g(x)}\,dx\,dt.$$

If we now set $s = \sqrt{t}$ *and introduce*

$$\tilde{f}(s) = \left(\frac{2}{\pi}\right)^{1/2}\int_0^\infty f(y)\sin sy\,dy,$$

then

$$([E_b - E_a]f, g) = \int_{\sqrt{a}}^{\sqrt{b}} \tilde{f}(s)\overline{\tilde{g}(s)}\,ds.$$

It can be shown ([59]) that $\sigma(A) \subset (0,\infty)$. *Consequently,* $E_a \to \Theta$ *as* $a \to 0$. *Hence*

$$(E_\lambda f, g) = \int_0^{\sqrt{\lambda}} \tilde{f}(s)\overline{\tilde{g}(s)}\,ds, \tag{4.41}$$

which in turn implies (write $g(s)$ *in full and interchange the order of integration)*

$$(E_\lambda f)(x) = \left(\frac{2}{\pi}\right)^{1/2}\int_0^{\sqrt{\lambda}} \tilde{f}(s)\sin sx\,ds =: \int_0^{\sqrt{\lambda}} \tilde{f}(s)\theta_x(s)\,ds,$$

where $\theta_x(s)$ *has been introduced for ease of presentation.*

The spectral theorem indicates that for $f \in D(A)$,

$$(Af)(x) = \int_0^\infty \lambda\,dE_\lambda f(x).$$

In the present case we obtain from (4.41)

$$\begin{aligned} dE_\lambda f(x) &= \frac{d}{d\lambda}\left(\int_0^{\sqrt{\lambda}} \tilde{f}(s)\theta_x(s)\,ds\right)d\lambda \\ &= \frac{1}{2\sqrt{\lambda}}\tilde{f}(\sqrt{\lambda})\theta_x(\sqrt{\lambda})\,d\lambda \\ &= \tilde{f}(s)\theta_x(s)\,ds, \qquad s=\sqrt{\lambda}, \end{aligned}$$

and hence

$$\begin{aligned} (Af)(x) &= \int_0^\infty \lambda\, dE_\lambda f(x) = \int_0^\infty s^2 \tilde{f}(s)\theta_x(s)\,ds \\ &= \left(\frac{2}{\pi}\right)^{1/2}\int_0^\infty s^2\tilde{f}(s)\sin sx\,ds, \end{aligned}$$

which we notice involves the Fourier sine transform of f. *This is to be expected bearing in mind the results of a classical analysis of the boundary value problem (4.39).*

4.6 ON FUNCTIONS OF AN OPERATOR

The spectral theorem (theorem 4.23) tells us that a self-adjoint operator A has a spectral representation in terms of its associated spectral family $\{E_\lambda\}$ given by (4.35). We will want to form functions of such an operator in later chapters. If $A \in B(H)$, where H, as usual, denotes a complex separable Hilbert space, then it is easy to see that a natural definition for $\exp(A)$, for example, is obtained by using the familiar expansion for the exponential function, namely,

$$\exp(A) = \sum_{n=0}^{\infty} \frac{A^n}{n!}.$$

This relation is well defined since the right-hand side of the expression converges in the operator norm. However, there are many more complicated operators than the one we have just considered that arise in applications. Nevertheless, the spectral theorem allows us to form a large class of functions of a self-adjoint operator A. This we do in the following manner.

Let φ be a complex-valued continuous function of the real variable λ. We can define an operator $\varphi(A)$, where A is a self-adjoint operator on a Hilbert space H, by writing

$$D(\varphi(A)) = \{f \in H : \int_{-\infty}^{\infty} |\varphi(\lambda)|^2\, d(E_\lambda f, f) < \infty\}, \qquad (4.42)$$

and for $f \in D(\varphi(A))$,

$$(\varphi(A)f, g) = \int_{-\infty}^{\infty} \varphi(\lambda)\, d(E_\lambda f, g), \qquad g \in H. \tag{4.43}$$

It then follows, formally at least, that we can write

$$\varphi(A) = \int_{-\infty}^{\infty} \varphi(\lambda)\, dE_\lambda. \tag{4.44}$$

For $\varphi(\lambda) = \lambda$ we recover the operator A as expected (see (4.35)).

Some of the basic properties of $\varphi(A)$ are contained in the following exercise.

Exercise 4.33 *Let H be a complex separable Hilbert space and let $A : H \to H$ be a bounded self-adjoint operator. Let $\varphi, \varphi_1, \varphi_2$, be complex-valued continuous functions defined on the support of $\{E_\lambda\}$. Then the following results are valid.*

(i) $\varphi(A)^* = \bar{\varphi}(A)$, where $\bar{\varphi}(\lambda) = \overline{\varphi(\lambda)}$.

(ii) If $\varphi(\lambda) = \varphi_1(\lambda)\varphi_2(\lambda)$, then $\varphi(A) = \varphi_1(A)\varphi_2(A)$.

(iii) If $\varphi(\lambda) = a_1\varphi_1(\lambda) + a_2\varphi_2(\lambda)$, then $\varphi(A) = a_1\varphi_1(A) + a_2\varphi_2(A)$.

(iv) $\varphi(A)$ is normal; that is, $\varphi(A)^*\varphi(A) = \varphi(A)\varphi(A)^*$.

(v) $\varphi(A)$ commutes with all bounded operators that commute with A.

(vi) If A is reduced by a projection P, then $\varphi(A)/PH = \varphi(A/PH)$.

The details are left as an exercise for the reader.

In applications a particularly interesting function of a self-adjoint operator A is its resolvent. The above discussion indicates that if we introduce the function φ_z defined by

$$\varphi_z(\lambda) = (\lambda - z)^{-1},$$

then we can define

$$\varphi_z(A) = R_z(A) = \int_{-\infty}^{\infty} (\lambda - z)^{-1}\, dE_\lambda. \tag{4.45}$$

The next example gives some properties of φ_z.

Example 4.34 *Let H be a complex separable Hilbert space and let $A : H \to H$ be a bounded self-adjoint operator. Also, let $z \in \mathbf{C}$ be such that* $\operatorname{Im} z \neq 0$. *Then*

(i) $\varphi_z(A) = (A - zI)^{-1} \in B(H)$,

(ii) $\|\varphi_z(A)\| \leq 1/\operatorname{Im} z$.

Proof. Since A is self-adjoint, $\sigma(A) \subset \mathbf{R}$. Hence for $\operatorname{Im} z \neq 0$ it follows that $z \in \rho(A)$, and part (i) follows by definition of the resolvent.

We now show that $\varphi_z(A)$ is bounded as in (ii). Indeed, for any

$$g \in D((A - zI)^{-1}) = R(A - zI) = (A - zI)D(A),$$

where $R(A - zI)$ denotes the range of $(A - zI)$, we have $f = (A - zI)^{-1} g \in D(A)$.

Consequently, writing $z = \alpha + i\beta$, we obtain

$$\begin{aligned}\|g\|^2 &= ((A - zI)f, (A - zI)f) \\ &= \|(A - \alpha I)f\|^2 + |\beta|^2 \|f\|^2 \geqslant |\beta|^2 \|f\|^2 \\ &= |\beta|^2 \left\|(A - zI)^{-1}g\right\|^2.\end{aligned}$$

We conclude that

$$\left\|(A - zI)^{-1}g\right\| \leq \frac{\|g\|}{|\beta|} \quad \text{for all } g \in D((A - zI)^{-1}),$$

and on recalling the definition of an operator norm, we establish part (ii). □

Finally in this subsection, we give some useful consequences of the functional calculus generated by the relations (4.43) and (4.44).

Example 4.35 *(i) For $\varphi(x) = x$, we have $\varphi(A) = A$. This follows from (4.44) and the spectral decomposition theorem, theorem 4.23, since for all $f \in D(A)$, $g \in H$,*

$$(\varphi(A)f, g) = \int_{-\infty}^{\infty} \varphi(\lambda)\, d(E_\lambda f, g) = \int_{-\infty}^{\infty} \lambda\, d(E_\lambda f, g) = (Af, g).$$

Arguing as in (i), the next results follow almost immediately.

(ii) For $\varphi(x) = 1$, we have $\varphi(A) = I$ because

$$(\varphi(A)f, g) = \int_{-\infty}^{\infty} d(E_\lambda f, g) = (f, g).$$

(iii) If f, g are continuous complex-valued functions of a real variable x and if $(fg)(x) = f(x)g(x)$, then for any $\varphi, \psi \in H$,

$$\begin{aligned}(f(A)g(A)\varphi, \psi) &= \int_{-\infty}^{\infty} f(\lambda) d_\lambda (E_\lambda g(A), \psi) \\ &= \int_{-\infty}^{\infty} f(\lambda)\, d_\lambda (g(A)\varphi, E_\lambda \psi) \quad \textit{since } E_\lambda = E_\lambda^2 = E_\lambda^* \\ &= \int_{-\infty}^{\infty} f(\lambda)\, d_\lambda \int_{-\infty}^{\infty} g(\mu) d_\mu (E_\mu \varphi, E_\lambda \psi) \quad \textit{by (4.43)} \\ &= \int_{-\infty}^{\infty} f(\lambda)\, d_\lambda \int_{-\infty}^{\lambda} g(\mu) d_\mu (\varphi, E_\mu \psi) \quad \textit{since} E_\lambda E_\mu = E_\mu,\ \mu < \lambda \\ &= \int_{-\infty}^{\infty} f(\lambda) g(\lambda)\, d_\lambda (\varphi, E_\lambda \psi) \\ &= \int_{-\infty}^{\infty} (fg)(\lambda)\, d_\lambda (E_\lambda \varphi, \psi).\end{aligned}$$

Hence $(fg)(A) = f(A)g(A)$.

(iv) For any $\varphi \in H$,

$$(f(A)\varphi, \varphi) = \int_{-\infty}^{\infty} f(\lambda)\, d(E_\lambda \varphi, \varphi) = \int_{-\infty}^{\infty} f(\lambda)\, d(\| E_\lambda \varphi \|^2).$$

4.7 SPECTRAL DECOMPOSITIONS OF HILBERT SPACES

There are relations involving a self-adjoint operator $A : H \supseteq D(A) \to H$ on a complex separable Hilbert space H that can be used to provide decompositions into simpler parts of not only a given operator such as A but also the underlying Hilbert space, the associated spectrum of A and other related quantities.

4.7.1 An Illustration

Recall that for any self-adjoint operator $A : H \to H$ the point spectrum $\sigma_p(A)$ of A is (definition 4.2)

$$\sigma_p(A) := \{\lambda \in \mathbf{R} : \exists\, \theta \neq u \in H \text{ s.t } Au = \lambda u\}.$$

We define (see definition 4.13) $H_p(A)$ to be the linear span of all eigenfunctions of A. As we have seen, $H_p(A)$ is a subspace of H and is called the *point subspace* of H with respect to A. Hence by means of the projection theorem we can write

$$H = H_p(A) \oplus H_c(A), \quad \text{where } H_c(A) = H_p^{\perp}(A). \tag{4.46}$$

We refer to $H_c(A)$ as the *subspace of continuity* of H with respect to A.

Let

$$P_p : H \to H_p(A)$$

denote the projection onto $H_p(A)$. Then for any $f \in H$,

$$f = P_p f + (I - P_p) f = P_p f + P_c f, \tag{4.47}$$

where

$$P_c := (I - P_p) : H \to H_c(A)$$

is a projection orthogonal to P_p.

Thus in (4.46) and (4.47) we have a decomposition of H with respect to A. Furthermore, a decomposition of A is also available. Using (4.47), we obtain

$$Af = AP_p f + AP_c f =: A_p f + A_c f, \tag{4.48}$$

where A_p, regarded as an operator on $H_p(A)$, is called the *discontinuous part* of A, whilst A_c, regarded as an operator on $H_c(A)$, is called the *continuous part* of A.

Also, using (4.47) together with the associated spectral family $\{E_\lambda\}$, we obtain

$$(E_\lambda f, f) = (E_\lambda P_p f, f) + (E_\lambda P_c f, f). \tag{4.49}$$

This result provides a means of decomposing integrals such as those appearing in (4.21)–(4.23). To see how this can be achieved, we need to return to the notion of a Riemann-Stieltjes integral introduced in (4.19).

4.7.2 A Little More About Riemann-Stieltjes Integrals

Let f and g denote real-valued functions defined on a closed interval $J : [a, b] \subset \mathbf{R}$. We shall assume throughout this subsection, unless otherwise stated, that f and g are bounded on J.

A *partition* π of J is a finite collection of nonoverlapping intervals whose union is J. We describe π by specifying a finite set of real numbers $(x_1, x_2, ..., x_n)$ such that

$$a = x_0 \leq x_1 \leq \cdots \leq x_n = b.$$

The subintervals occurring in the partition π are the intervals $[x_{k-1}, x_k]$, $k = 1, 2, \ldots, n$. The end points x_k, $k = 1, 2, ..., n$, are the *partition points* of π.

If π_1, π_2 are two partitions of J, then we say that π_2 is a *refinement* of π_1, or equivalently, that π_2 is *finer* than π_1, if every subinterval of π_2 is contained in some subinterval of π_1. Equivalently, every partition point of π_1 is also a partition point of π_2. When this is the case, we write $\pi_1 \subseteq \pi_2$.

DEFINITION 4.36 *If π is a partition of $J : [a, b] \subset \mathbf{R}$, then the Riemann-Stieltjes sum of f with respect to g corresponding to the partition $\pi = (x_1, x_2, ..., x_n)$ is a real number $S(\pi; f, g)$ defined by*

$$S(\pi;\, f, g) = \sum_{k=1}^{n} f(\xi_k)\{g(x_k) - g(x_{k-1})\}, \tag{4.50}$$

where $x_{k-1} \leq \xi_k \leq x_k$ for $k = 1, 2, ..., n$.

We notice that if $g(x) = x$, then the right-hand side of (4.50) reduces to

$$\sum_{k=1}^{n} f(\xi_k)\{x_k - x_{k-1}\}, \tag{4.51}$$

which is the usual *Riemann sum* of f corresponding to the partition π.

The Riemann sum (4.51) can be interpreted as the union of rectangles of height $f(\xi_k)$ and base $[x_k - x_{k-1}]$. As such, it offers an approximation to the area under the graph of f on the interval $[a, b]$. This approximation improves as π is refined.

The Riemann-Stieltjes sum (4.50) is similar to the Riemann sum (4.51) except that, instead of considering the *length* $(x_k - x_{k-1})$ of the subinterval $[x_{k-1}, x_k]$, we consider *some other measure of magnitude* of this subinterval, namely, the difference $(g(x_k) - g(x_{k-1}))$. The idea is that we want to be able to consider measures of magnitude of an interval other than the usual length.

DEFINITION 4.37 *A function f is Riemann-Stieltjes–integrable with respect to g on J if there exists a real number I such that for every $\varepsilon > 0$ there is a partition π_ε of J such that if π is any refinement of π_ε and if $S(\pi;\, f, g)$ is any Riemann-Stieltjes sum corresponding to π, then*

$$|S(\pi;\, f, g) - I| < \varepsilon.$$

In this case the number I is uniquely determined and is denoted by

$$I = \int_a^b f\,dg = \int_a^b f(t)\,dg(t), \tag{4.52}$$

which is called the Riemann-Stieltjes integral of f with respect to g over $J = [a, b]$. The function f is called the integrand, and the function g the integrator. In the particular case when $g(x) = x$ and f is Riemann-Stieltjes–integrable with respect to g, we say that f is Riemann-integrable and (4.52) reduces to

$$\int_a^b f(t)\,dt,$$

which is the familiar Riemann integral of f over $J = [a, b]$.

Example 4.38

(i) If g is constant on the interval $J = [a, b]$, then f is Riemann-Stieltjes–integrable with respect to g on J and the value of the integral is zero.

(ii) Let g be defined on $J = [a, b]$ by

$$g(x) = \begin{cases} 0, & x = a, \\ 1, & a < x < b. \end{cases} \tag{4.53}$$

The function f is Riemann-Stieltjes–integrable with respect to g if and only if f is continuous at $x = a$, in which case the value of the integral is $f(a)$.

The details of these two examples are left as exercises.

The function g in the integral in (4.52) provides a measure of the magnitude we associate with subintervals $[x_{k-1}, x_k] \subset [a, b]$, $k = 1, 2, \ldots, n$, other than simply the usual length of the interval. We say that g *generates* a measure on J that we shall refer to as a *Stieltjes measure*. In the particular case when $g(x) = x$, we recover the familiar integral and say that g in this case generates a *Lebesgue measure* on J.

If we now compare (4.52) and the results (4.29), (4.30), then we see that the integrals in (4.29), (4.30) are Stieltjes integrals in which integration is with respect to the numerically valued measure (integrator) generated by the function w defined by

$$w : \lambda \to w(\lambda) = (E_\lambda \varphi, \psi), \qquad \forall \varphi, \psi \in H.$$

Here w is a numerically valued function of λ and as such generates a measure on the real line. More particularly, since the action of the function w involves $\{E_\lambda\}$, the spectral family of a self-adjoint operator $A : H \to H$, we see that w generates a measure on $\sigma(A)$, the spectrum of A. We make these aspects more precise in the next subsection. We shall also indicate how we can choose (generate) a Stieltjes measure that is particularly useful in the context of scattering theory. As we shall see, it leads quite quickly to the types of decompositions we mentioned in the previous subsection.

4.7.3 Spectral Measure

Let H be a complex separable Hilbert space and let $A : H \to H$ be a bounded self-adjoint operator with associated spectral family $\{E_\lambda\}$.

For any interval $\Delta := (\lambda', \lambda''] \subset \mathbf{R}$ we define

$$E_\Delta = E_{\lambda''} - E_{\lambda'}. \tag{4.54}$$

The definition of the spectral family (definition 4.22) together with (4.54) indicate the following.

(i) E_Δ is a projection onto the subspace $M_{\lambda''} \ominus M_{\lambda'}$, that is, onto the orthogonal complement of $M_{\lambda'}$ in $M_{\lambda''}$.

(ii) If Δ_1, Δ_2 are disjoint intervals on $\mathbf{R}$, then

$$E_{\Delta_1} E_{\Delta_2} = E_{\Delta_2} E_{\Delta_1} = 0;$$

that is, the ranges of the E_{Δ_1} and E_{Δ_2} are orthogonal. This follows directly by using (4.54), writing the various products out in full and using definition 4.22.

(iii) $E_{\Delta_1} E_{\Delta_2} = E_{\Delta_1 \cap \Delta_2}$. Again, as in (ii), this follows by direct calculation.

The family $\{E_\lambda\}$ defined in this manner is called a *spectral measure* on the class of all sets Δ of the form indicated above.

The definition of E_Δ in (4.54) can be extended to closed and open intervals. To do this, we first recall the notation

$$E_{\lambda \pm 0} = s - \lim_{\eta \to 0^+} E_{\lambda \pm \eta}. \tag{4.55}$$

A spectral family $\{E_\lambda\}$ is said to be *right-continuous* if $E_{\lambda+0} = E_\lambda$ and *left-continuous* if $E_{\lambda-0} = E_\lambda$. We shall assume in the following that spectral families are right-continuous.

We now define

$$E\{\lambda\} = E_\lambda - E_{\lambda-0}. \tag{4.56}$$

The required extensions of (4.55) are given as follows.

(i) $\Delta_1 = [\lambda', \lambda'']$; then we set

$$E_{\Delta_1} = E_{d_1} + E\{\lambda'\}, \qquad d_1 = (\lambda', \lambda''].$$

(ii) $\Delta_2 = (\lambda', \lambda'')$; then we set

$$E_{\Delta_2} = s - \lim_{n \to \infty} E_{d_2}, \qquad d_2 = (\lambda', \lambda'' - 1/n].$$

This is meaningful since $\Delta_2 = \cup_n d_2$.

A more general introduction and treatment of spectral measures can be found in the texts cited in chapter 11. We would mention in particular [4, 102]. In this monograph we shall be mainly concerned with integration with respect to numerical measures generated by a function of the form

$$w : \lambda \to w(\lambda) = (E_\lambda \varphi, \psi), \qquad \forall \varphi, \psi \in H, \quad \lambda \in \mathbf{R}.$$

In this connection the following theorem is instructive.

THEOREM 4.39 *Let H be a complex separable Hilbert space and let $A : H \supseteq D(A) \to H$ be a self-adjoint operator with spectral family $\{E_\lambda\}$. For any $\varphi, \psi \in H$, the complex-valued function $\lambda \to (E_\lambda \varphi, \psi)$ is of bounded variation.*

Proof. Let $\Delta := (\lambda', \lambda''] \subset \mathbf{R}$ and let

$$\lambda' = \lambda_0 < \lambda_1 < \cdots < \lambda_n = \lambda''$$

be a partition of Δ.

If we set $\Delta_k := (\lambda_{k-1}, \lambda_k]$, then Δ is the union of all the $\{\Delta_k\}$, where the Δ_k are mutually disjoint. The properties of inner products and spectral families together with the Schwarz inequality (definition 3.67) allow us to write

$$\begin{aligned}
\sum_{k=1}^{n} \left|(E_{\lambda_k}\varphi, \psi) - (E_{\lambda_{k-1}}\varphi, \psi)\right| &= \sum_{k=1}^{n} |(E_{\Delta_k}\varphi, \psi)| \\
&= \sum_{k=1}^{n} |(E_{\Delta_k}\varphi, E_{\Delta_k}\psi)| \\
&\le \sum_{k=1}^{n} \|E_{\Delta_k}\varphi\| \|E_{\Delta_k}\psi\| \\
&\le \left(\sum_{k=1}^{n} \|E_{\Delta_k}\varphi\|^2\right)^{1/2} \left(\sum_{k=1}^{n} \|E_{\Delta_k}\psi\|^2\right)^{1/2} \\
&= \left(\sum_{k=1}^{n} (E_{\Delta_k}\varphi, \varphi)\right)^{1/2} \left(\sum_{k=1}^{n} (E_{\Delta_k}\psi, \psi)\right)^{1/2} \\
&= \|E_\Delta \varphi\| \, \|E_\Delta \psi\| \le \|\varphi\| \, \|\psi\|
\end{aligned}$$

Hence the total variation of $(E_\lambda \varphi, \psi)$ does not exceed $\|\varphi\| \, \|\psi\|$. □

Using standard methods ([102]), we can now use theorem 4.39 to define Riemann-Stieltjes integrals over any finite interval with respect to the measure generated by the numerically valued function w defined by

$$w : \lambda \to w(\lambda) = (E_\lambda \varphi, \psi), \qquad \varphi, \psi \in H.$$

As a consequence, if f is a complex-valued continuous function of λ, then the integral

$$\int_a^b f(\lambda)\, d(E_\lambda \varphi, \psi)$$

defined as the limit of Riemann-Stieltjes sums (definition 4.36) exists for every finite a, b and for all $\varphi, \psi \in H$. Improper integrals over $\mathbf{R}$ are defined, as in the case of Riemann integrals, as

$$\int_{-\infty}^{\infty} f(\lambda)\, d(E_\lambda\varphi, \psi) = \lim \int_a^b f(\lambda)\, d(E_\lambda\varphi, \psi) \tag{4.57}$$

as $a \to -\infty$ and $b \to \infty$ whenever they exist.

4.7.4 On Spectral Subspaces of a Hilbert Space

In this subsection we reintroduce some already familiar concepts and notation. This we do in a slightly different way that will have advantages later.

Let H be a complex separable Hilbert space and let $A : H \supseteq D(A) \to H$ be a self-adjoint operator with associated spectral family $\{E_\lambda\}$. We have seen in theorems 4.26 and 4.27 and in the notation introduced in (4.55) that $P_\lambda = E_\lambda - E_{\lambda-0}$ is nonzero if and only if λ is an eigenvalue of A and P_λ is the orthogonal projection onto the associated eigenspace. The set of all eigenvalues of A we have denoted by $\sigma_p(A)$ and have referred to it as the *point spectrum* of A. Let H_p denote the subspaces spanned by all the eigenvectors of A, that is, spanned by all the eigenvectors $P_\lambda H$. If $H_p = H$, then A is said to have a *pure point spectrum.*

In general $\sigma_p(A)$ is *not* a closed set. To see that this is the case, consider an operator A that has a pure point spectrum. The spectrum of A, denoted $\sigma(A)$, is the point spectrum of A together with all its points of accumulation; that is, $\sigma(A) := \overline{\sigma_p(A)}$. Furthermore, H_p reduces A since $P_\lambda H$ does, and A_p, the part of A in H_p, has a pure point spectrum. The proof of these statements is left as an exercise.

In the case when A has no eigenvalues, $H_p = \{\theta\}$ and A is said to have a *purely continuous spectrum.* We define $H_c = H_p^{\perp}$. In general, the part A_c of A defined in H_c has a purely continuous spectrum, which we denote by $\sigma(A_c)$. The spectrum of A_c, denoted $\sigma(A_c)$, is called the *continuous spectrum* of A and will be denoted in the future by $\sigma_c(A)$.

In the above we have offered a decomposition of $\sigma(A)$, the spectrum of A, in terms of a decomposition of H rather than a decomposition of $\mathbf{C}$ as previously.

The subspaces $H_p \equiv H_p(A)$ and $H_c \equiv H_c(A)$ are called the *subspace of discontinuity* and the *subspace of continuity*, respectively. When there is no danger of confusion, we shall simply write H_p and H_c.

The following result characterises H_c.

THEOREM 4.40 *$f \in H_c$ if and only if $(E_\lambda f, f)$ is a continuous function of λ.*

Proof. Let $w(\lambda) = (E_\lambda f, f)$. If w is a continuous function of λ, then $w(\lambda) \to w(\mu)$ as $\lambda \to \mu$. This implies that $((E_\lambda - E_\mu) f, f) \to 0$ as $\lambda \to \mu$. Hence, recalling (4.56), we can conclude that $(P_\lambda f, f) = 0$ for all $\lambda \in \mathbf{R}$.

Since P_λ is a projection and recalling the Schwarz inequality $|(a, b)| \le \|a\|\,\|b\|$, we obtain for all $g \in H$,

$$|(E_\lambda g, f)|^2 = |(E_\lambda g, E_\lambda f)|^2 \le (E_\lambda g, g)(E_\lambda f, f) = 0.$$

Hence f is orthogonal to the ranges of all P_λ and therefore to H_p. Hence $f \in H_c$.

Conversely, if $f \in H_c = H_p^{\perp}$, then f is orthogonal to P_λ for all λ so that, again recalling (4.56), $(E_\lambda f, f)$ is continuous. □

We shall find it useful when developing a scattering theory to further subdivide (decompose) H_c.

We have seen in subsection 4.7.3 that the spectral family $\{E_\lambda\}$ generates a spectral measure E_Δ. Thus, for any fixed $f \in H$, we can construct a non-negative measure by defining

$$m_f(\Delta) = (E_\Delta f, f) = \|E_\Delta f\|^2. \tag{4.58}$$

We now introduce two further subspaces of H.

DEFINITION 4.41 *An element $f \in H$ is said to be absolutely continuous with respect to A if m_f is absolutely continuous with respect to the Lebesgue measure $|.|$ on* **R**. *That is, $|\Delta| = 0$ implies $m_f(\Delta) = \|E_\Delta f\|^2 = 0$.*

DEFINITION 4.42 *A measure m_f is singular with respect to Lebesgue measure on* **R** *if there is a set Δ_0 with $|\Delta_0| = 0$ such that $m_f(\Delta) = m_f(\Delta \cap \Delta_0)$ for all sets $\Delta \subseteq \mathbf{R}$. In which case, f is said to be singular with respect to A.*

The set of all elements in H that are absolutely continuous (singular) with respect to A, denoted $H_{ac}(H_s)$, is called the subspace of absolute continuity (singularity) with respect to A.

THEOREM 4.43 *H_{ac} and H_s are subspaces of H, are orthogonal complements of each other and reduce A.*

A proof of this theorem can be found in [59] and [95].

Since the point set $\{\lambda\}$ has Lebesgue measure zero, we have $(E_\lambda f, f) = (E_{\lambda-0} f, f)$ for all $f \in H_{ac}$ and all $\lambda \in \mathbf{R}$. Therefore, by theorem 4.40, we have $H_{ac} \subseteq H_c$ and $H_p \subseteq H_s$. If we now set

$$H_{sc} := H_c \ominus H_{ac}, \tag{4.59}$$

then we obtain, for each self-adjoint operator $A : H \to H$, the following decompositions of H.

$$H = H_{ac} \oplus H_s = H_{ac} \oplus H_{sc} \oplus H_p. \tag{4.60}$$

If $H_{ac} = H$, then A is said to be (*spectrally*) *absolutely continuous*. If $H_s = H$, then A is said to be (*spectrally*) *singularly continuous*.

The parts of A on these various subspaces are denoted A_{ac}, A_s, A_{sc}, respectively. We write $\sigma_{ac}(A), \sigma_s(A), \sigma_{sc}(A)$ to denote the absolutely continuous, the singular and the singularly continuous spectra of A, respectively. The associated components of the spectrum of A are given by $\sigma(A_{ac})$, $\sigma(A_s)$ and $\sigma(A_{sc})$, respectively.

The physical relevance of the decomposition (4.60) can be considerable, particularly in some areas of quantum scattering ([95]). For the moment, we simply state that $H_p(A)$ usually contains the *bound states* of A, whilst $H_c(A)$, and more especially $H_{ac}(A)$, consists of *scattering states* of A. For most self-adjoint operators arising in applications, it turns out that $H_{sc} = \{\theta\}$, which implies that $H_c = H_{ac}$.

Further detailed discussions along these lines can be found in the texts cited in chapter 11.

Finally in this subsection, we mention yet another way of decomposing the spectrum of a self-adjoint operator $A : H \to H$, where H is a Hilbert space.

DEFINITION 4.44 *The set of all $\lambda \in \sigma(A)$ with the range of P_λ finite-dimensional forms $\sigma_d(A)$ the discrete spectrum of A.*

The set complementary to $\sigma_d(A)$ in $\sigma(A)$ constitutes the essential spectrum of A and is denoted $\sigma_e(A)$.

The sets $\sigma_d(A)$ and $\sigma_e(A)$ are disjoint. The set $\sigma_e(A)$ consists of the continuous spectrum of A, the accumulation points of the point spectrum of A and eigenvalues of A of infinite multiplicities.

Finally, we would remark that in this chapter we have not worked explicitly with unbounded operators. If we wish to do this, then more care must be taken when handling the domain of the operator. In this connection see chapter 3 and [53, 59].

Chapter Five

On Nonautonomous Problems

5.1 INTRODUCTION

In Chapter 1 we saw that in the acoustic case an IBVP could be reduced, formally at least, to an IVP. The IBVP was defined in $\mathbf{R}^n \times \mathbf{R}$ in terms of a partial differential equation that was of second order in time, whilst the IVP was a Cauchy problem for a system of ordinary differential equations that was first order in time and defined on an appropriate energy space. It turns out that for these first-order equations, results concerning existence, uniqueness and stability of solutions can be obtained in an efficient and elegant manner using results from the theory of semigroups and the theory of Volterra integral equations. Furthermore, such an approach, which will be used for both APs and NAPs, will be seen to offer good prospects for developing constructive methods of solution.

We would point out that, in keeping with the spirit of this monograph, most of the results offered in this chapter are simply stated without proof. This we have done because in most practical problems the main aim is centred on making progress by the application of analytical mathematical results rather than by working through their proofs, which are often quite lengthy. However, references will be given either here in the text or in chapter 11 indicating where detailed proofs are to be found.

We remark that we shall work here in a Hilbert space setting. However, a similar analysis can be conducted in a more general Banach space setting ([105, 83]).

5.2 CONCERNING SEMIGROUP METHODS

We shall confine our attention initially to a discussion of APs and then go on to show how the various results and techniques have to be modified when we come to deal with NAPs.

Let H denote a Hilbert space and consider the IVP

$$\left\{\frac{d}{dt} - G\right\} w(t) = 0, \qquad t \in \mathbf{R}^+ =: (0, \infty), \qquad w(0) = w_0, \tag{5.1}$$

where $w \in C(\mathbf{R}^+, H)$ and $G : H \supseteq D(G) \to H$. We remark that any boundary conditions imposed on the originating problem are accommodated in the definition of $D(G)$, the domain of G.

An IVP of the form (5.1) governs the manner in which a system evolves from an initial state $w(0) = w_0$ to another state $w(t)$ at some other time $t \neq 0$. The operator

G characterises the particular class of problem being considered. The presentation in this section will allow a wide range of specific forms to be accommodated.

When faced with a problem such as (5.1), the first requirement is to clarify the meaning of the defining equation. The definition of an ordinary derivative indicates that (5.1) should be interpreted to mean that

(i) $w(t) \in D(G)$,
(ii) $\lim_{h \to 0} \|h^{-1}\{w(t+h) - w(t)\} - Gw(t)\| = 0$, where $\|.\|$ denotes the norm in H.

When the problem (5.1) models a physical evolutionary system, ideally it should be well posed in the following sense.

Definition 5.1 *A problem is said to be well posed if it has a unique solution that depends continuously on the given data.*

This definition implies that small changes in the given data produce only small changes in the solution.

Let the evolutionary problem (5.1) be well posed and let $U(t)$ denote the transformation that maps $w(s)$, the solution at time s, onto $w(s+t)$, the solution at time $(s+t)$; that is,

$$w(s+t) = U(t)w(s).$$

In particular we have

$$w(t) = U(t)w(0) = U(t)w_0.$$

Therefore, since (5.1) is assumed to be well posed, a solution of (5.1) is unique and we have

$$U(s+t)w_0 = w(s+t) = U(t)w(s) = U(t)U(s)w_0,$$

which implies the so-called *semigroup properties*

$$U(s+t) = U(s)U(t), \qquad s, t \in \mathbf{R}^+, \qquad U(0) = I. \tag{5.2}$$

We are thus led to the consideration of a family of operators $\{U(t)\}_{t \geq 0}$. Our first aim is to determine the family $\{U(t)\}_{t \geq 0}$. With this in mind we notice that (5.2) is reminiscent of the properties of the familiar exponential function. In support of this remark we recall the following result obtained when investigating *Cauchy's functional equation*. Specifically, if $f : [0, \infty) \to \mathbf{R}$ is such that

(i) $f(s+t) = f(s)f(t)$ for all $s, t \geq 0$,
(ii) $f(0) = 1$,
(iii) f is continuous on $[0, \infty)$ (on the right at the origin), then f is defined by

$$f(t) = \exp\{tA\}$$

for some *constant* $A \in \mathbf{R}$.

Furthermore, if we apply an integrating factor technique to (5.1), then, formally at least, we have

$$w(t) = \exp\{tG\}w_0$$

provided that G is regarded as a constant. With these results in mind it is natural to conjecture that

$$U(t) = \exp\{tG\} \tag{5.3}$$

for some *operator* G. Of course, this conjecture has to be proved if it is to be of any use. In this connection we introduce the following notion.

DEFINITION 5.2 *A family $U := \{U(t)\}_{t\geq 0}$ of bounded linear operators on a Hilbert space into itself is called a strongly continuous, one-parameter semigroup, denoted a C_0-semigroup, provided*
(i) $U(s+t)f = U(s)U(t)f = U(t)U(s)f$ for all $f \in H$ and $s,t \geq 0$,
(ii) $U(0)f = f$ for all $f \in H$,
(iii) the mapping $t \to U(t)$ is continuous for $t \geq 0$ and for all $f \in H$.
If, in addition we have
(iv) $\|U(t)f\| \leq \|f\|$ for all $f \in H$,
then U is called a C_0-contraction semigroup.

We remark that the restriction $U(t) \in B(H)$ enables us to compare different solutions by means of the relation

$$\|U(t)u_0 - U(t)v_0\| \leq \|U(t)\| \, \|u_0 - v_0\| ,$$

which only makes sense if $U(t) \in B(H)$.

Once we have introduced the notion of a C_0-semigroup, three questions are immediate.

1. Given a C_0-semigroup $\{U(t)\}_{t\geq 0}$, how can we obtain the operator G whose existence was conjectured in (5.3)?
2. What types of operators G can appear in (5.3)?
3. Given a suitable operator G, how can we construct an associated semigroup $\{\exp(tG)\}_{t\geq 0}$?

These three questions are investigated in great detail in the general theory of C_0-semigroups [83, 94].Whilst we will always be aware of questions 1and 2, our main interest in this monograph is centred on question 3.

If we recall the interpretation given to (5.1), then the following definition appears quite natural.

DEFINITION 5.3 *The infinitesimal generator of the C_0-semigroup $\{U(t)\}_{t\geq 0}$ is the linear operator*

$$G : H \supseteq D(G) \to H,$$

defined by

$$Gf := \lim_{h\to 0}\left\{h^{-1}(U(t)f - f)\right\} \quad \textit{for } f \in D(G),$$

$$D(G) = f \in H : \lim_{h\to 0}\left\{h^{-1}(U(t)f - f)\right\} \quad \textit{exists in } H.$$

Example 5.4 *The defining properties of a C_0-semigroup suggest that, formally at least, for the family $U = \{U(t)\}_{t\geq 0}$ defined by*

$$U(t) = \exp(tG),$$

we have

$$G = \left.\frac{dU(t)}{dt}\right|_{t=0} = U'(0),$$

which indicates that U is the semigroup generated by the operator G. Furthermore, the familiar integrating factor technique indicates that the solution of (5.1) can be expressed in the form

$$w(t) = \exp\{tG\}.w_0 = U(t)w_0$$

provided $w_0 \in D(G)$. To see this last point, notice that (5.1) implies that we must have

$$Gw_0 - \left.\frac{dU(t)w}{dt}\right|_{t=0} = Gw_0 - Gw_0 = 0$$

and that this result is only valid if $w_0 \in D(G)$.

We now collect some well-known facts from the theory of semigroups. The presentation is essentially informal and is intended to provide a reference source rather than a comprehensive, self-contained account. More details can be found in the references cited here and in chapter 11.

Theorem 5.5 *[83, 105] Let $G \in B(H)$. The family $\{U(t)\}_{t\geq 0}$ defined by*

$$U := \left\{ U(t) = \exp(tG) = \sum_{n=0}^{\infty} \frac{(tG)^n}{n!} : t \in \mathbf{R}^+ \right\}$$

is a C_0-semigroup that satisfies

$$\|U(t) - I\| \to 0 \quad \textit{as } t \to 0.$$

Moreover, G is the generator of U.

Conversely, if U is a C_0-semigroup satisfying the above relation, then the generator of U is an element $G \in B(H)$.

COROLLARY 5.6 *If $U(t) = \exp(tG)$, defined as theorem 5.5, then*
(i) $\|U(t)\| = \|\exp(tG)\| \le \exp\{t\, \|G\|\}$, $t \in \mathbf{R}^+$,
(ii) $U(t) : \mathbf{R}^+ \to H$ *continuously for all* $t \in \mathbf{R}^+$,
(iii) $\frac{d^n}{dt^n}\{U(t)\} = G^n U(t) = U(t) G^n$.

In many cases of practical interest the operator G appearing in (5.1) can be unbounded. For such an operator some of the quantities used above such as $\|G\|$ and $\exp(tG)$, defined as in theorem 5.5, are meaningless. Consequently, results such as theorem 5.5 have to be modified. With this in mind the following results are available ([83, 105]).

THEOREM 5.7 *Let H be a Hilbert space and let $U := \{U(t), t \ge 0\}$ be a C_0-semigroup with generator G.*
(i) $U(t)U(s) = U(s)U(t)$ *for all* $t, s \ge 0$.
(ii) U *is exponentially bounded in the sense that there exist constants* $M > 0$ *and* $\omega \in \mathbf{R}$ *such that* $\|U(t)\| \le M \exp(\omega t)$.
(iii) $\overline{D(G)} = H$, *and the operator* G *is closed.*
(iv) $\frac{d}{dt}\{U(t)f\} = U(t)Gf = GU(t)f$ *for all* $f \in D(G)$.
(v) For all $\lambda \in \mathbf{C}$ *such that* $\mathrm{Re}\{\lambda\} > \omega$, *there exists* $R\{G, \lambda\} := (\lambda I - G)^{-1}$ *and*

$$R\{G, \lambda\} f = \int_0^\infty e^{-\lambda t} U(t) f, \qquad f \in H.$$

In the next subsections we give conditions that ensure that problems of the form (5.1) are well posed. We indicate when the operator G in (5.1) actually generates a C_0-semigroup suitable for ensuring that (5.1) is well posed.

5.2.1 On the Well-posedness of Problems

Let H be a Hilbert space and let G be a densely defined operator on H. Consider the IVP

$$\left\{ \frac{d}{dt} - G \right\} w(t) = 0, \qquad t \in \mathbf{R}^+, \, w(0) = w_0. \tag{5.4}$$

A more precise definition than that given earlier of the well-posedness of problems such as (5.4) is as follows.

DEFINITION 5.8 *The problem (5.4) is well posed if the resolvent set* $\rho(G) \neq \phi$ *and if for all* $w_0 \in D(G)$ *there exists a unique solution* $w : \mathbf{R}^+ \to D(G)$ *of (5.4) with* $w \in C^1((0, \infty), H) \cap C([0, \infty], H)$.

Results along the following lines can now be established.

THEOREM 5.9 *The problem (5.4) is well posed if G generates a C_0-semigroup U on H. In this case the solution of (5.4) is given by* $w(t) = U(t) w_0$, $t \in \mathbf{R}^+$.

Proof. Let G generate a C_0-semigroup $U = \{U(t) : t \in \mathbf{R}^+\}$. If $w_0 \in D(G)$, then by theorem 5.7 we see that $w(.) = U(.)w_0 \in C^1(\mathbf{R}^+, H)$ is $D(G)$-valued and (5.4) holds. To prove well-posedness, it remains to establish uniqueness. To this end, let φ be any solution of (5.4). Then, for $0 \leq s \leq t < \infty$, we have

$$\frac{d}{ds}\{U(t-s)\varphi(s)\} = U(t-s)G\varphi(s) - U(t-s)G\varphi(s) = 0.$$

Hence $U(t-s)\varphi(s)$ is independent of s. Consequently, since $U(0)=I$, this independence allows us to write

$$\varphi(t) = U(t-s)\varphi(s) = U(t)\varphi(0) = U(t)w_0.$$

The last equality follows from the assumption that $\varphi(t)$ is any solution of (5.4) and must therefore satisfy the imposed initial condition. The required uniqueness now follows since the right-hand side is $w(t)$. □

The converse of this theorem also holds. The details can be found in [83] and in the titles cited in chapter 11.

We will also be interested in nonhomogeneous forms of the problem (5.1). Specifically, we will want to discuss problems of the form

$$\left\{\frac{d}{dt} - G\right\} v(t) = f(t), \qquad t \in \mathbf{R}^+, \qquad v(0) = v_0, \tag{5.5}$$

where f and v_0 are given data functions. In this connection the following result holds.

Theorem 5.10 *Let H be a Hilbert space and let $G : H \supseteq D(G) \to H$ be the generator of a C_0-semigroup $U = \{U(t) : t \geq 0\} \subseteq B(H)$. If $v_0 \in D(G)$ and $f \in C^1(\mathbf{R}^+, H)$, then (5.5) has a unique solution $v \in C^1(\mathbf{R}^+, H)$ with values in $D(G)$.*

A proof of this theorem can be obtained by first noticing that a formal application to (5.5) of the familiar integrating factor technique yields

$$v(t) = U(t)v_0 + \int_0^t U(t-s)f(s)\,ds. \tag{5.6}$$

It now remains to prove that (5.6) is indeed a solution of (5.5) and, moreover, that v has all the properties indicated in the statement of theorem 5.10. This is a straightforward matter. The details are left as an exercise but can be found in [105] if required.

5.2.2 On Generators of Semigroups

We have seen that the well-posedness of an IVP can be established provided there is associated with the IVP a C_0-semigroup. The following results help to characterise the linear operators that actually generate C_0 -semigroups. The results are simply listed for our convenience in this monograph. Detailed proofs can be found in the texts cited in chapter 11. We would particularly mention [83, 105] as starting texts. We would remark that now the linear operators we will be dealing with are not necessarily bounded.

THEOREM 5.11 *Let $U = \{U(t) : t \geq 0\} \subseteq B(H)$ be a C_0-semigroup with generator G. Then $D(G)$ is dense in H. Furthermore, $G : H \supseteq D(G) \to H$ is a closed linear operator.*

THEOREM 5.12 *A C_0-semigroup is uniquely determined by its generator.*

Theorem 5.11 is a valuable result whenever we know that G is the generator of a C_0-semigroup. We also want to know when it is that an operator G is indeed the generator of a C_0-semigroup. The answer is given by the celebrated Hille-Yosida theorem [59, 83].

It will be convenient at this stage to introduce the following notation.

For real numbers $M > 0$ and $\omega > 0$, the set of all generators of C_0-semigroups $\{U(t)\}_{t \geq 0}$ that satisfy on a Hilbert space H the relation

$$\|U(t)\| \leq M \exp(\omega t)$$

will be denoted $\mathcal{G}(M, \omega, H)$.We remark that this notation is also used in general Banach spaces.

Necessary and sufficient conditions for an operator G to belong to $\mathcal{G}(M, \omega, H)$ are provided by the following theorem.

THEOREM 5.13 *(Hille-Yosida theorem) A linear operator $G : H \supseteq D(G) \to H$ is an element of $\mathcal{G}(M, \omega, H)$ if and only if*

(i) G is a closed linear operator with domain $D(G)$ dense in H,

(ii) a real number $\lambda > \omega$ is such that $\lambda \in \rho(G)$, the resolvent set of G,

(iii) $R(G, \lambda) := (\lambda I - G)^{-1}$ is such that

$$\left\| [R(G, \lambda)]^n \right\| \leq \frac{M}{(\lambda - \omega)^n}, \qquad n = 1, 2, \ldots.$$

We now turn to question 3: Specifically, if a given operator G satisfies the conditions of the Hille-Yosida theorem, then how can the C_0-semigroup $\{U(t)\}_{t \geq 0}$ generated by G be constructed?

We have seen in theorem 5.5 that when $G \in B(H)$, the semigroup $\{U(t)\}_{t \geq 0}$ generated by G is defined by

$$U := \left\{ U(t) = \exp(tG) = \sum_{n=0}^{\infty} \frac{(tG)^n}{n!} : t \in \mathbf{R}^+ \right\}. \tag{5.7}$$

However, in many applications the given operator G is not necessarily bounded. Consequently, for a not necessarily bounded operator, we try to find a family of *bounded* operators that approximate G in some sense. As a first attempt in this direction, let $\{G_k\}_{k=1}^{\infty}$ be a family of bounded operators the elements of which generate the associated semigroups $U_k := \{U_k(t)\}_{t \geq 0}$ of the form given in (5.7). We would like to have a result of the form

$$\exp\{tG\} = \lim_{k \to \infty} (\exp\{tG_k\}).$$

This would then enable us to write

$$U(t)g = \lim_{k\to\infty} U_k(t)g, \qquad t \geq 0, \qquad g \in H.$$

THEOREM 5.14 *Let G be the generator of a C_0-semigroup $\{U(t)\}_{t\geq 0} \subset B(H)$. Then*

$$U(t)g = \lim_{h\to 0^+} (\exp\{tG_h\})g, \qquad t \geq 0, \quad g \in H,$$

where G_h is defined by

$$G_h g = \{U(h)g - g\}/h, \qquad g \in H,$$

and the convergence is uniform with respect to t for $0 \leq t \leq t_0$ with $t_0 \geq 0$ arbitrary [83, 94].

There is a major practical difficulty associated with this result that is centred on the approximations $G_h = \{U(h) - I\}/h$. These quantities are only known if $U(h)$ is known, and this is what we are trying to find!

A way around the above difficulty can be obtained by first recalling that when G is a real or complex number,

$$\exp\{tG\} = [\exp\{-tG\}]^{-1} = \lim_{n\to\infty}\{[I - tG/n]^n\}^{-1} = \lim_{n\to\infty}\{[I - tG/n]^{-1}\}^n.$$

When G is an operator in H, then the analogue of the result seems to be given by

$$(\exp\{tG\})g = \left(\lim_{n\to\infty}\{[I - tG/n]^{-1}\}^n\right) g =: \lim_{n\to\infty}\{[V(t/n)]^n g\}.$$

This indeed proves to be the case, as the following result can be obtained [83, 94].

THEOREM 5.15 *Let G be the generator of a C_0-semigroup $\{U(t)\}_{t\geq 0} \subset B(H)$. Then for all $g \in H$,*

$$U(t)g = \lim_{n\to\infty}\{\{[V(t/n)]^n g\}, \qquad t \geq 0 \quad \text{for all } g \in H,$$

where

$$V(t/n) = [I - tG/n]^{-1}$$

The convergence is uniform with respect to t for $0 \leq t \leq t_0$.

Finally in this subsection, we give Stone's theorem, which will be particularly useful when developing scattering theories. In order to do this, we require some preparation.

DEFINITION 5.16 *A C_0-group on a Hilbert space H is a family of operators $U := \{U(t) : t \in \mathbf{R}\} \subset B(H)$ satisfying all the requirements of definition 5.2 but with $s, t \in \mathbf{R}$. The generator G of a C_0-group U on H is defined by*

$$Gf = \lim_{h \to 0} h^{-1}\{U(h)f - f\},$$

where $D(G)$ is the domain of definition G, which is the set of all $f \in H$ for which the above limit exists. This limit is two-sided in the sense that $t \to 0$ and not just $t \to 0^+$.

We would point out that G is the generator of a C_0-group U if and only if $G_\pm$, defined as above but with $t \to 0^\pm$, respectively, generate C_0-semigroups $U_\pm$ where

$$U(t) = \begin{cases} U_+(t), & t \geq 0, \\ U_-(t), & t \leq 0. \end{cases}$$

DEFINITION 5.17 *Let H be a Hilbert space with structure $(.,.)$ and $\|.\|$.*

(i) An operator $A : H \supseteq D(A) \to H$ is a symmetric operator on H if it is densely defined on H and if

$$(Af, g) = (f, Ag) \quad \text{for all } f, g \in D(A).$$

(ii) The operator A is skew-symmetric if $A \subset -A^*$.

(iii) The operator A is self-adjoint if $A = A^*$.

(iv) The operator A is skew-adjoint if $A = -A^*$.

(v) When H is a complex Hilbert space, A is skew-adjoint if and only if (iA) is self-adjoint.

An instructive exercise is to prove the following.

THEOREM 5.18 *Let H be a Hilbert space and let $G : H \supseteq D(G) \to H$ generate a C_0-semigroup $U := \{U(t) : t \in \mathbf{R}^+\} \subset B(H)$. Then $U^* := \{U^*(t) : t \in \mathbf{R}^+\}$ is a semigroup with generator G^*.*

The semigroup U is a self-adjoint C_0-semigroup; that is, $U(t)$ is self-adjoint for all $t \in \mathbf{R}^+$ if and only if its generator is self-adjoint.

DEFINITION 5.19 *A C_0-unitary group is a C_0-group of unitary operators.*

We now state the following.

THEOREM 5.20 *(Stone's theorem) [49] Let H be a Hilbert space. An operator $G : H \supseteq D(G) \to H$ is the generator of a C_0-unitary group U on H if and only if G is skew-adjoint.*

In the remaining sections of this chapter we shall be concerned with the development for nonautonomous problems of similar results to those introduced in the preceding sections.

5.3 THE PROPAGATOR AND ITS PROPERTIES

When we come to consider NAPs, we will find that we will have to consider, for some fixed number $T > 0$, families of operators $\{G(t) : 0 \leq t \leq T\}$. In the first instance we make the following enabling assumptions.

ASSUMPTION 5.21 *(i) There exist Hilbert spaces H and H_1 with structure $(\|.\|, (.,.))$ and $(\|.\|_1, (.,.)_1)$, respectively, such that*

$$H_1 \overset{d}{\hookrightarrow} H;$$

that is, H_1 is continuously and densely embedded in H.

(ii) The operators $G(t)$, $t \in [0, T]$, are closed linear operators in H with $D(G(t)) = H_1$ for all $t \in [0, T]$.

(iii) For each $t \in [0, T]$, the operator $G(t)$ generates a semigroup.

When dealing with NAPs, we will find that we have to consider IVPs of the form

$$\left\{\frac{d}{dt} - G(t)\right\} u(t) = f(t), \qquad t \in (s, t), \qquad u(s) = u_s, \tag{5.8}$$

where $s \in [0, T)$ is fixed with $u_s \in H$ and $f \in C([s, T], H)$ are given data functions.

DEFINITION 5.22 *By a solution of (5.8) we mean an element $u \in C([s, T], H) \cap C^1((s, T], H)$ such that $u(t) \in H_1$ for $t \in (s, T]$ and $u(s) = u_s$ that satisfies*

$$\{d_t - G(t)\}u(t) = f(t), \qquad t \in (s, t), \qquad d_t := \frac{d}{dt}$$

REMARK 5.23 *All that we have said so far has been conducted in a Hilbert space setting. This is in anticipation of use in later chapters. The whole matter can be conducted in general Banach spaces if required [59, 142].*

For APs, the operator $G(t) = G$ is independent of t, and we find that we come to work with the semigroup $U := \{U(t) = \exp\{tG\}\}$. In arriving at this result we would have used an integrating factor technique. The use of the same technique when investigating (5.8) leads, formally at least, to a representation of the required solution in the form

$$u(t) = U(t, s)u_s + \int_s^t U(t, \tau) f(\tau)\, d\tau, \tag{5.9}$$

where

$$U(t, s) := \exp\left\{\int_s^t G(\eta)\, d\eta\right\}. \tag{5.10}$$

With this in mind it seems to be convenient to introduce, when dealing with NAPs, an operator $U(t, s)$ with the following properties [83].

Definition 5.24 *Let*

$$\overline{\Omega(T)} := \{(t, s) : 0 \leq s \leq t \leq T\}, \qquad \Omega(T) := \{(t, s) : 0 \leq s < t \leq T\}.$$

A propagator (evolution operator, fundamental solution) for the family $\{G(t)\}_{0 \leq t \leq T}$ *is a mapping* $U : \overline{\Omega(T)} \to B(H)$ *with the properties*

(i) $U \in C(\overline{\Omega(T)}, B(H)) \cap C(\overline{\Omega(T)}, B(H_1)) \cap C(\Omega(T), B(H, H_1))$,

(ii) $U(t, t) = I$, *where* I *is the identity operator on* H,

(iii) $U(t, s) = U(t, \tau)U(\tau, s)$ *for all* $0 \leq s \leq \tau \leq t \leq T$,

(iv) The mapping $(t, s) \to G(t)U(t, s)$ *is an element in* $C(\Omega(T), B(H))$ *and*

$$\sup_{(t,s) \in \Omega(T)} (t - s) \, \|G(t)U(t, s)\| < \infty,$$

(v) $U(., s) \in C^1((s, T], B(H))$ *for each* $s \in [0, T)$ *and for all* $t \in (s, T]$ *such that* $\partial_t U(t, s) \equiv \partial_1 U(t, s) = G(t)U(t, s)$,

(vi) $U(t, .) \in C^1([0, t), B(H_1, H))$ *for each* $t \in (0, T]$ *and for all* $s \in [0, t)$ *such that* $\partial_s U(t, s) \equiv \partial_2 U(t, s) = -U(t, s)G(s)$ *for all elements in* H_1.

In general, the relations in (vi) involve unbounded operators. We assume, therefore, that they hold on a suitable dense subspace of H, which has to be determined for each equation being considered.

We notice that when $G(t) = G$ is independent of t,

$$U(t, s) = U(t - s) = \exp\{(t - s)G\}$$

is the associated propagator, as might have been expected, and that (iv) in definition 5.24 holds on $D(G)$.

It is clear that, formally, $U(t, s)$ in (5.10) satisfies the requirements of definition 5.24. However, we emphasise that the validity of (5.9), (5.10) depends crucially on the existence of $U(t, s)$.

In the general theory of propagators a number of additional assumptions are introduced. We list in the following those that will be useful to us later (see [59, 83]) and chapter 11.

Assumptions 5.25

(i) For all $t \in [0, T]$, *we have*

$$[\mathrm{Re}\,\mu \geq 0] := \{\mu \in \mathbf{C} : \mathrm{Re}\,\mu \geq 0\} \subset \rho(G(t)).$$

(ii) There exists a constant $M \geq 1$, *independent of* $t \in [0, T]$, *such that*

$$\|R(G(t), \lambda)\| := \left\|(\lambda I - G(t))^{-1}\right\| \leq \frac{M}{1 + |\lambda|}$$

for all $(\lambda, t) \in [\mathrm{Re}\,\mu \geq 0] \times [0, T]$.

(iii) There exists a constant $b \in (0, 1)$ *such that*

$$G(.) \in C^b([0, T], B(H, H_1)).$$

Here C^b *denotes Hölder continuity of index* b; *see chapter 11, appendix A1.*

The usefulness of the propagator is illustrated by the following technical results, which are important in the study of the solvability of (5.8).

THEOREM 5.26 *(i) There exists at most one propagator $U(t, s)$ for the family $\{G(t): 0 \leq t \leq T\}$. If u is any solution of (5.8), then the representation*

$$u(t) = U(t, s)u_s + \int_s^t U(t, \tau) f(\tau)\, d\tau$$

holds for all $t \in [0, T]$.

(ii) There exists a unique propagator for the family $\{G(t): 0 \leq t \leq T\}$. Furthermore,

$$\|U(t, s)\|_* \quad \text{and} \quad (t - s)\, \|G(t)U(t, s)\|,$$

where $\| \|_*$ denotes the norm in either $B(H_1)$ or $B(H)$ as appropriate, are uniformly bounded for $(t, s) \in \{(t, s): 0 \leq s \leq t \leq T\}$ with a constant of continuity depending only on M, the Hölder norm of $G(.)$, and a bound for $\left\|G(t)G^{-1}(s)\right\|$.

Proof. (i) Let u be any solution of (5.8). Then, by definition 5.24(iv), we have for any $s < r < t \leq T$,

$$\partial_2\{U(t, r)u(r)\} = U(t, r)u'(r) - U(t, r)G(r)u(r) = U(t, r) f(r).$$

Now integrate both sides from $\varepsilon \in (s, t)$ to t and let $\varepsilon \to 0$ to obtain the required result. The formula (5.10) then clearly indicates the uniqueness.

The representation formula (5.9) is used frequently in the following chapters. It is referred to as the *variation of parameters formula.*

(ii) This theorem was also proved independently by Sobolevski [130] and by Tanabe [141]. We do not give a proof here since it is long and technical. Complete proofs appear in the above references and also in [142] and [94]. □

In applications the actual determination of the propagator is of crucial importance. A move in this direction is to write $U(t, s)$ as a perturbation of the semigroup $U(s)$, which is generated by $G(s)$, where s is *fixed* in $[0, T)$, as follows.

$$\begin{aligned} U(t, s) &= \exp\{(t - s)G(s)\} + P(t, s) \\ &= U(t - s) + P(t, s), \qquad 0 \leq s \leq t \leq T. \end{aligned} \tag{5.11}$$

It is clear that $P(t, t) = 0$ for all t.

Using definition 5.24(iv) and (5.11), we obtain

$$\partial_1 U(t, s) = G(t)U(t, s) = G(t)\{U(t - s) + P(t, s)\}$$

and

$$\begin{aligned} \partial_1 U(t, s) &= \partial_1\{U(t - s) + P(t, s)\} \\ &= G(s)U(t - s) + \partial_1 P(t, s). \end{aligned}$$

These last two relations taken together imply that the perturbation $P(t, s)$ must satisfy the IVP

$$\{\partial_t - G(t)\}P(t, s) = (G(t) - G(s))U(t - s) =: R_1(t, s),$$

$$P(s, s) = 0.$$

As in (5.11), we require $0 \le s \le t \le T$.

The variation of parameters formula (5.9) applied to this IVP indicates that

$$P(t, s) = \int_s^t U(t, \tau)R_1(\tau, s)\, d\tau,$$

and hence, recalling (5.11), we obtain

$$U(t, s) = U(t - s) + \int_s^t U(t, \tau)R_1(\tau, s)\, d\tau, \tag{5.12}$$

which is a Volterra integral equation of the second kind for $U(t, s)$.

We gather together in the final section of this chapter a number of results from the general theory of integral equations that can be specialised to deal with (5.12). Specifically, we can obtain

$$U(t, s) = U(t - s) + \int_s^t U(t - \tau)R(\tau, s)\, d\tau, \tag{5.13}$$

where
$U(t - s) = \exp\{(t - s)G(s)\}$, $s \in \mathbf{R}$ fixed
$R(t, s) = \sum_{m=1}^{\infty} R_m(t, s)$
$R_m(t, s) = \int_s^t R_{m-1}(t, \tau)R_1(\tau, s)\, d\tau$
$R_1(t, s) = \{G(t) - G(s)\}U(t - s)$.

This result taken together with theorem 5.26 forms a basis from which to develop approximate solutions of NAPs.

Finally in this section, we point out that the solvability of (5.8) can be settled using the following result.

Theorem 5.27 *For any $s \in [0, T]$, $u_0 \in H$ and $f \in C^{\nu}([s, T], H)$ with $\nu \in (0, 1)$, there exists a unique solution of (5.8), and the solution is given by the variation of parameters formula.*

The proof of this theorem has been provided independently in [130] and in [141]. The proofs will be seen to depend on estimates for and differentiability properties of the perturbation $W(t, s)$. For a detailed proof we also refer the reader to [94] and [142].

5.4 ON THE SOLUTION OF A NONAUTONOMOUS WAVE PROBLEM

In this section we indicate the use of the above material by considering an acoustic wave problem that involves a boundary condition. The analysis given here will act as a prototype for much of the work in later chapters.

The central problem with which we shall be concerned can be simply stated as follows.

A system consists of a medium containing a transmitter and a receiver. The transmitter emits a signal that is eventually detected at the receiver, possibly after it has been perturbed by some inhomogeneity in the medium. We are interested in the manner in which the emitted signal evolves through the medium and the form it assumes at the receiver.

The study of NAPs is, as we have already mentioned, motivated by problems arising in such areas as, for example, radar, sonar, nondestructive testing and ultrasonic medical diagnosis. In all these areas a powerful diagnostic is the dynamical response of the media to the emitted signal. Mathematically, many of these problems can be conveniently modelled in terms of an initial boundary value problem. To fix ideas, we shall confine our attention here to acoustic problems and to IBVPs for the classical wave equation. Specifically, we shall be interested in an IBVP outlined in the next subsection.

5.4.1 A Mathematical Model

Let Q be an open region in $\mathbf{R}^n \times \mathbf{R}$ and define

$$\Omega(t) = \{x \in \mathbf{R}^n : (x,t) \in Q\}, \qquad B(t) = \{x \in \mathbf{R}^n : (x,t) \notin Q\}.$$

The region $\Omega(t)$ denotes the exterior, at time t, of a scattering target $B(t)$. For each value of t, the domain $\Omega(t)$ is open in $(\mathbf{R}^n \times \mathbf{R})$ and is assumed to have a smooth boundary $\partial\Omega(t)$. The lateral surface of Q, denoted ∂Q, is defined by $\partial Q := \cup_{t\in\mathcal{I}} \partial\Omega(t)$, where for fixed $T \geq 0$ we denote $\mathcal{I} := \{t \in \mathbf{R} : 0 \leq t \leq T\}$. We shall assume throughout that $B(t)$ remains in a fixed ball in $\mathbf{R}^n$ and that ∂Q is timelike in the sense that the boundary speed is less than the speed of propagation.

The basic mathematical problem we investigate is the following.

Determine a quantity $u(x,t)$ satisfying the IBVP

$$\{\partial_t^2 + L(x,t)\}u(x,t) = f(x,t), \qquad (x,t) \in Q, \tag{5.14}$$

$$u(x,s) = \varphi(x,s), \qquad u_t(x,s) = \psi(x,s), \qquad x \in \Omega(s), \qquad s \in \mathbf{R}, \tag{5.15}$$

$$u(x,t) \in (bc)(t), \qquad (x,t) \in \partial\Omega(t) \times \mathbf{R}, \tag{5.16}$$

where

$L(x,t) =$ a differential expression characterising the wave field

$f, \varphi(.,s), \psi(.,s) =$ given data functions

$s \in \mathbf{R} =$ a fixed initial time

$(bc)(t) =$ the boundary condition to be satisfied by $u(.,.)$.

We remark that in (5.14) the inhomogeneous term f characterises the transmitter and the signals it emits.

To fix ideas, we shall confine our attention here to an acoustic Dirichlet problem for which we take

$$L(x,t) = -\Delta, \quad \text{for all } (x,t), \tag{5.17}$$

$$(bc)(t) = \{u(.,.) : u(x,t) = 0, \ (x,t) \in \partial\Omega(t) \times \mathbf{R})\}. \tag{5.18}$$

Our aim here is to discuss and determine solutions of the acoustic Dirichlet problem defined above. This is most conveniently done by representing the problem in an energy space setting.

5.4.2 Energy Space Setting and Solution Concepts

We shall use the notation

$$\mathbf{f} = \begin{bmatrix} f_1 \\ f_2 \end{bmatrix} = \langle f_1, \ f_2 \rangle. \tag{5.19}$$

We introduce the energy norm

$$\|\mathbf{f}\|^2 = \|\langle f_1, \ f_2\rangle\|^2 = \frac{1}{2}\int_{\mathbf{R}^n} \{|\nabla f_1(x)|^2 + |f_2(x)|^2\}\, dx, \tag{5.20}$$

where we assume $f_1, f_2 \in C_0^\infty(\mathbf{R}^n)$.

Associated with the norm in (5.20) is the inner product

$$(\mathbf{f}, \mathbf{g}) = (\nabla f_1, \nabla g_1) + (f_2, g_2), \tag{5.21}$$

where on the right-hand side of (5.21) the notation $(.,.)$ denotes the usual $L_2(\mathbf{R}^n)$ inner product.

We shall write $H_0 = H_0(\mathbf{R}^n)$ to denote the completion of $C_0^\infty(\mathbf{R}^n) \times C_0^\infty(\mathbf{R}^n)$ with respect to the energy norm (5.20) and introduce

$$H(t) = \{\mathbf{f} \in H_0 : \mathbf{f} = 0 \text{ on } B(t)\}, \tag{5.22}$$

$$H^{\text{loc}}(t) = \{\mathbf{f} = \langle f_1, \ f_2\rangle : \zeta\mathbf{f} \in H(t) \ \forall \zeta \in C_0^\infty(\mathbf{R}^n)\}. \tag{5.23}$$

We notice that

(i) the first component of $\mathbf{f} \in H^{\text{loc}}(t)$ must vanish on $\partial\Omega(t)$,
(ii) the first component of $\mathbf{f} \in H^{\text{loc}}(t)$ must vanish on $B(t)$.

It is convenient at this stage to emphasise the following notation.

(iii) $h = h(.,.) : (x,t) \to h(x,t) \in \mathbf{K} = \mathbf{R}$ or $\mathbf{C}$,
(iv) $h = h(.,.) : t \to h(.,t) =: h(t) \in H_0$ (say).

In this latter case we refer to h as an H_0-valued function of t.

We are now in a position to say what we mean by a solution of our problem.

Definition 5.28 *A function $u = u(.,.)$ is a solution of locally finite energy of the IBVP (5.14)–(5.18) if*

(i) $\mathbf{u} = \langle u_1, u_2 \rangle \in C(\mathbf{R}, H^{\text{loc}}(t))$,

(ii) $\{\partial_t^2 + L(x,t)\}u(x,t) = f(x,t)$, $(x,t) \in Q$

(in the sense of distributions).

We say that u is a solution of finite energy if $\mathbf{u} \in C(\mathbf{R}, H_0)$ and $\mathbf{u}(t) \in H(t)$ for each t. A function u is a free solution of finite energy if $\mathbf{u} \in C(\mathbf{R}, H_0)$.

If $\mathbf{u} \in H_0$, then u has finite energy and we write

$$\|\mathbf{u}(t)\|^2 = \|u(t)\|^2 = \frac{1}{2}\int_{\mathbf{R}^n} \{|\nabla u_1(x,t)|^2 + |u_2(x,t)|^2\}\, dx \tag{5.24}$$

for the total energy of u at time t. (This apparent abuse of notation will be clarified when we come to deal with specific problems, as in the next subsection.)

The wave energy in a sphere is obtained from (5.24) by restricting the range of integration appropriately.

5.4.3 Reduction to a First-Order System

By giving the IBVP (5.14)–(5.18) an energy space setting, we are led, as in chapter 1 and section 2.6, to the consideration of an initial value problem, of the following form.

$$\{\partial_t + \mathbf{N}(t)\}\mathbf{u}(x,t) = \mathbf{F}(x,t), \qquad (x,t) \in Q, \tag{5.25}$$

$$\mathbf{u}(x,s) = \langle \varphi(.,s), \psi(.,s) \rangle(x), \qquad x \in \Omega(s), \tag{5.26}$$

where we will now use boldface symbols to indicate vectors and matrices; specifically,

$$\mathbf{u}(x,t) = \langle u, u_t \rangle(x,t), \qquad \mathbf{F}(x,t) = \langle 0, f \rangle(x,t),$$

$$\mathbf{N}(t) := \begin{bmatrix} 0 & -I \\ A(t) & 0 \end{bmatrix},$$

and on denoting by $L_2^D(\Omega(t))$ the completion of $C_0^\infty(\Omega(t))$ in $L_2(\Omega(t))$, we have introduced

$$A(t) : L_2^D(\Omega(t)) \to L_2^D(\Omega(t)),$$

$$A(t)u(.,t) = L(.,t)u(.,t), \qquad u(.,t) \in D(A(t)),$$

$$D(A(t)) = \{u \in L_2^D(\Omega(t)) : L(.,t)u(.,t) \in L_2^D(\Omega(t))\}.$$

Throughout we assume that the receiver and the transmitter are in the far field of $B(t)$ and, furthermore, that $\operatorname{supp} f \subset \{(x,t), t_0 \le t \le T, |x - x_0| \le \delta_0\}$, where x_0 denotes the position of the transmitter and t_0, T, δ_0 are constants.

With the understanding that

$$\mathbf{u} = \mathbf{u}(.,.) : t \to \mathbf{u}(.,t) =: \mathbf{u}(t) \in H(t),$$

we now introduce

$$\mathbf{G}(t) : H(t) \to H(t), \tag{5.27}$$

$$\mathbf{G}(t)\mathbf{u}(t) = i\mathbf{N}(t)\mathbf{u}(t), \qquad \mathbf{u}(t) \in D(\mathbf{G}(t)),$$

$$D(\mathbf{G}(t)) = \{\mathbf{u}(t) \in H(t) : \mathbf{N}(t)\mathbf{u}(t) \in H(t)\}.$$

Then the IVP (5.25) can be realised as a first-order system in $H(t)$ in the form

$$\{d_t - i\mathbf{G}(t)\}\mathbf{u}(t) = \mathbf{F}(t), \qquad t \in \mathbf{R},$$
$$\mathbf{u}(s) = \mathbf{u}_s. \tag{5.28}$$

A simple integrating factor technique applied to (5.28) yields

$$\mathbf{u}(t) = \mathbf{U}(t,s)\mathbf{u}_s + \int_s^t \mathbf{U}(t,\tau)\mathbf{F}(\tau)\,d\tau =: \mathbf{U}(t,s)\{\mathbf{u}_s + \mathbf{h}(s,t)\}, \tag{5.29}$$

where

$$\mathbf{U}(t,s) = \exp\left\{i\int_s^t \mathbf{G}(\eta)\,d\eta\right\}. \tag{5.30}$$

The quantity $\mathbf{U}(t,s)$ is the propagator for the family $\{iG(t)\}_{0\le s\le T}$ and is the operator $\mathbf{u(s)} \to \mathbf{u(t)}$ that maps $H(s)$ onto $H(t)$. The results indicated in section 5.3 provide conditions under which the family $\{iG(t)\}_{0\le s\le T}$ has at most one associated propagator.

The relation (5.29) is the familiar variation of parameter formula, and the integral involved is a corresponding Duhamel-type integral.

5.4.4 On the Construction of the Propagator and the Solution

We have seen in the previous sections that $\mathbf{U}(t,s)$ can be obtained as the solution of an integral equation of the form (5.12). In the present case this equation has a solution with a form that can be obtained by replacing G in (5.13) by $i\mathbf{G}(t)$. Specifically, we obtain

$$\mathbf{U}(t,s) = \mathbf{U}(t-s) + \int_s^t \mathbf{U}(t-\tau)\mathbf{R}(\tau,s)\,d\tau, \tag{5.31}$$

where

$\mathbf{R}(t,s) = \sum_{m=1}^{\infty} \mathbf{R}_m(t,s)$

$\mathbf{R}_m(t,s) = \int_s^t \mathbf{R}_{m-1}(t,\tau)\mathbf{R}_1(\tau,s)\,d\tau,\ m \ge 2,\ 0 \le s \le t \le T$

$\mathbf{R}_1(t,s) = i\{\mathbf{G}(t) - \mathbf{G}(s)\}\mathbf{U}(t-s)$

and the kernel can be estimated by [83]

$$\|\mathbf{R}(t, s)\| \leq c(t-s)^{b-1}$$

and

$$i\mathbf{G}(.) \in C^b([0, T], B(H(.))).$$

It then follows that an exact representation of the required solution is

$$\mathbf{u}(t) = \left\{ \mathbf{U}(t-s) + \sum_{m=1}^{\infty} \int_s^t \mathbf{U}(t-\tau)\mathbf{R}_m(\tau, s)\, d\tau \right\} \mathbf{h}(s). \tag{5.32}$$

The results (5.31) and (5.32) provide a firm base from which to develop approximation theories.

The required solutions of problems of the form (5.14)–(5.18) are given by the first component of $\mathbf{u}$ in (5.32). This is obtained by making use of (2.72), which indicates that

$$\begin{aligned} \mathbf{U}(t) &= \exp(it\mathbf{G}(s)) \\ &= \begin{bmatrix} \cos tA^{1/2}(s) & A^{-1/2}(s)\sin tA^{1/2}(s) \\ -A^{1/2}(s)\sin tA^{1/2}(s) & \cos tA^{1/2}(s) \end{bmatrix}. \end{aligned} \tag{5.33}$$

Since s is fixed in time, the various entries in the above matrix can be interpreted by means of either the familiar spectral theorem or some appropriate generalised eigenfunction expansion theorem. It will be at this stage that the full influence of the boundary conditions and any other perturbations will come into play. We shall return to this aspect when we come to discuss scattering processes in more detail.

5.5 SOME RESULTS FROM THE THEORY OF INTEGRAL EQUATIONS

We have seen that solutions of the IVP (5.8) have the representation (5.9). For this representation to be of practical use, we must be able to determine the propagator $U(t, s)$ that, as we have shown, is the solution of the integral equation (5.12).

In this section we gather together a number of results from the general theory of integral equations that can be specialised to deal with integral equations of the form (5.12).

Throughout this section we assume that the reader is familiar with the material introduced in chapter 3 and in particular in subsection 3.6.3.

We shall consider linear integral equations that have the typical form

$$\varphi(x) = f(x) + \lambda \int K(x, y)\varphi(y)\, dy \tag{5.34}$$

The function f is a given data function, λ is a scalar, φ is the required solution function and $K(x, y)$ is the kernel of the integral expression. An equation of the form (5.34) is known as an *integral equation of the second kind*.

An integral equation with the typical form

$$\lambda \int K(x, y)\varphi(y)\,dy = f(x)$$

is known as an *integral equation of the first kind*. When the given term $f(x)$ is identically zero, the equations are called *homogeneous equations*.

When the limits on the integral are constants, the integral equations are called *Fredholm integral equations*. If one or the other of the limits on the integral in (5.34) are functions of the variable x, then the integral equations are called *Volterra integral equations*.

We remark that by working with integral equations, we can include both the salient features of a differential equation and its associated boundary and initial condition in a single equation [104].

In this section we shall be concerned almost entirely with L_2-kernels (square-integrable kernels). We first consider Fredholm integral equations.

DEFINITION 5.29 *Let $K(x, y)$ satisfy the following.*

(i) $K(.,.)$ is a measurable function of (x, y) in the square $a \leq x \leq b$, $a \leq y \leq b$,

$$\int_a^b \int_a^b |K(x, y)|^2\,dx\,dy < \infty.$$

(ii) For each value of x, the quantity $K(x, y)$ defines a measurable function of y such that

$$\int_a^b |K(x, y)|^2\,dy < \infty.$$

(iii) For each value of y, the quantity $K(x, y)$ defines a measurable function of x such that

$$\int_a^b |K(x, y)|^2\,dx < \infty.$$

A quantity $K(x, y)$ that satisfies (i)–(iii) is called an L_2-kernel. [104, 86, 50, 145].

We shall analyse (5.34) in an $L_2(a, b) =: H$ setting. With this in mind we introduce the linear operator K defined by

$$K : H \to H, \tag{5.35}$$

$$K\varphi = \int_a^b K(., y)\varphi(y)\,dy, \qquad \varphi \in D(K),$$

$$D(K) = \left\{\varphi \in H : \int_a^b K(., y)\varphi(y)dy \in H\right\}.$$

We refer to K as the integral operator with kernel $K(x, y)$.

The integral equation (5.34) can now be written in an $L_2(a, b) =: H$ setting as

$$(I - \lambda K)\varphi = f. \tag{5.36}$$

If (5.36) has a solution $\varphi \in H$, then it is natural to enquire whether φ can be expressed in terms of f by means of a formula similar to (5.36). That is, can we write

$$\varphi = (I + \lambda L) f = f + \lambda L f, \tag{5.37}$$

where L is a linear integral operator with kernel $L(x, y) \equiv L(x, y, \lambda)$ that is also an L_2-kernel?

We notice first that if there exists a $\varphi \in H$ that satisfies both (5.36) and (5.37), then we must have

$$(I - \lambda K)(I + \lambda L) f = f = If.$$

This relation suggests that L should satisfy the operator equation

$$I - \lambda K + \lambda L - \lambda^2 KL = I,$$

and if $\lambda \neq 0$, then we obtain

$$L - K = \lambda KL. \tag{5.38}$$

Also, if $\varphi = (I + \lambda L) f \in H$ is the unique solution of (5.36), then (5.37) indicates that we must have

$$\varphi = (I + \lambda L) f = (I + \lambda L)(I - \lambda K).$$

This relation suggests that L must also satisfy

$$L - K = \lambda LK. \tag{5.39}$$

We now introduce the following definitions.

DEFINITION 5.30 *If, for a given value of λ, there is an L_2-kernel $L(x, y, \lambda)$ that ensures that (5.38) and (5.39) hold, that is, if on writing (5.38) and (5.39) in terms of kernels we have*

$$\begin{aligned} L(x, y, \lambda) - K(x, y) &= \lambda \int_a^b K(x, z) L(z, y, \lambda)\, dz \\ &= \lambda \int_a^b L(x, z, \lambda) K(z, y)\, dz, \end{aligned}$$

then

(i) $L(x, y, \lambda)$ is traditionally called the resolvent kernel of $K(x, y)$ for the value of λ involved, in which case λ is called a regular value of the kernel $K(x, y)$ or, equivalently, of the associated linear integral operator K.

(ii) In this section only the equation

$$L - K = \lambda KL = \lambda LK \tag{5.40}$$

is called the resolvent equation of K.

We notice from (5.39) and (5.40) that if $\lambda = 0$, then we have $L = K$, and we can conclude that λ is a regular value of K.

Two immediately useful results from the general theory of integral equations are the following.

THEOREM 5.31 *If for a given value of λ a resolvent kernel $L(x, y, \lambda)$ for the L_2-kernel $K(x, y)$ exists, then it is unique.*

THEOREM 5.32 *Let λ be a regular value of the L_2-kernel $K(x, y)$ and let $L(x, y, \lambda)$ be the associated resolvent kernel for the kernel $K(x, y)$.*

If $K : H \to H$ is the linear integral operator with kernel $K(x, y)$ and if $f \in H$, then the equation

$$\varphi = f + \lambda K\varphi \tag{5.41}$$

has a unique solution $\varphi \in H$ given by

$$\varphi = f + \lambda Lf, \tag{5.42}$$

where $L : H \to H$ is a linear integral operator with kernel $L(x, y, \lambda)$.

If $K(x, y)$ is an L_2-kernel, then so is the kernel $K^*(x, y) = \overline{K(y, x)}$, where the overbar denotes a complex conjugate. The kernel $K^*(x, y)$ is known as the adjoint of the kernel $K(x, y)$. The corresponding linear integral operators, denoted K and K^*, respectively, obey, as in chapter 3, the rules

$$(K\varphi, \psi) = (\varphi, K^*\psi), \quad \text{where } (.,.) \text{ is the inner product in } H,$$

$$(K^*)^* = K, \|K^*\| = \|K\|, (\lambda K)^* = \overline{\lambda}K^*, (K + L)^* = K^* + L^*,$$

$$(KL)^* = L^*K^*.$$

Given the integral equation

$$\varphi = f + \lambda K\varphi, \tag{5.43}$$

the associated adjoint equation has the form

$$\psi = g + \overline{\lambda}K^*\psi. \tag{5.44}$$

Theorem 5.33 *If $K : H \to H$ is a linear integral operator with an L_2-kernel, then λ is a regular value of K if and only if $\overline{\lambda}$ is a regular value of K^*.*

If the resolvent of K for λ is L, then the resolvent of K^* for $\overline{\lambda}$ is L^*.

Writing the integral equation (5.47) in the operator notation introduced above, we have

$$\varphi = f + \lambda K\varphi.$$

If in this equation the function f is identically zero, then we have the homogeneous linear integral equation

$$\varphi = \lambda K\varphi. \tag{5.45}$$

This equation has the trivial solution $\varphi \equiv 0$. If (5.45) has a solution $\varphi \in H$ other than the trivial solution, then λ is called a characteristic value of K and φ is a characteristic function of K belonging to the characteristic value λ.

We notice that a characteristic value of K is the reciprocal of the notion of an eigenvalue introduced in chapter 4.

Theorem 5.34 *Let K denote a linear integral operator with an L_2-kernel $K(x, y)$. No regular value of K can be a characteristic value of K.*

Proof. If λ is a regular value of K, then there is a unique $\varphi \in H$ satisfying

$$\varphi = f + \lambda K\varphi. \tag{5.46}$$

If λ is also a characteristic value of K, then there is a nontrivial φ such that

$$\varphi = \lambda K\varphi.$$

These two results combine to yield $f \equiv 0$. However, (5.37) shows that the unique solution φ of (5.46) can be written

$$\varphi = f + \lambda Lf.$$

Hence, since $f \equiv 0$, it follows that $\varphi = 0$, and the required result follows. □

As an aid in determining the resolvent of an operator, we introduce the following notion.

Definition 5.35 *[102] A sequence $\{\varphi_n\}_{n=1}^{\infty} \subset H = L_2(a, b)$ is said to be relatively uniformly convergent to a limiting function $\varphi \in H$ if there exists a non-negative function $p \in H$ such that, given $\varepsilon > 0$, there exists a positive integer $n_0(\varepsilon)$ for which*

$$\left|\varphi_n(x) - \varphi(x)\right| \leq \varepsilon p(x), \qquad n \geq n_0, \qquad a \leq x \leq b.$$

Equivalently, an infinite series of functions in H is said to be relatively uniformly convergent if the sequence formed from its partial sums is relatively uniformly convergent.

An infinite series $\sum \varphi_n(x)$ of elements $\varphi_n \in H$ is relatively uniformly absolutely convergent if the series $\sum |\varphi_n(x)|$ is relatively uniformly convergent.

The main properties of relatively uniform convergence that will be useful in our discussions of integral equations are summarised in the following theorem.

Theorem 5.36 *(i) The statements in definition 5.35 apply to L_2-kernels.*

(ii) If $\varphi_n \to \varphi \in H = L_2(a,b)$ as $n \to \infty$ relatively uniformly and if $f \in H$, then

$$(\varphi_n, f) \to (\varphi, f), \qquad n \to \infty,$$

where $(.,.)$ is the H inner product.

(iii) If $K_n(x,y) \to K(x,y)$ as $n \to \infty$ is a relatively uniformly convergent sequence of L_2-kernels and if $f \in H$, then

$$\int_a^b K_n(x,y) f(y)\, dy \to \int_a^b K(x,y) f(y)\, dy, \qquad n \to \infty$$

relatively uniformly.

(iv) If $K_n(x,y) \to K(x,y)$ as $n \to \infty$ relatively uniformly and if $L(x,y)$ is an L_2-kernel, then

$$\begin{aligned} (K_n L)(x,y) &\to (KL)(x,y), \qquad n \to \infty, \\ (L K_n)(x,y) &\to (LK)(x,y), \qquad n \to \infty, \end{aligned}$$

where the convergence is relatively uniform in each case.

(v) If $\sum_{n=1}^{\infty} f_n(x) = f(x)$ is a relatively uniformly absolutely convergent series, then

$$\sum_{n=1}^{\infty} (f_n, g) \to (f, g), \qquad g \in H,$$

the series on the left being absolutely convergent.

(vi) If $\sum_{n=1}^{\infty} K_n(x,y) = K(x,y)$ is a relatively uniformly, absolutely convergent series of L_2-kernels and if $f \in H$, then

$$\sum_{n=1}^{\infty} \int_a^b K_n(x,y) f(y)\, dy = \int_a^b K(x,y) f(y)\, dy,$$

the series being relatively uniformly absolutely convergent.

(vii) If $\sum_{n=1}^{\infty} K_n(x,y) = K(x,y)$ is a relatively uniformly absolutely convergent series of L_2-kernels and if $L(x,y)$ is an L_2-kernel, then

$$\begin{aligned} \sum_{n=1}^{\infty} (K_n L)(x,y) &= (KL)(x,y), \qquad n \to \infty, \\ \sum_{n=1}^{\infty} (L K_n)(x,y) &= (LK)(x,y), \qquad n \to \infty, \end{aligned}$$

the series being relatively uniformly absolutely convergent.

We have seen that if λ is a regular value of the linear integral operator $K: H \to H$, which has an L_2-kernel $K(x, y)$, the equation

$$\varphi = f + \lambda K\varphi \tag{5.47}$$

has the unique solution

$$\varphi = f + \lambda Lf, \tag{5.48}$$

where L is the resolvent of K. In order to solve (5.47), we need an explicit form for L. There are a number of equivalent ways of doing this. First, we can use the resolvent equation (5.40),

$$L - K = \lambda LK = \lambda KL,$$

which can be rewritten in the form

$$L(I - \lambda K) = (I - \lambda K)L = K.$$

We now proceed formally to obtain

$$L = K(I - \lambda K)^{-1} = K + \lambda K^2 + \lambda^2 K^3 + \cdots = \sum_{n=1}^{\infty} \lambda^{n-1} K^n, \tag{5.49}$$

where $(I - \lambda K)^{-1}$ has been expanded by the binomial theorem.

A second means of obtaining an explicit expression for L is to use successive approximations. As a first approximation to the solution of (5.47), we write

$$\varphi_1 = f.$$

Substituting φ_1 for φ in the right-hand side of (5.47), we get the second approximation

$$\varphi_2 = f + \lambda Kf.$$

We obtain, similarly,

$$\varphi_3 = f + \lambda Kf + \lambda^2 K^2 f$$

and

$$\varphi_n = f + \lambda Kf + \cdots + \lambda^{n-1} K^{n-1} f.$$

This suggests that the series

$$\varphi = \sum_{n=0}^{\infty} .\lambda^n K^n f \tag{5.50}$$

may be a solution of (5.47). This is indeed the case since (5.50) can be written in the form

$$\varphi = f + \lambda \sum_{n=1}^{\infty} \lambda^{n-1} K^n f, \tag{5.51}$$

and we have recovered the series (5.49) and the equation (5.48).

An alternative method of obtaining successive approximations to the solution of (5.47) is to assume from the outset that φ can be written as a power series in λ. Specifically, we assume

$$\varphi(x) = \sum_{m=0}^{\infty} \lambda^m \varphi_m(x), \tag{5.52}$$

where the φ_m have to be determined. With this in mind substitute (5.52) into (5.47) and equate powers of λ to obtain

$$\varphi_0(x) = f(x),$$

$$\varphi_n(x) = \int_a^b K_n(x, y) f(y) dy, \qquad n = 1, 2, \ldots,$$

where

$$K_1(x, y) = K(x, y),$$

$$K_{n+1}(x, y) = \int K(x, z) K_n(z, y)\, dz,$$

It is a straightforward matter to show that, under the conditions placed on f and K, the series is uniformly convergent with respect to x and λ, and moreover, that the sum of the series is a solution of (5.47).

Using the quantities just determined above, we can write (5.65) in the form

$$\varphi(x) = f(x) + \sum_{m=1}^{\infty} \lambda^m \int_a^b K_m(x, z) f(z)\, dz. \tag{5.53}$$

This result expresses the resolvent operator L in terms of the associated kernels.

Two theorems that make precise the nonrigorous arguments used above are the following [50, 86, 145].

Theorem 5.37 *If $K(x, y)$ is an L_2-kernel and K is the associated linear integral operator, then the series*

$$\sum_{m=1}^{\infty} \lambda^{m-1} K_m(x, y)$$

is relatively uniformly absolutely convergent for all λ such that $|\lambda| . \|K\| < 1$, and its sum $L(x, y)$ is the resolvent kernel of $K(x, y)$. Furthermore, every value of λ such that $|\lambda| . \|K\| < 1$ is a regular value of $L(x, y)$.

Theorem 5.38 *If $K(x, y)$ is an L_2-kernel and $K : H \to H$ is the associated linear integral operator and if $f \in H$ and $|\lambda| \,.\, \|K\| < 1$, then the integral equation*

$$\varphi(x) = f(x) + \lambda \int_a^b K(x, y)\varphi(y)\, dy$$

has a unique solution $\varphi \in H$ given by

$$\varphi(x) = f(x) + \sum_{m=1}^{\infty} \lambda^m f_m(x), \tag{5.54}$$

where

$$f_m(x) = \int_a^b K_m(x, y) f(y)\, dy, \qquad m \geq 1,$$

the series in (5.54) being relatively uniformly absolutely convergent. The homogeneous integral equation

$$\varphi(x) = \lambda \int K(x, y)\varphi(y)\, dy$$

has the unique L_2-solution defined by $\varphi(x) = 0$.

The series (5.54) is known as the Neumann series for the integral equation.

In the event that the kernels and given data functions are assumed to be continuous, all the proceeding results hold with relatively uniform convergence replaced by uniform convergence.

To illustrate how Volterra integral equations are dealt with, we consider an equation of the form

$$\varphi(x) = f(x) + \lambda \int_a^x K(x, y)\varphi(y)\, dy, \qquad a \leq x \leq b. \tag{5.55}$$

If we extend the domain of definition of $K(x, y)$ by defining it to be zero when $a \leq x < y \leq b$, then (5.55) assumes the form we have been dealing with so far, namely,

$$\varphi(x) = f(x) + \lambda \int_a^b K(x, y)\varphi(y)\, dy, \qquad a \leq x \leq b. \tag{5.56}$$

When dealing with Volterra integral equations of the form (5.55), we shall assume that $K(x, y)$ is defined in the square $a \leq x \leq b$, $a \leq y \leq b$, that it vanishes when $a \leq x < y \leq b$ and that it is an L_2-kernel. A kernel of this type is known as a *Volterra kernel*. We shall also assume that in (5.55) the given function $f \in H = L_2(a, b)$.

One of the main results that emerges from a general analysis of Volterra integral equations of the form (5.55) is that the associated Neumann series converges for all complex coefficients λ. This implies that if a linear integral operator $K : H \to H$ is an operator with an L_2-Volterra kernel, then $\rho(K)$, the resolvent set of K, is the entire complex plane. This result is established by means of the following two estimates.

THEOREM 5.39 *[102] Let $K(x, y)$ be an L_2-Volterra kernel and let $\varphi \in H = L_2(a, b)$.*
If

$$\varphi_n(x) = \int_a^x K^n(x, y)\varphi(y)\, dy, \qquad n = 1, 2, \ldots,$$

then

$$\left|\varphi_n(x)\right| \le \frac{k_1(x)\, \|\varphi\|}{\sqrt{(n-1)!}} \left\{ \int_a^x (k_1(y))^2\, dy \right\}^{(n-1)/2},$$

where

$$k_1(x) = \left\{ \int_a^x |K(x, y)|^2\, dy \right\}^{1/2}.$$

THEOREM 5.40 *If $K : H \to H$ is a linear integral operator with an L_2-Volterra kernel, then*

$$\left|K^{n+1}(x, y)\right| \le \frac{\|K\|^{n-1}}{\sqrt{(n-1)!}} k_1(x) k_2(y),$$

where $k_1(x)$ is defined as in theorem 5.39 and

$$k_2(y) = \left\{ \int_y^b |K(x, y)|^2\, dx \right\}^{1/2}$$

On the basis of these two results the following can be established.

THEOREM 5.41 *If $K(x, y)$ is an L_2-Volterra kernel, then the series*

$$\sum_{m=0}^{\infty} \lambda^m K^{m+1}(x, y)$$

is relatively uniformly absolutely convergent for every complex number λ. The sum of the series is $L(x, y, \lambda)$, the resolvent kernel of $K(x, y)$.

Finally in this section, we summarise the previous results and discussions in a form that will be particularly useful in later sections and chapters.

For $T > 0$, introduce the sets

$$\overline{\Omega(T)} = \{(t, s) : 0 \le s \le t \le T\},$$

$$\Omega(T) = \{(t, s) : 0 \le s < t \le T\}.$$

Let H be a Hilbert space. We shall consider the Volterra integral equations

$$w(t, s) = a(t, s) + \int_s^t w(t, \tau) k_1(\tau, s)\, d\tau, \tag{5.57}$$

$$v(t, s) = b(t, s) + \int_s^t h_1(t, \tau) v(\tau, t)\, d\tau, \tag{5.58}$$

defined in the space $C(\overline{\Omega(T)}, B(H))$. We shall assume that

(i) $a(.,.)$ and $b(.,.)$ belong to $C(\overline{\Omega(T)}, B(H))$,
(ii) $k_1(.,.)$ and $h_1(.,.)$ belong to $C(\Omega(T), B(H))$.

Furthermore, we shall assume that there exist constants $c_0 \geq 0$ and $\alpha \in (0, 1]$ such that

(iii) $\|k_1(t, s)\| \leq c_0(t-s)^{\alpha-1}$,
(iv) $\|h_1(t, s)\| \leq c_0(t-s)^{\alpha-1}$,

where (iii) and (iv) hold for all $(t, s) \in \Omega(T)$.

We shall use the notations

$$k_m(t, s) = \int_s^t k_{m-1}(t, \tau) k_1(\tau, s)\, d\tau, \tag{5.59}$$

$$h_m(t, s) = \int_s^t h_{m-1}(t, \tau) h_1(\tau, s)\, d\tau, \tag{5.60}$$

where $m \geq 2$ and $(t, s) \in \Omega(T)$. Furthermore, we shall write

$$k(t, s) = \sum_{m=1}^{\infty} k_m(t, s), \qquad h(t, s) = \sum_{m=1}^{\infty} h_m(t, s). \tag{5.61}$$

With this preparation we can summarise our previous discussions in the following.

Theorem 5.42 *The Volterra integral equations (5.57), (5.58) have unique solutions in $C(\overline{\Omega(T)}, B(H))$. In terms of the resolvent kernels $k(t, s)$ and $h(t, s)$, these solutions can be represented in the form*

$$w(t, s) = a(t, s) + \int_s^t a(t, \tau) k(\tau, s)\, d\tau, \tag{5.62}$$

$$v(t, s) = b(t, s) + \int_s^t h(t, \tau) b(\tau, s)\, d\tau. \tag{5.63}$$

If $a(.,.)$ and $b(.,.)$ are elements of $C(\overline{\Omega(T)}, B(H))$, then so also are $w(.,.)$ and $v(.,.)$. Furthermore, the resolvent kernels are estimated according to

$$\|k(t, s)\| \leq C(t-s)^{\alpha-1} \quad \text{and} \quad \|h(t, s)\| \leq C(t-s)^{\alpha-1}$$

for all $(t, s) \in \Omega(T)$, where c is a constant depending on c_0, α and T.

Chapter Six

On Scattering Theory Strategies

6.1 INTRODUCTION

In this chapter we first recall salient features of scattering theories that have been developed for APs. Some of these have already been indicated in chapter 1, however, here we provide a rather more detailed account. Topics to be covered include propagation aspects, solution decay, scattering states, solutions with finite energy, representations of solutions, expansion theorems and construction of solutions. The comparison of solutions for large time is discussed, as is the evolution operator for a wave equation and the asymptotic equality of solutions. Results are recalled concerning the existence, uniqueness and completeness of wave and scattering operators, and mention is also made of the principle of limiting absorption. Finally, in the first section, a method is outlined for the construction of wave and scattering operators. Much of the philosophy introduced in the first section will be acknowledged when dealing with NAPs in the final section.

6.2 ON SCATTERING PROCESSES IN AUTONOMOUS PROBLEMS

Scattering theory is the study of the way in which interacting systems behave for large values of time and/or distance.In this monograph we are particularly interested in systems that support wave motions.

Scattering processes describe the effects of a perturbation of a system about which everything is known in the absence of that perturbation. Such processes can be studied in three stages: generation, interaction and measurement. In the generation stage an incident wave, a signal, is generated far away from the perturbation in both space and time. At this stage the interaction between the incident wave and the perturbation is negligible, and the system evolves as though it were a free system, that is, a system in which there are no perturbations. Eventually, the incident wave and the perturbation interact and exert considerable influences on each other. At this stage the resulting effects, the scattered waves, often have a very complicated structure. After the interaction during which the scattering has occurred, the now scattered wave and the perturbation can once more become quite distant from each other, and the interaction effects again become negligible. Consequently, any measurement of the scattered wave at this stage would indicate that the system is once again beginning to evolve as a free system.

In practical situations measurements of a wave are usually made far away from any perturbation and as such are really the only data available. This suggests that one of the fundamental questions to be addressed is of the following type. If an observer far distant from any perturbation measures the scattered wave (signal), then what was the incident wave (signal)? Furthermore, we would like to be able to answer this question without having to investigate, in too much detail, the actual interaction stage. For this reason we shall be particularly interested in the asymptotic behaviour of solutions to wave equations in the distant future (i.e., as $t \to +\infty$) and in the distant past (i.e., as $t \to -\infty$) and especially in the asymptotic equality of systems as outlined in chapter 1.

However, even more basic than the above question is the assumption we have made in chapter 1, namely, that the scattered wave can indeed be characterised in terms of quantities associated with a free system. This leads to the *asymptotic condition* and the notion of *asymptotic completeness*. We shall discuss these concepts in later subsections. Our first concern is to determine whether or not the systems of interest actually have solutions and that these solutions represent propagating waves.

6.2.1 Propagation Aspects

We shall fix ideas by considering the IVPs

$$\{\partial_t^2 + A_j\}u_j(x,t) = 0, \qquad (x,t) \in \mathbf{R}^n \times \mathbf{R}, \qquad j = 0, 1, \tag{6.1}$$

$$u_j(x,0) = \varphi_j(x), \qquad u_{jt}(x,0) = \psi_j(x), \qquad j = 0, 1, \tag{6.2}$$

where $j = 0$ represents a FP and $j = 1$ a PP. We shall assume that

$$A_j : H(\mathbf{R}^n) \to H(\mathbf{R}^n) \equiv L_2(\mathbf{R}^n), \qquad j = 0, 1,$$

and therefore $H(\mathbf{R}^n)$ is a Hilbert space. We remark that here, for ease of presentation, we have assumed that both IVPs are defined in the same space. It should be noticed that we will not always be able to assume this. This assumption could well hold when A_0 is perturbed by additional terms, as in potential scattering. It is unlikely to hold, without further assumptions, when $D(A_0)$ is perturbed, as would be the case for target scattering. Furthermore, for the sake of illustration we shall assume here that A_0 is a realisation in $H(\mathbf{R}^n)$ of the *negative* Laplacian and that A_1 is some perturbation of A_0.

An analysis of the given IVPs (6.1), (6.2) can begin by interpreting them as IVPs for ordinary differential equations rather than for partial differential equations. This can be achieved in the following manner. Let X be a Hilbert space. Furthermore, let $\Lambda \subseteq \mathbf{R}$ be a Lebesgue-measurable subset of $\mathbf{R}$ and let f denote a function of $x \in \mathbf{R}^n$ and $t \in \mathbf{R}$ that has the action

$$f \equiv f(.,.) : t \to f(.,t) =: f(t) \in X, \qquad t \in \Lambda; \tag{6.3}$$

that is, f is interpreted as an X-valued function of $t \in \Lambda$.

We shall denote by $L_2(\Lambda, X) =: H$ the set of all equivalence classes of measurable functions defined on Λ with values in X satisfying

$$\|f\|_H^2 := \int_\Lambda \|f(t)\|_X^2 \, dt < \infty, \tag{6.4}$$

where $\|.\|_H$ denotes a norm on H and $\|.\|_X$ the norm on X. It is an easy matter to show that H is a Hilbert space with inner product

$$(f, g) := \int_\Lambda (f(t), g(t))_X \, dt. \tag{6.5}$$

Therefore, with this notation and understanding, we can interpret u in (6.1), (6.2) as

$$u \equiv u(.,.) : t \to u(., t) =: u(t) \in X.$$

The IVP (6.1), (6.2) can now be interpreted as an IVP for an ordinary differential equation, *defined in* H, of the form

$$\{d_t^2 + A_j\}u_j(t) = 0, \qquad u_j(0) = \varphi_j, \qquad u_{jt}(0) = \psi_j, \qquad j = 0, 1. \tag{6.6}$$

When these IVPs are known to have solutions, they can be represented in the form

$$u_j(t) = (\cos t A_j^{1/2})\varphi_j + A_j^{-1/2}(\sin t A_j^{1/2})\psi_j, \qquad j = 0, 1. \tag{6.7}$$

Hence the solution of the given problem (6.1), (6.2) can be written in the form

$$u_j(x, t) = (\cos t A_j^{1/2})\varphi_j(x) + A_j^{-1/2}(\sin t A_j^{1/2})\psi_j(x), \qquad j = 0, 1. \tag{6.8}$$

From (6.8), provided that the spectral theorem is available, it then follows that, for $j = 0, 1$,

$$u_j(x, t) = \int_{\sigma(A_j)} \{\cos t\sqrt{\lambda}\} \, dE_j(\lambda)\varphi_j(x) + \int_{\sigma(A_j)} \left\{\frac{\sin t\sqrt{\lambda}}{\sqrt{\lambda}}\right\} dE_j(\lambda)\psi_j(x), \tag{6.9}$$

where $\sigma(A_j)$ denotes the spectrum of A_j and $\{E_j(\lambda)\}_{\lambda \in \sigma(A_j)}$ is the spectral family of A_j.

The representation (6.8) is only meaningful if we know that the problems (6.1), (6.2) actually have solutions that, moreover, are known to be unique. We return to this point a little later. Furthermore, the practical usefulness of the representation (6.9) depends crucially on how readily the spectral family $\{E_j(\lambda)\}_{\lambda \in \sigma(A_j)}$ can be determined.

An alternative approach frequently adopted when discussing wave motions governed by an IVP of the generic form (6.1), (6.2) is to replace the given IVP by an equivalent system of equations that are of first order in time. We have already given an indication of how this can be done in chapters 1 and 5. This approach has a number of advantages, not the least being that it can provide a straightforward means of including energy considerations. Results governing the existence and uniqueness of solutions with finite energy (wfe) can then be quite readily obtained. To develop this approach, we introduce an "energy space" $H_E(\mathbf{R}^n)$, which is defined as the completion of $C_0^\infty(\mathbf{R}^n) \times C_0^\infty(\mathbf{R}^n)$ with respect to the energy norm $\|.\|_E$, where for

$$\mathbf{f} = \begin{bmatrix} f_1 \\ f_2 \end{bmatrix} =: \langle f_1, f_2 \rangle \in C_0^\infty(\mathbf{R}^n) \times C_0^\infty(\mathbf{R}^n), \tag{6.10}$$

we define

$$\|\mathbf{f}\|_E^2 := \int_{\mathbf{R}^n} \{|\nabla f_1(x)|^2 + |f_2(x)|^2\}\, dx. \tag{6.11}$$

We notice that $H_E(\mathbf{R}^n)$ has the decomposition

$$H_E(\mathbf{R}^n) = H_D(\mathbf{R}^n) \oplus L_2(\mathbf{R}^n), \tag{6.12}$$

where $H_D(\mathbf{R}^n)$ is the completion of $C_0^\infty(\mathbf{R}^n)$ with respect to the norm defined by

$$\|f\|_E^2 := \int_{\mathbf{R}^n} \{|\nabla f(x)|^2\} dx, \qquad f \in C_0^\infty(\mathbf{R}^n). \tag{6.13}$$

Furthermore, $H_E(\mathbf{R}^n)$ is readily seen to be a Hilbert space with respect to the inner product $(.,.)_E$ defined by

$$(\mathbf{f}, \mathbf{g})_E := (\nabla f_1, \nabla g_1) + (f_2, g_2), \tag{6.14}$$

where $\mathbf{f} =: \langle f_1, f_2 \rangle$, $\mathbf{g} =: \langle g_1, g_2 \rangle$ are elements of $H_E(\mathbf{R}^n)$ and $(.,.)$ denotes the usual $L_2(\mathbf{R}^n)$ inner product.

We now write the IVP (6.1), (6.2) in the form

$$\begin{bmatrix} u_j \\ u_{jt} \end{bmatrix}_t (x,t) + \begin{bmatrix} 0 & -I \\ A_j & 0 \end{bmatrix} \begin{bmatrix} u_j \\ u_{jt} \end{bmatrix} (x,t) = \begin{bmatrix} 0 \\ 0 \end{bmatrix}, \qquad j = 0, 1, \tag{6.15}$$

$$\begin{bmatrix} u_j \\ u_{jt} \end{bmatrix} (x,0) = \begin{bmatrix} \varphi_j \\ \psi_j \end{bmatrix}, \qquad j = 0, 1. \tag{6.16}$$

This array can be conveniently written in the form

$$(\partial_t - i\mathbf{G}_j)\mathbf{u}_j(x,t) = 0, \qquad \mathbf{u}_j(x,0) = \mathbf{u}_j^0(x), \qquad j = 0, 1, \tag{6.17}$$

where

$$\mathbf{u}_j(x,t) = \begin{bmatrix} u_j \\ u_{jt} \end{bmatrix} (x,t), \qquad \mathbf{u}_j^0(x) = \begin{bmatrix} \varphi_j \\ \psi_j \end{bmatrix} (x), \tag{6.18}$$

$$-i\mathbf{G}_j = \begin{bmatrix} 0 & -I \\ A_j & 0 \end{bmatrix}. \tag{6.19}$$

We now interpret $\mathbf{u}_j$, $j = 0, 1$, as H_E-valued functions of t in the sense that

$$\mathbf{u}_j \equiv \mathbf{u}_j(.,.) : t \to \mathbf{u}_j(.,t) =: \mathbf{u}_j(t) \in H_E(\mathbf{R}^n), \qquad j = 0, 1. \tag{6.20}$$

In this case (6.17) can be reformulated in $H_E(\mathbf{R}^n)$ as an IVP for an ordinary differential equation of the form

$$\{d_t - i\mathbf{G}_j\}\mathbf{u}_j(t) = 0, \qquad \mathbf{u}_j(0) = \mathbf{u}_j^0, \tag{6.21}$$

where for $j = 0, 1$,

$$\mathbf{G}_j : H_E(\mathbf{R}^n) \supseteq D(\mathbf{G}_j) \to H_E(\mathbf{R}^n),$$

$$\mathbf{G}_j \boldsymbol{\xi} = i \begin{bmatrix} 0 & -I \\ A_j & 0 \end{bmatrix} \begin{bmatrix} \xi_1 \\ \xi_2 \end{bmatrix}, \qquad \boldsymbol{\xi} = \langle \xi_1, \xi_2 \rangle \in D(\mathbf{G}_j),$$

$$D(\mathbf{G}_j) = \{\boldsymbol{\xi} = \langle \xi_1, \xi_2 \rangle \in H_E(\mathbf{R}^n) : A_j \xi_1 \in L_2(\mathbf{R}^n), \xi_2 \in H_D(\mathbf{R}^n)\}.$$

Once we have obtained the representations (6.21) of the given IVPs (6.1), (6.2), the following questions are immediate.

1. Are the problems (6.21) and (6.1), (6.2) well posed?
2. How can the solutions of the problems (6.21) be represented whenever they exist?
3. How can solutions of (6.21) yield the required solutions to (6.1), (6.2)?

It is clear that if the problems (6.21) are well posed, then it will follow that the problems (6.1), (6.2) are also well posed. To establish the well-posedness of (6.21), we use results from the theory of semigroups introduced in chapter 5. For our later convenience we gather together here the relevant results. For ease of presentation we shall consider, for the moment, the IVP

$$\left\{ \frac{d}{dt} - B \right\} w(t) = 0, \qquad t \in \mathbf{R}^+, \qquad w(0) = w_0. \tag{6.22}$$

Theorem 5.9 *The problem (6.22) is well posed if B generates a C_0-semigroup U on H. In this case the solution of (6.22) is given by $w(t) = U(t)w_0$, $t \in \mathbf{R}^+$.*

We will also want to discuss nonhomogeneous problems of the form

$$\left\{ \frac{d}{dt} - B \right\} v(t) = f(t), \qquad t \in \mathbf{R}^+, \quad v(0) = v_0, \tag{6.23}$$

where f and v_0 are given data functions. In this connection the following result holds.

Theorem 5.10 *Let H be a Hilbert space and let $B : H \supseteq D(B) \to H$ be the generator of a C_0-semigroup $U = \{U(t) : t \geq 0\} \subseteq B(H)$. If $v_0 \in D(B)$ and $f \in C^1(\mathbf{R}^+, H)$, then (6.23) has a unique solution $v \in C^1(\mathbf{R}^+, H)$ with values in $D(B)$.*

A formal application to (6.23) of the familiar integrating factor technique yields the solution form

$$v(t) = U(t)v_0 + \int_0^t U(t-s) f(s)\, ds.$$

These results settle the well-posedness of the IVPs concerned provided we can show that B is the generator of a suitable semigroup. With this in mind and bearing in mind definition 5.17, we recall the following.

Theorem 5.18 *(Stone's theorem) [49] Let H be a Hilbert space. An operator $B : H \supseteq D(B) \to H$ is the generator of a C_0-unitary group U on H if and only if B is skew-adjoint.*

Returning now to our original notation, we remark that in most cases of practical interest it can be shown [105] that the $\mathbf{G}_j$, $j=0, 1$, are positive self-adjoint operators on $H_E(\mathbf{R}^n)$. Furthermore, since the well-posedness of the problem (6.22) will imply the well-posedness of the problem (6.1), (6.2) for each $j=0, 1$, we can summarise the use of the above results as follows.

Theorem 6.1 *Let H be a Hilbert space and let $A_j : H \supseteq D(A_j) \to H$, $j=0, 1$, be positive self-adjoint operators on H. Let H_E denote an energy space associated with H. If for $j=0, 1$ the operators $\mathbf{G}_j : H_E \supseteq D(\mathbf{G}_j) \to H_E$ of the form*

$$i\mathbf{G}_j = \begin{bmatrix} 0 & I \\ -A_j & 0 \end{bmatrix}$$

are self-adjoint on H_E, then $(i\mathbf{G}_j)$ generates a C_0-group $\{\mathbf{U}_j(t), t \in \mathbf{R}\}$ defined by

$$\mathbf{U}_j(t) = \exp(it\mathbf{G}_j) = \cos tA_j^{1/2} \begin{bmatrix} I & 0 \\ 0 & I \end{bmatrix} - A_j^{-1/2} \sin tA_j^{1/2} \begin{bmatrix} 0 & -I \\ A_j & 0 \end{bmatrix} \tag{6.24}$$

$$= \begin{bmatrix} \cos tA_j^{1/2} & A_j^{-1/2} \sin tA_j^{1/2} \\ -A_j^{1/2} \sin tA_j^{1/2} & \cos tA_j^{1/2} \end{bmatrix}. \tag{6.25}$$

Therefore, recalling the material in chapter 5, in particular subsection 5.2.1, it follows that the IVPs

$$\{d_t^2 + A_j\}u_j(t) = 0, \tag{6.26}$$

$$u_j(0) = \varphi_j \in D(A_j), \qquad u_{jt}(0) = \psi_j \in D(A_j^{1/2}) \tag{6.27}$$

are also well posed.

The solution of (6.21) can be obtained, using an integrating factor technique, in the form

$$\mathbf{u}_j(t) = \exp\{it\mathbf{G}_j\}\mathbf{u}_j^0 = \mathbf{U}_j(t)\mathbf{u}_j^0. \tag{6.28}$$

Consequently, provided we ensure that the $(i\mathbf{G}_j)$, $j=0, 1$, generate C_0-groups and that they are of the form (6.24), then it is clear that (6.1), (6.2) are well-posed problems and, moreover, *the first component of* (6.28) *yields the same solution as* (6.8). These observations will enable us to settle propagation problems associated with (6.1), (6.2).

In the next section we shall indicate the influence of the initial conditions on the solution forms obtained above. First, we recall that a practical interpretation of these solution forms relies on a detailed knowledge of the spectra $\sigma(A_j)$, $j=0, 1$, and the spectral families $\{E_j(\lambda)\}_{\lambda \in \sigma(A_j)}$, $j=0, 1$. The spectral families can be determined by means of Stone's formula [71, 105], which for $j=0, 1$ has the form

$$((E_j(\lambda)-E_j(\mu))f, g) = \lim_{\delta\downarrow 0, \varepsilon\downarrow 0} \int_{\mu+\delta}^{\lambda+\delta} ([R_j(t+i\varepsilon)-R_j(t-i\varepsilon)]f, g)\, dt, \quad (6.29)$$

where

$$R_j(t\pm i\varepsilon) = (A_j - (t\pm i\varepsilon))^{-1}. \quad (6.30)$$

Hence for $j=0, 1$ the spectral families $\{E_j(\lambda)\}_{\lambda\in\sigma(A_j)}$, $j=0, 1$, can be obtained via an investigation of the resolvent $R_j(\lambda)$ of A_j. This in turn yields details of the underlying spectral properties of A_j.

From a practical point of view determination of the spectral families is quite demanding, and detailed investigations are more profitably left to specific cases. However, we notice that since, to begin with, an investigation of $R_j(\lambda)$ will always be required, it seems that an alternative approach based more directly on the theory of eigenfunction expansions offers good prospects for developing constructive methods. We shall tend to concentrate on this approach in the following chapters.

6.2.2 Solutions with Finite Energy and Scattering States

Bearing in mind (6.12) and the notion of energy in a wave [11], we introduce

$$E(u, t) = \int_{R^n} \{|\nabla u(x, t)|^2 + |u_t(x, t)|^2\}\, dx = \|u(t)\|_E^2 \quad (6.31)$$

and

$$E(B, u, t) = \int_B \{|\nabla u(x, t)|^2 + |u_t(x, t)|^2\}\, dx = \|u(t)\|_{B,E}^2, \quad (6.32)$$

where $B \subseteq \mathbf{R}$ is any bounded set. $E(u, t)$ denotes the global energy of the wave at time t, whilst $E(B, u, t)$ denotes the energy of the wave in B at time t.

In this section we shall only be interested in those systems in which the global energy is conserved, that is,

$$E(u, t) = E(u, 0) = \text{constant}. \quad (6.33)$$

Therefore, if we are dealing with wave equation problems of the typical form (6.1), (6.2), then the energy integrals (6.31) associated with these problems have the form

$$E(u_j, t) = \int_{R^n} \{|\nabla\varphi_j(x)|^2 + |\psi_j(x)|^2\}\, dx, \qquad j=0, 1. \quad (6.34)$$

For the FP, that is, for the case $j=0$, the following result can be obtained [154].

THEOREM 6.2

(i) A_0 is a self-adjoint non-negative operator on $L_2(\mathbf{R}^n)$.

(ii) A_0 has a unique non-negative square root $A_0^{1/2}$ with domain

$$D(A_0^{1/2}) = \{u \in L_2(\mathbf{R}^n) : D^\alpha u \in L_2(\mathbf{R}^n), |\alpha| \le 1\} =: L_2^1(\mathbf{R}^n),$$

where α is a multi-index of the form $\alpha=(\alpha_1,\alpha_2,\ldots,\alpha_n)$ and the α_k are non-negative integers for $k=1,2,\ldots,n$ and $|\alpha|=\sum_{k=1}^{n}\alpha_k$. Further, we define $D^\alpha := D_1^{\alpha_1}D_2^{\alpha_2}\ldots D_n^{\alpha_n}$, where $D_k=\partial/\partial x_k$, $k=1,2,\ldots,n$.

We remark that $L_2^m(\mathbf{R}^n)$, $m=0,1,\ldots,$ are the usual Sobolev-Hilbert spaces [1].

Consequently, using (6.12), we see that if $\varphi_0\in D(A_0^{1/2})=L_2^1(\mathbf{R}^n)$ and $\psi_0\in L_2(\mathbf{R}^n)$, then the representation

$$u_0(t)=(\cos tA_0^{1/2})\varphi_0+A_0^{-1/2}(\sin tA_0^{1/2})\psi_0 \tag{6.35}$$

implies that $u_0(t)\in L_2^1(\mathbf{R}^n)$ and $u_{0t}(t)\in L_2(\mathbf{R}^n)$. In this case the energy integral $E(u_0,t)$ is finite and u_0 is called a solution with finite energy (wfe).

When we come to deal with the PP, we will require a similar result to theorem 6.2 for the operator A_1 in (6.1).

Although we have assumed that the global wave energies $E(u_j,t)$, $j=0,1$, remain constant, this is not necessarily the case for the local energies $E(B,u_j,t)$, $j=0,1$. As a consequence, it is natural to say that the $u_j(x,t)$, $j=0,1$, represent *scattering waves* if for every bounded measurable set $B\subseteq\mathbf{R}^n$,

$$\lim_{t\to\infty}E(B,u_j,t)=0. \tag{6.36}$$

If we assume that φ_0 and ψ_0 are real-valued functions such that $\varphi_0\in H(\mathbf{R}^n)=: L_2(\mathbf{R}^n)$ and $\psi_0\in D(A_0^{-1/2})$ and if we define

$$h_0=\varphi_0+iA_0^{-1/2}\psi_0, \tag{6.37}$$

then (6.35) can be expressed in the form

$$u_0(t)=u_0(.,t)=\mathrm{Re}(v_0(.,t)), \tag{6.38}$$

where

$$v_0(t)\equiv v_0(.,t)=\exp\{-itA_0^{1/2}\}h_0=: U_0(t)h_0 \tag{6.39}$$

is the complex-valued solution in $H(\mathbf{R}^n)$ of (6.1), (6.2) with $j=0$. The representation (6.38), (6.39) implies that the evolution and asymptotic behaviour of $u_0(x,t)$ are determined by that of $v_0(x,t)$.

If, with (6.39) in mind, a wave system of interest evolves according to

$$v(x,t)=U_0(t)h(x), \tag{6.40}$$

then it is natural to say that $h\in H(\mathbf{R}^n)$ is a *scattering state* if and only if (6.36) holds.

If we introduce the mapping

$$Q_q: H_E(\mathbf{R}^n)\to H_E(\mathbf{R}^n), \tag{6.41}$$

where

$$Q_q w(x)=\chi_q(x)w(x) \quad \text{for all } x\in\mathbf{R}^n$$

and χ_q is the characteristic function for

$$B(q):=\{x\in\mathbf{R}^n:|x|\le q\},$$

then we notice that

$$E(B(q), v_0, t) = \|Q_q U_0(t) h_0\|_E^2.$$

Hence (6.36) is equivalent to

$$\lim_{t\to\infty} \|Q_q U_0(t) h_0\|_E = 0 \quad \text{for every } q > 0. \tag{6.42}$$

It is a straightforward matter to verify, for $0 \le q < \infty$, that Q_q is an orthogonal projection on H_E and that

$$s - \lim_{q\to\infty} Q_q = I.$$

This leads to the following definition [153].

Definition 6.3

(i) A family of orthogonal projections on H_E, denoted by $\{Q_q : 0 \le q \le \infty\}$, is called a family of localising operators on H_E if $Q_q : H_E \to H_E$ satisfies $s - \lim_{q\to\infty} Q_q = I$.

(ii) An element $h_0 \in H_E$ is a scattering state for A_0 and $\{Q_q\}$ if and only if (6.42) holds. The set of all such scattering states will be denoted by H^s.

6.2.3 On the Construction of Solutions

Once questions of existence and uniqueness of solution have been settled then we can turn our attention to methods for actually determining such solutions.

We consider the IVP (6.1), (6.2) for the case $j = 0$, 1, and with future applications in mind we shall take $n = 3$.

We first notice that theorem 6.2 indicates that the spectral theorem is available for interpreting the solution forms (6.35) and (6.39). Specifically, if $\{E_0(\lambda)\}_{\lambda\in\sigma(A_0)}$ denotes the spectral family of A_0, then we have the spectral representations

$$A_0 = \int_0^\infty \lambda\, dE_0(\lambda), \tag{6.43}$$

$$\Phi(A_0) = \int_0^\infty \Phi(\lambda)\, dE_0(\lambda), \tag{6.44}$$

where Φ is a bounded Lebesgue-measurable function of λ. However, as we have already mentioned, a difficulty associated with the results (6.43) and (6.44) concerns the practical determination of the spectral family $\{E_0(\lambda)\}_{\lambda\in\sigma(A_0)}$. For the case of the FP that we are concerned with, the situation can be eased by introducing results for Fourier transforms in $L_2(\mathbf{R}^3) := H(\mathbf{R}^3)$. The Plancherel theory indicates that for any $f \in H(\mathbf{R}^3)$ the following limits exist.

$$(F_0 f)(p) = \hat{f}(p) := \lim_{r\to\infty} \frac{1}{(2\pi)^{3/2}} \int_{|x|\le r} \exp(-ix.p) f(x)\, dx, \tag{6.45}$$

$$f(x) = (F_0^* \hat{f})(x) := \lim_{r\to\infty} \frac{1}{(2\pi)^{3/2}} \int_{|p|\le r} \exp(ix.p)\, \hat{f}(p)\, dp, \tag{6.46}$$

where $x, p \in \mathbf{R}^3$. It can also be shown that for any bounded Lebesgue-measurable function Φ we have

$$(\Phi(A_0)f)(x) = \lim_{r\to\infty} \frac{1}{(2\pi)^{3/2}} \int_{|p|\le r} \exp(ix.p)\Phi(|p|^2)\hat{f}(p)\,dp. \tag{6.47}$$

We would emphasise that the limits in (6.45)–(6.47) have to be taken in the $H(\mathbf{R}^3)$ sense. Furthermore, the theory of Fourier transform indicates that $F_0 : H(\mathbf{R}^3) \to H(\mathbf{R}^3)$ and, moreover, that it is a unitary operator. Consequently, we have $F_0^{-1} = F_0^*$.

We notice that

$$w_0(x, p) = \frac{1}{(2\pi)^{3/2}} \exp(ix.p), \qquad x, p \in \mathbf{R}^3 \tag{6.48}$$

satisfies the Helmholtz equation

$$(\Delta + |p|^2)w_0(x, p) = 0, \qquad x, p \in \mathbf{R}^3. \tag{6.49}$$

Thus w_0 might be thought to be an eigenfunction of $A_0 = -\Delta$ with associated eigenvalue $|p|^2$. However, a direct calculation shows that $w_0 \notin H(\mathbf{R}^3)$, and so w_0 must be a generalised eigenfunction of A_0. Nevertheless, the Fourier-Plancherel theory, *which has been developed independently of any scattering aspects*, indicates that all the limits (6.45)–(6.47) exist. Consequently, the spectral decomposition of A_0 can be written as a generalised eigenfunction expansion in the form

$$(F_0 f)(p) = \hat{f}(p) := \lim_{r\to\infty} \int_{|x|\le r} \overline{w_0(x, p)}\, f(x)\,dx, \tag{6.50}$$

$$f(x) = (F_0^* \hat{f})(x) := \lim_{r\to\infty} \int_{|p|\le r} w_0(x, p)\,\hat{f}(p)\,dp, \tag{6.51}$$

$$(\Phi(A_0)f)(x) = \lim_{r\to\infty} \int_{|p|\le r} w_0(x, p)\Phi(|p|^2)\,\hat{f}(p)\,dp, \tag{6.52}$$

where, as before, all limits are taken in the $H(\mathbf{R}^3)$ sense. It will be useful later on to bear in mind that (6.52) can also be written in the form

$$F(\Phi(A_0)f)(p) = \Phi(|p|^2)\,\hat{f}(p). \tag{6.53}$$

These various results imply that the wave function v_0, introduced in (6.39), can be interpreted in the form

$$v_0(x, t) = \int_{\mathbf{R}^3} w_0(x, p) \exp(-it\,|p|)\,\overset{\wedge}{h_0}(p)\,dp. \tag{6.54}$$

We remark that the improper integral in (6.54) must be interpreted in the $H(\mathbf{R}^3)$ limit sense as in (6.50)–(6.52). With this understanding we should also notice that in (6.54),

$$w_0(x, p)\exp(-it\,|p|) = \frac{1}{(2\pi)^{3/2}} \exp(i(x.p - t\,|p|)) \tag{6.55}$$

are solutions of (6.1) with $j=0$ and as such represent plane waves propagating in the direction of the vector p. Therefore the wave function given by (6.54) is a representation of a wave (acoustic) in terms of elementary plane waves (6.55).

We now turn our attention to the PP given by (6.1), (6.2) with $j=1$. As we have already mentioned, for ease of presentation at this stage we shall assume that the FP and the PP are both defined in the same Hilbert space. We have seen that the complex-valued solution of the FP is given by (6.39). Consequently, arguing as for the FP, we find that provided $\varphi_1 \in H(\mathbf{R}^3)$ and $\psi_1 \in D(A_1^{-1/2})$, the complex-valued solution of the PP is given by

$$v_1(x,t) \equiv v_1(x,t) = \exp\{-itA_1^{1/2}\}h_1(x) =: U_1(t)h_1(x), \tag{6.56}$$

where

$$h_1 = \varphi_1 + iA_1^{-1/2}\psi_1. \tag{6.57}$$

For the FP the Fourier-Plancherel theory provides us with the generalised eigenfunction expansion (6.50)–(6.52). As a consequence, we could interpret (6.39) in the form (6.54). We would like to have a similar result for the PP. Specifically, associated with A_1 we want a generalised eigenfunction expansion theorem of the form

$$(F_1 f)(p) = \tilde{f}(p) := \lim_{r\to\infty} \int_{|x|\le r} \overline{w_1(x,p)} f(x)\, dx, \tag{6.58}$$

$$f(x) = (F_1^* \tilde{f})(x) := \lim_{r\to\infty} \int_{|p|\le r} w_1(x,p)\tilde{f}(p)\, dp, \tag{6.59}$$

$$(\Phi(A_1)f)(x) = \lim_{r\to\infty} \int_{|p|\le r} w_1(x,p)\Phi(|p|^2)\tilde{f}(p)\, dp, \tag{6.60}$$

where, as previously, the above limits have to be taken in the $H(\mathbf{R}^3)$ sense. The kernels $w_1(x,p)$ are taken to be solutions of

$$(A_1 - |p|^2)w_1(x,p) = 0, \qquad x, p \in \mathbf{R}^3 \tag{6.61}$$

and as such are to be generalised eigenfunctions of A_1.

We would emphasise that for any specific perturbed problem it has to be *proved* that a generalised eigenfunction expansion (spectral decomposition) such as (6.58)–(6.60) is indeed available for use. For specific physical problems this can often involve a great deal of work. A full spectral analysis of A_1 is required, and functions such as w_1, which are intimately connected with the particular problem being considered, have to be determined. We shall return to these various aspects in later chapters when we come to deal with specific scattering problems. For the remainder of this chapter we shall assume that such generalised eigenfunction expansions are available. Consequently, we will then be able to write (6.56) in the following form

$$v_1(x,t) = \int_{\mathbf{R}^3} w_1(x,p)\exp(-it\,|p|)\tilde{h}_1(p)\, dp, \tag{6.62}$$

which is interpreted in the same way as (6.54).

We remark that in (6.54) and (6.62) the p need not be the same for both. It is associated with eigenvalues of A_0 in (6.54) and with eigenvalues of A_1 in (6.62).

From (6.54) and (6.62) it is a straightforward matter to obtain the representations

$$u_0(x,t)=\int_{R^3} w_0(x,p)\left\{\hat{\varphi}_0(p)\cos t\,|p|+\hat{\psi}_0(p)\frac{\sin t\,|p|}{|p|}\right\}dp, \tag{6.63}$$

$$u_1(x,t)=\int_{R^3} w_1(x,p)\left\{\tilde{\varphi}_1(p)\cos t\,|p|+\tilde{\psi}_1(p)\frac{\sin t\,|p|}{|p|}\right\}dp. \tag{6.64}$$

Hence, provided we can establish an eigenfunction expansion theorem of the form (6.58)–(6.60), since all the terms in (6.63) and (6.64) are computable, we have available in (6.63) and (6.64) a practical means of constructing solutions to the FP and the PP, respectively.

For the purpose of developing a scattering theory, it remains to investigate whether or not these solutions can be considered asymptotically equal, in some sense, as $t \to \pm\infty$. We shall begin to investigate this aspect in the next subsection.

6.2.4 Wave Operators and Their Construction

In chapter 1 we introduced the notions of asymptotic equality and wave operators. Specifically, we say that v_j, $j=0, 1$, the complex solutions of (6.1), (6.2), are AE as $t \to \pm\infty$ if

$$\lim_{t\to\pm\infty} \|v_1(t)-v_0(t)\|=0, \tag{6.65}$$

where $\|.\|$ denotes the norm on $H(\mathbf{R}^3)$.

Using (6.39) and (6.56), we find

$$\begin{aligned}\|v_1(t)-v_0(t)\| &= \|U_1(t)h_1-U_0(t)h_0\|\\ &= \|U_0^*(t)U_1(t)h_1-h_0\|\\ &=: \|W(t)h_1-h_0\|\,.\end{aligned}$$

Hence

$$\lim_{t\to\pm\infty} \|v_1(t)-v_0(t)\|=\|W_\pm h_1-h_0\|, \tag{6.66}$$

where

$$W_\pm := \lim_{t\to\pm\infty} W(t)=\lim_{t\to\pm\infty} U_0^*(t)U_1(t)=\lim_{t\to\pm\infty} \exp(itA_0^{1/2})\exp(-itA_1^{1/2}) \tag{6.67}$$

are the WOs associated with A_0 and A_1.

The manipulations leading to (6.66) have always to be justified but are certainly valid when the $U_j(t)$, $j=0, 1$, are unitary operators. In practice we endeavour to ensure that this is the case. We notice that if the A_j, $j=0, 1$, are self-adjoint operators, then Stone's theorem [105] ensures that the $U_j(t)$, $j=0, 1$, are indeed unitary operators.

Thus we see that the limit on the left-hand side of (6.66) will be zero, and so the FP and the PP will be AE provided that the initial data for the FP and the PP are related according to

$$h_{\pm} = W_{\pm} h_1, \tag{6.68}$$

where, for the sake of clarity, h_0 has been replaced by $h_{\pm}$ to indicate that different initial values might have to be considered for the FP when investigating $t \to +\infty$ and $t \to -\infty$.

Before indicating a means of constructing the wave operators $W_{\pm}$, we first recall some features of waves on a semi-infinite string.

Example 6.4 *The wave motion of a semi-infinite string is governed by an equation of the form*

$$(\partial_t^2 - \partial_x^2)u(x, t) = 0, \qquad (x, t) \in \Gamma \times \mathbf{R}, \tag{6.69}$$

where $\Gamma = (0, \infty)$.

Since Γ *is an unbounded region, then any solution of (6.69) will in practice be required to satisfy certain growth conditions, called radiation conditions, as* $|x| \to \infty$. *To indicate the nature of these conditions, we recall that equations like (6.69) have solutions that can be written in the form*

$$u(x, t) = f(x - t) + g(x + t), \tag{6.70}$$

where f *and* g *are arbitrary functions characterising a wave of constant profile travelling with unit velocity from left to right and from right to left, respectively. The precise form of* f *and* g *is settled in terms of the initial and boundary conditions that are imposed on solutions of (6.69). In the particular case when both waves can be assumed to have the same time dependency,* $\exp(-i\omega t)$, *we can expect to be able to write (6.70) in the form*

$$u(x, t) = e^{-i\omega t} u_+(x) + e^{-i\omega t} u_-(x). \tag{6.71}$$

Direct substitution of (6.71) into (6.69) shows that the two quantities u_+ *and* u_- *must satisfy*

$$(d_x^2 + \omega^2) u_{\pm}(x) = 0. \tag{6.72}$$

Now (6.72) does not imply that the $u_{\pm}$ *are necessarily the same. Indeed,*

$$u_+(x) = e^{i\omega x} \quad \text{and} \quad u_-(x) = e^{-i\omega x} \tag{6.73}$$

both satisfy (6.72). Combining (6.71) and (6.73), we obtain

$$u(x, t) = \exp(-i\omega(t - x)) + \exp(-i\omega(t + x)). \tag{6.74}$$

Thus, on recalling (6.70), we see that u_+ *characterises a wave moving from left to right and* u_- *a wave moving from right to left, both having the same time dependency* $\exp(-i\omega t)$. *Equivalently, we can say that* u_+ *is an outgoing wave since it is* *moving away from the origin, whilst u_- is an incoming wave as it is moving towards the origin. This particular feature of wave motion can be neatly encapsulated as follows. (See chapter 11 and the references cited there.)*

DEFINITION 6.5 *Solutions $u_\pm$ of the equation*

$$(\Delta+\omega^2)u_\pm(x)=f(x), \qquad x\in R^n,$$

are said to satisfy the Sommerfeld radiation conditions if and only if

$$\left\{\frac{\partial}{\partial r}\mp i\omega\right\}u_\pm(x)=\mathrm{o}\left(\frac{1}{r^{(n-1)/2}}\right) \quad \text{as } r=|x|\to\infty, \tag{6.75}$$

$$u_\pm(x)=\mathrm{O}\left(\frac{1}{r^{(n-1)/2}}\right) \quad \text{as } r=|x|\to\infty. \tag{6.76}$$

The estimates in (6.75) and (6.76) are considered to hold uniformly with respect to the direction $x/|x|$.

The estimate (6.75) taken with a minus (plus) sign is called the Sommerfeld outgoing (incoming) radiation condition.

With $u_\pm$ defined as in (6.73) it is clear that u_+ is outgoing, whilst u_- is incoming. From the practical point of view this is entirely to be expected. Furthermore, it will often be convenient to think of u_- as an incident wave and u_+ as a scattered wave.

Since we are dealing with perturbation processes, it is reasonable to assume that w_1, the kernel function in the generalised eigenfunction expansion theorem (6.58)–(6.60), is a perturbation of w_0, the kernel function in the generalised eigenfunction expansion theorem (6.50)–(6.52). Since w_0 characterises a plane wave, we shall refer to w_1, a perturbation of w_0, as a *distorted plane wave*.

DEFINITION 6.6 *An outgoing (incoming) distorted plane wave $w_+(x, p)(w_-(x, p))$ satisfies*

(i) $(\Delta+\omega^2)w_\pm(x)=0, x, p\in \mathbf{R}^n$,

(ii) $w_+(x, p)-w_0(x, p)$ satisfies the outgoing radiation condition (or $w_-(x, p) - w_0(x, p)$ satisfies the incoming radiation condition).

Consequently, we shall assume here that the kernel $w_1(x, p)$ is either an outgoing or an incoming distorted plane wave and we shall write

$$w_1(x, p)\equiv w_\pm(x, p)=w_0(x, p)+w'_\pm(x, p), \tag{6.77}$$

where $w'_+(w'_-)$ behaves like an outgoing (incoming) wave.

Of course, when dealing with specific physical problems, the existence and structure of the distorted plane waves must be established.

The existence of the distorted plane waves can be established by means of the *limiting absorption* (LAB) *principle* [35, 36], which is based on noticing that if A is a self-adjoint linear operator in a Hilbert space H and if $\lambda=\mu+i\nu\in\mathbb{C}$, $\nu\neq 0$, then the equation

$$(A-\lambda I)u(x, \lambda)=f(x)$$

has a solution $u(.,\lambda)\in H$ for each $f\in H$ because $\lambda\notin\sigma(A)$. In the LAB method we look for solutions in the form

$$u_{\pm}(x,\mu)=\lim_{\nu\to 0^{\pm}} u(x,\lambda).$$

The difficulty with this approach is centred on the interpretation of this limit. In general, it can only be understood in the sense of convergence in a Hilbert space $H(\varpi)$, where ϖ is an arbitrary subdomain of the region over which functions in H are defined. Physically, the quantity $u(x,\lambda)$, $\nu\neq 0$, describes a steady-state wave in an energy-absorbing medium with an absorption coefficient proportional to ν [154]. We shall deal with this method in more detail when we come to consider the specific problems discussed in later chapters.

If we assume the existence of the $w_{\pm}$ and, moreover, that they form two complete sets of generalised eigenfunctions for A_1, then on substituting (6.77) into (6.58)–(6.60), we obtain

$$\tilde{f}_{\pm}(p)=(F_{\pm}f)(p)=\lim_{r\to\infty}\int_{|x|\le r}\overline{w_{\pm}(x,p)}f(x)\,dx, \tag{6.78}$$

$$f(x)=(F_{\pm}^{*}\tilde{f})(x)=\lim_{r\to\infty}\int_{|p|\le r}w_{\pm}(x,p)\tilde{f}_{\pm}(p)\,dp, \tag{6.79}$$

$$(\Phi(A_1)f)(x)=\lim_{r\to\infty}\int_{|p|\le r}w_{\pm}(x,p)\Phi(|p|^2)\tilde{f}_{\pm}(p)\,dp, \tag{6.80}$$

provided these limits exist. We refer to F_+ as an *outgoing generalised Fourier transform* and to F_- as an *incoming generalised Fourier transform.*

On the basis of these various assumptions we see that the solution $v_1(x,t)$ given in (6.62) has two spectral representations depending on whether w_+ or w_- is used in the expansion theorem (6.78)–(6.80). Specifically, we have

$$v_1(x,t)=\lim_{r\to\infty}\int_{|p|\le r}w_{+}(x,p)\exp(-it\,|p|)\tilde{h}_{+}(p)\,dp \tag{6.81}$$

and

$$v_1(x,t)=\lim_{r\to\infty}\int_{|p|\le r}w_{-}(x,p)\exp(-it\,|p|)\tilde{h}_{-}(p)\,dp, \tag{6.82}$$

where

$$\tilde{h}_{\pm}(p)=\lim_{r\to\infty}\int_{|x|\le r}\overline{w_{\pm}(x,p)}h(x)\,dx. \tag{6.83}$$

Since w_+(resp. w_-) is an outgoing (resp. incoming) distorted plane wave, we refer to (6.81) (or (6.82)) as the outgoing (or incoming) spectral representations of v_1.

We are now in a position to construct a useful form for the wave operators $W_{\pm}$. If we substitute the decomposition (6.77) for w_- into (6.82), then we obtain

$$v_1(x,t)=v_0^{-}(x,t)+v^{-}(x,t), \tag{6.84}$$

where

$$v_0^-(x,t) = \lim_{r\to\infty} \int_{|p|\le r} w_0(x,p)\exp(-it\,|p|)\tilde{h}_-(p)\,dp, \tag{6.85}$$

$$v^-(x,t) = \lim_{r\to\infty} \int_{|p|\le r} w'_-(x,p)\exp(-it\,|p|)\tilde{h}_-(p)\,dp. \tag{6.86}$$

We now notice that since the kernel function in the integral (6.85) is w_0, it follows that v_0^- represents a free wave. Therefore we can write

$$v_0^-(x,t) = U_0(t)h_0^-(x) = \exp(-itA_0^{1/2})h_0^-(x), \tag{6.87}$$

where

$$h_0^-(x) = v_0^-(x,0). \tag{6.88}$$

Hence, bearing in mind (6.85), (6.68) and (6.58), we find

$$h_0^-(x) = v_0^-(x,0) = (F_0^*\tilde{h}_-)(x) = (F_0^*F_-h_1)(x). \tag{6.89}$$

Now (6.89) relates the initial data for a FP and the initial data for an associated PP. Therefore we conclude that, as $t \to -\infty$, we might expect that

$$h_0^-(x) = (F_0^*F_-h)(x) = W_-h(x); \tag{6.90}$$

that is, we might expect that

$$W_- = F_0^*F_-. \tag{6.91}$$

It turns out that this is indeed the case provided we have local energy decay of the form

$$\lim_{t\to-\infty} v^-(.,t) = 0. \tag{6.92}$$

Using (6.84), we see that (6.92) is equivalent to

$$\lim_{t\to-\infty} \|v_1(.,t) - v_0^-(.,t)\| = 0, \tag{6.93}$$

where $\|.\|$ is the $H(\mathbf{R}^3)$ norm. It now follows that

$$\begin{aligned}\|v_1(.,t) - v_0^-(.,t)\| &= \|\exp(-itA_1^{1/2})h_1 - \exp(-itA_0^{1/2})h_0^-\| \\ &= \|\{\exp(itA_0^{1/2})\exp(-itA_1^{1/2}) - F_0^*F_-\}h_1\|.\end{aligned} \tag{6.94}$$

Equation (6.94), together with (6.93) and the definition of the WO given in (6.67), implies that W_- exists and is given by

$$W_- = F_0^*F_- \tag{6.95}$$

If we substitute the decomposition (6.77) for w_+ into (6.81), then we obtain

$$v_1(x,t) = v_0^+(x,t) + v^+(x,t), \tag{6.96}$$

where

$$v_0^+(x,t) = \lim_{r\to\infty} \int_{|p|\leq r} w_0(x,p)\exp(-it\,|p|)\tilde{h}_+(p)\,dp, \tag{6.97}$$

$$v^+(x,t) = \lim_{r\to\infty} \int_{|p|\leq r} w'_+(x,p)\exp(-it\,|p|)\tilde{h}_+(p)\,dp. \tag{6.98}$$

Arguing as before, we see that $v_0^+(x,t)$ represents a free wave and that we can write

$$v_0^+(x,t) = U_0(t)h_0^+(x) = \exp(-itA_0^{1/2})h_0^+(x), \tag{6.99}$$

where

$$h_0^+(x) = v_0^+(x,0) = (F_0^*\tilde{h}_+)(x) = (F_0^*F_+h_1)(x). \tag{6.100}$$

This result implies that we might expect that

$$W_+ = F_0^*F_+. \tag{6.101}$$

We can show that W_+ exists and that (6.100) is indeed the case provided that we have local energy decay of the form

$$\lim_{t\to+\infty} v^+(.,t) = 0. \tag{6.102}$$

The proof follows as for the case of W_-, and the details are left as an exercise.

Once we have determined the existence and the form of the wave operators $W_\pm$, a scattering operator S that links the initial conditions $h_0^\pm$ can be introduced as follows.

The above results indicate that

$$h_0^\pm = W_\pm h_1 = F_0^* F_\pm h_1. \tag{6.103}$$

This in turn implies

$$F_0 h_0^\pm = \hat{h}_0^\pm = F_\pm h_1.$$

Hence

$$\hat{h}_0^+ = F_+h_1 = F_+F_-^*\hat{h}_0^- =: S\hat{h}_0^-, \tag{6.104}$$

and we see that

$$S := F_+F_-^* : \hat{h}_0^- \to \hat{h}_0^+. \tag{6.105}$$

This operator and the unitarily equivalent operator

$$F_0^*SF_0 := F_0^*F_+F_-^*F_0 : h_0^- \to h_0^+ \tag{6.106}$$

are particularly useful when discussing the theoretical and practical details of the asymptotic condition and the associated AE results.

6.2.5 More About Asymptotic Conditions

We introduced in chapter 1 the notion of AE. The aim in this subsection is to provide a more precise formulation of this asymptotic property.

The requirement that the scattered waves can be characterised, at large positive and large negative times, in terms of free waves that are totally unaffected by any scatterer, is called the *asymptotic condition*. To place this in a mathematical framework, we again consider a simple case. Specifically, we consider a FP characterised by an operator A_0 and an associated group $\{U_0(t)\}$, and a PP, describing the wave scattering, characterised in terms of an operator A_1 and a group $\{U_1(t)\}$. Here, as introduced earlier,

$$U_0(t) = \exp\{-itA_0^{1/2}\}, \qquad U_1(t) = \exp\{-itA_1^{1/2}\}. \tag{6.107}$$

Furthermore, we have seen that the FP has an initial state vector h_0 given by (6.37), whilst the PP has an initial state vector h_1 given by (6.57). We shall assume that A_0 and A_1 both act in the same Hilbert space H.

In a typical scattering situation the time evolution of the scattered wave that has initial state $h_1 \in H$ is governed by the group $\{U_1(t)\}$; that is, the state of the scattered wave at some other time t is $U_1(t)h_1$. In the absence of any scattering mechanism the time evolution of the free wave having a initial state $h_0 \in H$ is governed by the group $\{U_0(t)\}$. The state of the free wave at some other time t is then $U_0(t)h_0$. Our aim is to see if we can approximate, as $t \to \pm\infty$, the waves arising from the PP by waves arising from the free evolution of suitable FPs. This is most readily done by assuming that for $h_1 \in H$ there exist two initial states $h_\pm$ such that the wave $U_1(t)h_1$ converges to $U_0(t)h_\pm$ as $t \to \pm\infty$. Symbolically, we mean that we should be able to satisfy the requirements

$$\lim_{t\to-\infty} \|U_1(t)h_1 - U_0(t)h_-\| = 0, \qquad \lim_{t\to+\infty} \|U_1(t)h_1 - U_0(t)h_+\| = 0, \tag{6.108}$$

where $\|.\|$ is the norm on H. We refer to (6.108) as the *asymptotic conditions*. When these requirements are satisfied, it will mean that the scattered wave, characterised by $U_1(t)h_1$, is virtually indistinguishable from the wave $U_0(t)h_-$ in the remote past and from $U_0(t)h_+$ in the distant future. The requirement that vectors such as $h_\pm$ should exist and, moreover, exist as elements of H is really quite a severe restriction. However, we would remark that an indication of how this requirement can be met has already been given in the previous subsection.

The set of vectors $h_1 \in H$ for which (6.108) can be satisfied is called the *set of scattering states* for A_1 and will be denoted by $M(A_1)$. We shall assume that, for each self-adjoint operator A with which we shall be concerned, the set of scattering states $M(A)$ has the following properties.

(i) $M(A)$ is a subspace of H.
(ii) $M(A)$ is invariant under the group $\{U(t)\}$; that is, if $h \in M(A)$, then

$$U(t)h = \exp(-itA^{1/2})h \in M(A) \quad \text{for all } t \in \mathbf{R}.$$

We notice that if h is an eigenfunction of $A^{1/2}$, that is, for some $\mu \in H$,

$$A^{1/2}h = \mu h,$$

then the state $U(t)h=\exp(-itA^{1/2})h=\exp(-it\mu)h$ is simply a multiple of the state h and as such cannot define a scattering process. It defines a *bound state* of A. For this reason scattering states are expected to be associated with the continuous spectrum of A.

In this description of the asymptotic condition we are associating with each initial state $h_- \in M(A_0)$ another state vector $h_+ \in M(A_0)$, both state vectors being interpreted as initial states at time $t=0$. If no scattering is taking place, then $U_1(t)=U_0(t)$ and clearly we have $h_-=h_+$. However, when scattering does occur, the correspondence between h_- and h_+ is effected by means of a scattering operator S. Typical forms for the scattering operator have been indicated in (6.105) and (6.106).

When dealing with specific problems, it is sometimes convenient to alter the various requirements mentioned above. This is because h_1 is usually associated with the PP and as such is a given quantity. Our task then is to determine the states $h_\pm$ so that the asymptotic conditions (6.108) are satisfied. This is done by means of the WO, as in (6.68), and using a SO to determine the relation between h_- and h_+, as in (6.105) or (6.106). We can summarise the above discussion in the following manner.

Definition 6.7 *A wave evolving according to*

$$U_1(t)h_1(x)=\exp(-itA_1^{1/2})h_1(x)$$

is said to satisfy the asymptotic condition for $t \to +\infty$*, defined with respect to the family of operators*

$$\{U_0(t)=\exp(-itA_0^{1/2})\},$$

if there exists an element $h_+ \in M(A_0) \subseteq H$ *such that, as* $t \to +\infty$*, the wave* $U_1(t)$ $h_1(x)$ *is asymptotically indistinguishable from the wave* $U_0(t)h_+(x)$*.*

A similar definition holds as $t \to -\infty$*.*

This definition implies that an element $h_\pm \in H$ is such that the free evolution $U_0(t)h_\pm$ defines, as $t \to \pm\infty$, the asymptotes of some evolution $U_1(t)h_1$. However, we need to determine whether or not every $h_1 \in H$ evolving according to $U_1(t)h_1$ has asymptotes, as $t \to \pm\infty$, of the form $U_0(t)h_\pm$. To settle this question, we need some preparation involving the properties of the wave operators $W_\pm$.

First we show that the wave operators $W_\pm$ satisfy the *intertwining relation*

$$A_0W_\pm = W_\pm A_1. \tag{6.109}$$

To see that this is the case, we notice

$$\begin{aligned}\exp(-i\tau A_0^{1/2})W_\pm &= \exp(-i\tau A_0^{1/2}) \lim_{t\to\pm\infty}\{\exp(itA_0^{1/2})\exp(-itA_1^{1/2})\}\\ &= \lim_{t\to\pm\infty}\{\exp(i(t-\tau)A_0^{1/2})\exp(-itA_1^{1/2})\}\\ &= \lim_{t\to\pm\infty}\{\exp(i(t-\tau)A_0^{1/2})\exp(-i(t-\tau)A_1^{1/2})\}\exp(-i\tau A_1^{1/2})\\ &= W_\pm \exp(-i\tau A_1^{1/2}).\end{aligned}$$

Differentiate with respect to τ and set $\tau=0$ to obtain (6.109).

We next notice, using the properties of inner products on Hilbert spaces, that for all $f, g \in D(W_{\pm}^{*})$, we have

$$(W_{\pm}^{*} f, W_{\pm}^{*} g) = \lim_{t\to\pm\infty} (\exp(itA_0^{1/2})\exp(-itA_1^{1/2})f, \exp(itA_0^{1/2})\exp(-itA_1^{1/2})g)$$
$$= (f, g). \tag{6.110}$$

Hence the wave operators $W_{\pm}^{*}$ are isometries. Furthermore, we can use (6.110) to obtain

$$(W_{\pm}^{*} f, W_{\pm}^{*} g) = (f, W_{\pm} W_{\pm}^{*} g) = (f, g),$$

from which it follows that

$$W_{\pm} W_{\pm}^{*} = I. \tag{6.111}$$

We remark that from (6.111) we might expect that $W_{\pm}^{*}$ should behave like $W_{\pm}^{-1}$. We obtain conditions below that can ensure this.

We notice that for $f, g \in H$ related according to

$$g := W_{\pm}^{*} f \in R(W_{\pm}^{*}) = \text{ range of } W_{\pm}^{*},$$

we have

$$W_{+}^{*} W_{+} g = W_{+}^{*}(W_{+} W_{+}^{*} f) = W_{+}^{*} f = g. \tag{6.112}$$

If, however, an element $h \in H$ is orthogonal to $R(W_{+}^{*})$, then

$$0 = (h, W_{+}^{*} f) = (W_{+} h, f) \quad \text{for all } f \in H. \tag{6.113}$$

Hence it follows that

$$W_{+} h = 0, \qquad h \in R(W_{\pm}^{*})^{\perp}, \tag{6.114}$$

and that

$$W_{+}^{*} W_{+} h = W_{+}^{*}(W_{+} h) = 0.$$

Hence we have shown that $W_{+}^{*} W_{+}$ is a projection onto $R(W_{+}^{*})$. Similarly, $W_{-}^{*} W_{-}$ is a projection onto $R(W_{-}^{*})$.

With this preparation we return to the question of the availability of the initial states $h_{\pm}$ for the FP that will yield the required AE for the PP.

We emphasise that, as always in this monograph, the FP will be concerned with an incident wave (signal) in the absence of any perturbation, whilst the PP will be concerned with scattered waves that are a consequence of some perturbation of the incident wave.

We have seen that when developing a scattering theory, we try to relate the evolution of a given PP and the evolution of a rather simpler, associated FP. If the evolution of the PP is governed by the group $\{U_1(t)\}$ and that of the FP by the group $\{U_0(t)\}$ and if the PP has given initial state h_1, then the solutions of the PP and the FP will be AE as $t \to -\infty$ provided the FP has initial state h_{-} given by

$$h_{-} = W_{-} h_1, \tag{6.115}$$

where $W_{-} = \lim_{t\to-\infty} U_0^{*}(t) U_1(t)$ (see (6.67), (6.68)).

Once the initial state h_- is determined according to (6.115), it then remains to determine whether there exist some, possibly different, initial state h_+ for the FP that will ensure that we *also* have AE of the PP and the FP as $t \to +\infty$. With this in mind we first notice from (6.115) that to ensure AE as $t \to -\infty$, the initial state for the PP must satisfy

$$h_1 = W_-^* h_- \in R(W_-^*). \tag{6.116}$$

However, with AE as $t \to +\infty$ in mind, we see that we can also introduce the influence of the wave operator W_+ by expressing h_1 in the form

$$h_1 = h_+ + h^\perp, \qquad h_+ \in R(W_+^*), \qquad h^\perp \in R(W_+^*)^\perp. \tag{6.117}$$

Since $W_+^* W_+$ is a projection onto $R(W_+^*)$, the first component of (6.117) indicates, remembering (6.116), that the following must also hold.

$$R(W_+^*) \ni h_+ = W_+^* W_+ h_1 = W_+^* W_+ W_-^* h_- =: W_+^* S h_-, \tag{6.118}$$

where

$$S := W_+ W_-^* : h_- \to h_+ \tag{6.119}$$

is the scattering operator that, when known, enables us to determine the required initial state from the previously obtained state h_-.

As $t \to +\infty$, the evolution of the initial state h_1 will still be governed by the group $\{U_1(t)\}$, and from (6.117) we have

$$U_1(t) h_1 = U_1(t) h_+ + U_1(t) h^\perp. \tag{6.120}$$

Consequently,

$$W_+ U_1(t) h_1 = W_+ U_1(t) h_+ + W_+ U_1(t) h^\perp.$$

Now, using the intertwining relation (6.109) and (6.114), we find

$$W_+ U_1(t) h^\perp = U_0(t) W_+ h^\perp = 0.$$

Consequently, we see that, as $t \to +\infty$, the component $U_1(t) h^\perp$ of the state of the system remains orthogonal to the scattering subspace generated by h_+.

Finally in this section, we demonstrate some useful connections between the ranges of the WO and the properties of the SO.

Theorem 6.8
(i) $R(W_-^*) \subseteq R(W_+^*)$ if and only if S is isometric.
(ii) $R(W_+^*) \subseteq R(W_-^*)$ if and only if S^* is isometric.
(iii) $R(W_+^*) = R(W_-^*)$ if and only if S is unitary.

Proof. (i) Assume $R(W_-^*) \subseteq R(W_+^*)$. For any element $f \in H$ we then have $W_-^* f \in R(W_+^*)$. Since $W_+^* W_+$ is a projection onto $R(W_+^*)$, this implies

$$W_+^* W_+ W_-^* f = W_-^* f. \tag{6.121}$$

Furthermore, since we always have

$$\|W_\pm h\| = \lim_{t \to \pm\infty} \|\exp(itA_0^{1/2}) \exp(-itA_1^{1/2}) h\| = \|h\|, \tag{6.122}$$

then using (6.119), (6.122) and the properties of projection operators, we obtain

$$\|Sf\| = \|W_+^* Sf\| = \|W_+^* W_+ W_-^* f\| = \|W_-^* f\| = \|f\|. \tag{6.123}$$

Hence S is isometric.

Conversely, assume that S is isometric. Then (6.123) indicates that the projection of $W_-^* f$ onto $R(W_+^*)$ has a norm identical to $\|W_-^* f\|$. Hence $W_-^* f \in R(W_+)$. Hence $R(W_-^*) \subseteq R(W_+^*)$.

(ii) This follows by noticing that $S^* = W_- W_+^*$ and using the same argument as for (i) with plus and minus signs interchanged.

(iii) This follows by noticing that S is unitary if and only if $SS^* = S^* S = I$, that is, if and only if both S and S^* are isometric. Consequently, (ii) will follow from (i) and (ii). □

A physical interpretation of this theorem can be obtained, for instance, by considering the condition $R(W_-^*) \subseteq R(W_+^*)$. We have seen that if the initial states for the PP and the FP are related as $t \to -\infty$ according to

$$h_1 = W_-^* h_-, \tag{6.124}$$

then the PP and the FP are AE as $t \to -\infty$. In the above, theorem (6.121) implies that we also have

$$h_1 = W_+^* S h_-. \tag{6.125}$$

For AE of the PP and the FP as $t \to +\infty$, when h_- is already fixed to ensure AE as $t \to -\infty$, we consider

$$\begin{aligned}\lim_{t\to+\infty} \|U_1(t)h_1 - U_0(t)h_+\| &= \|W_+ h_1 - h_+\| \\ &= \|W_+ W_-^* h_- - h_+\| \\ &= \|Sh_- - h_+\|.\end{aligned}$$

The right-hand side vanishes by virtue of (6.125) and (6.68), which implies that we have the required AE as $t \to +\infty$. Therefore we see that the condition $R(W_-^*) \subseteq R(W_+^*)$ implies that if solutions of the PP can be shown to be asymptotically free as $t \to -\infty$, then they become asymptotically free again as $t \to +\infty$. The scattering operator S provides the transformation between the initial state for the asymptotically free state as $t \to -\infty$ and the initial state for the asymptotically free state as $t \to +\infty$. Similar interpretations for (ii) and (iii) can also be given.

In summary, we have seen that there are conditions that ensure that a given initial state h_1 evolving according to $U_1(t)h_1$ at some general time t will approach $U_0(t)h_-$ as $t \to -\infty$ and $U_0(t)Sh_- = U_0(t)h_+$ as $t \to +\infty$.

In many cases of practical interest the wave operators can be shown to have the following property.

Definition 6.9 *The wave operators $W_\pm$ defined as in (6.67) are said to be asymptotically complete if $R(W_+^*) = R(W_-^*)$.*

Asymptotic completeness of the wave operators is thus equivalent to the unitarity of the scattering operator. However, as we shall see in later chapters, to prove for a given problem that the associated SO is indeed unitary is not always a simple matter.

6.2.6 A Remark About Spectral Families

A comparison of (6.44) and (6.47), bearing in mind (6.48), seems to suggest that

$$dE_0(\lambda)f(x) = w_0(x, p)\,\hat{f}(p)\,dp. \tag{6.126}$$

To make this more precise, we first recall that $E_0(\mu)$ satisfies

$$E_0(\mu) = \begin{cases} I, & \mu \geq 0, \\ 0, & \mu < 0. \end{cases}$$

Consequently, $E_0(\mu)$ has the property of the Heaviside unit function $H(\tau)$. Thus if in (6.47) we take $\Phi(\lambda) = H(\mu - \lambda)$, then we obtain

$$E_0(\mu)f(x) = \lim_{R\to\infty} \int_{|p|\leq R} w_0(x, p)H(\mu - |p|^2)\,\hat{f}(p)\,dp, \qquad \mu \geq 0,$$

from which it follows that

$$E_0(\mu)f(x) = \begin{cases} \int_{|p|\leq\sqrt{\mu}} w_0(x, p)\,\hat{f}(p)\,dp, & \mu \geq 0, \\ 0, & \mu < 0. \end{cases} \tag{6.127}$$

Differentiating (6.127), we recover (6.126).

Even for the FP the practical determination of the spectral family $\{E_0(\lambda)\}$ is a difficult matter. It seems best left to abstract analytical discussions where it can be of considerable use [59, 95].

The results (6.45)–(6.47) are often referred to as a *Fourier inversion theorem.* We have already mentioned that these results can be obtained quite independently of any scattering considerations by means of the Plancherel theory of Fourier transforms. We shall endeavour, for both FPs and PPs, to use the Fourier inversion theorem approach rather than the spectral family approach.

6.2.7 Some Comparisons of the Two Approaches

In the last few subsections we have worked directly with wave equations and their solutions. However, we could have worked throughout in terms of the equivalent first-order systems introduced earlier to settle the well-posedness of the problems.

For convenience we gather together here the salient results from the treatment of wave equations and their associated first-order systems. We then indicate how they are related.

When working with IVPs of the form

$$\{\partial_t^2 + A_j\}u_j(x, t) = 0, \qquad (x, t) \in \mathbf{R}^n \times \mathbf{R}, \qquad j = 0, 1, \tag{6.1}$$

$$u_j(x, 0) = \varphi_j(x), \qquad u_{jt}(x, 0) = \psi_j(x), \qquad j = 0, 1, \tag{6.2}$$

we noticed that their solutions could be written in the form (see (6.37) and (6.57))

$$u_j(t) = (\cos tA_j^{1/2})\varphi_j + A_j^{-1/2}(\sin tA_j^{1/2})\psi_j, \qquad j = 0, 1.$$

We then defined

$$h_j(x) = \varphi_j(x) + iA_j^{-1/2}\psi_j(x), \qquad j = 0, 1,$$

and combined these results to obtain (see (6.37) and (6.57))

$$u_j(t) \equiv u_j(., t) = \mathrm{Re}(v_j(., t)), \qquad j = 0, 1, \tag{6.38}$$

where

$$v_j(t) \equiv v_j(., t) = \exp\{-itA_j^{1/2}\}h_j =: U_j(t)h_j, \qquad j = 0, 1. \tag{6.39}$$

The quantity v_j is referred to as the complex-valued solution of (6.1).

We remark that if the initial time is $t = s$ rather than zero, then in the above t has to be replaced by $(t - s)$.

We then went on to discuss the AE of the solutions $v_j(x, t)$ and, as a consequence, introduced the wave operators $W_\pm$ defined by

$$W_\pm := \lim_{t\to\pm\infty} W(t) = \lim_{t\to\pm\infty} U_0^*(t)U_1(t) = \lim_{t\to\pm\infty} \exp(itA_0^{1/2})\exp(-itA_1^{1/2}). \tag{6.67}$$

The wave operators were then determined in the form (see (6.91) and (6.101))

$$W_\pm = F_0^* F_\pm,$$

where F_0 is the Fourier transform defined in (6.45) and $F_\pm$ are the incoming and outgoing generalised Fourier transforms defined in (6.78), bearing in mind (6.77) and (6.58).

The IVPs (6.1) and (6.2) can be written as first-order systems of the form

$$\begin{bmatrix} u_j \\ u_{jt} \end{bmatrix}_t (x, t) + \begin{bmatrix} 0 & -I \\ A_j & 0 \end{bmatrix} \begin{bmatrix} u_j \\ u_{jt} \end{bmatrix} (x, t) = \begin{bmatrix} 0 \\ 0 \end{bmatrix}, \qquad j = 0, 1, \tag{6.15}$$

$$\begin{bmatrix} u_j \\ u_{jt} \end{bmatrix} (x, 0) = \begin{bmatrix} \varphi_j \\ \psi_j \end{bmatrix} (x), \qquad j = 0, 1. \tag{6.16}$$

This array can be conveniently written as an IVP for an ordinary differential equation in H_E of the form

$$\{d_t - i\mathbf{G}_j\}\mathbf{u}_j(t) = 0, \qquad \mathbf{u}_j(0) = \mathbf{u}_j^0, \tag{6.21}$$

where, for $j = 0, 1$,

$$\mathbf{G}_j : H_E(\mathbf{R}^n) \supseteq D(\mathbf{G}_j) \to H_E(\mathbf{R}^n),$$

$$\mathbf{G}_j\boldsymbol{\xi} = i\begin{bmatrix} 0 & -I \\ A_j & 0 \end{bmatrix} \begin{bmatrix} \xi_1 \\ \xi_2 \end{bmatrix}, \qquad \boldsymbol{\xi} = \langle \xi_1, \xi_2 \rangle \in D(\mathbf{G}_j),$$

$$D(\mathbf{G}_j) = \{\boldsymbol{\xi} = \langle \xi_1, \xi_2 \rangle \in H_E(\mathbf{R}^n) : A_j\xi_1 \in L_2(\mathbf{R}^n), \xi_2 \in H_D(\mathbf{R}^n)\}$$

and where we understand

$$\mathbf{u}_j \equiv \mathbf{u}_j(.,.) : t \to \mathbf{u}_j(.,t) =: \mathbf{u}_j(t) \in H_E(\mathbf{R}^n), \qquad j = 0, 1.$$

The IVPs (6.21) have solutions of the form

$$\mathbf{u}_j(t) = \exp\{it\mathbf{G}_j\}\mathbf{u}_j^0 = \mathbf{U}_j(t)\mathbf{u}_j^0, \tag{6.28}$$

where

$$\mathbf{U}_j(t) = \exp(it\mathbf{G}_j) = \cos tA_j^{1/2}\begin{bmatrix} I & 0 \\ 0 & I \end{bmatrix} - A_j^{-1/2}\sin tA_j^{1/2}\begin{bmatrix} 0 & -I \\ A_j & 0 \end{bmatrix} \tag{6.24}$$

$$= \begin{bmatrix} \cos tA_j^{1/2} & A_j^{-1/2}\sin tA_j^{1/2} \\ -A_j^{1/2}\sin tA_j^{1/2} & \cos tA_j^{1/2} \end{bmatrix}. \tag{6.25}$$

Acknowledging (6.37) and (6.57), we obtain, for $j = 0, 1$,

$$\mathbf{u}_j(x,t) = \mathbf{U}_j(t)\mathbf{u}_j^0(x) = \mathrm{Re}\begin{bmatrix} (\exp\{-itA_j^{1/2}\})h_j(x) \\ (-iA_j^{1/2})(\exp\{-itA_j^{1/2}\})h_j(x) \end{bmatrix}$$
$$= \mathrm{Re}\{\mathbf{v}_j(x,t)\},$$

where the complex-valued solutions of (6.21) can be written in the form

$$\mathbf{v}_j(x,t) := U_j(t)\begin{bmatrix} I \\ -iA_j^{1/2} \end{bmatrix} h_j(x) =: U_j(t)\mathbf{g}_j(x), \qquad j = 0, 1,$$

with

$$\mathbf{g}_j(x) := \begin{bmatrix} I \\ -iA_j^{1/2} \end{bmatrix} h_j(x), \qquad j = 0, 1,$$

and the $U_j(t)$ are defined as in (6.39) and (6.56).

In a similar manner to that outlined above when discussing the AE of solutions $v_j(x,t)$, we investigate the AE of the solutions $\mathbf{v}_j(x,t)$ by requiring

$$0 = \lim_{t\to\pm\infty} \|\mathbf{v}_1(.,t) - \mathbf{v}_0(.,t)\|_{H_E}$$
$$= \lim_{t\to\pm\infty} \|U_1(t)\mathbf{g}_1(.) - U_0(t)\mathbf{g}_0(.)\|_{H_E}$$
$$= \|W_\pm\mathbf{g}_1(.) - \mathbf{g}_0(.)\|_{H_E},$$

where $\|.\|_{H_E}$ denotes the norm in the energy space $H_E(\mathbf{R}^n)$, (6.11), and, as in (6.67),

$$W_\pm = \lim_{t\to\pm\infty} U_0^*(t)U_1(t).$$

Thus we see that we will have the required AE if

$$\mathbf{g}_0(x) = W_\pm\mathbf{g}_1(x),$$

that is, if

$$\begin{bmatrix} I \\ -iA_0^{1/2} \end{bmatrix} h_0(x) = W_\pm \begin{bmatrix} I \\ -iA_1^{1/2} \end{bmatrix} h_1(x).$$

We see that the first component of this result yields, as might have been expected, the same result as that obtained in subsection 6.2.4.

6.2.8 Summary

In the above subsections we have outlined in a reasonably precise way a strategy, already hinted at in chapter 1, for the analysis of wave scattering problems in the time domain. We now see that this strategy, written more compactly, consists of the following fundamental problems.

- Settle the existence and uniqueness of solutions to (6.1), (6.2) and determine their propagation properties. We have seen that this can be achieved by using Stone's theorem to show that the solution forms (6.28) are valid.
- Establish the existence and uniqueness of the wave operators $W_\pm$ defined in (6.67). We have seen in the above subsections that this can be achieved using generalised eigenfunction (generalised Fourier transform) techniques in conjunction with certain energy decay requirements.
- Provide a spectral analysis of the associated spatial operators A_j, $j=0, 1$, in order to be able to generate appropriate generalised eigenfunction expansion theorems.
- Prove the limiting absorption principle for the operators A_j, $j=0, 1$, and as a consequence settle the existence and uniqueness of appropriate distorted plane waves.
- Investigate the completeness of the wave operators. This means determining whether or not all solutions of the PP are asymptotically free as $t \to \pm\infty$. This is closely related to establishing the existence of the quantities $h_\pm$ introduced in (6.68).

The literature devoted to these, often very difficult, problems is now extensive in the autonomous problem case; see, for instance, [6, 71, 95, 101, 105].

Once these several problems have been satisfactorily resolved, we will be well placed to develop the required scattering theory. In this connection see definition 1.1 and theorem 1.2 together with definition 1.3 and theorem 1.4 with the remarks that follow it. Indeed, by working through this programme of problems we will be able to develop for autonomous problems promising methods for the practical construction of solutions, wave operators and the scattering operator.

6.3 ON SCATTERING PROCESSES IN NONAUTONOMOUS PROBLEMS

When we come to analysing NAPs and, in particular, developing an associated scattering theory, we would like to be able to mimic as much as possible the strategies adopted when dealing with APs. However, we will not have available all the structure outlined above. The essential differences and sources of difficulty when dealing with NAPs are centred on the following points.

- The separation of variables technique is, in general, no longer available. Hence analysis in the frequency domain is not immediately possible.
- Fourier transform methods are not immediately available.
- Operators in the defining equations are now time-dependent. Consequently, solutions of the form (6.7) and (6.28) are no longer available.

Nevertheless, a profitable way of investigating NAPs is again to reduce them to first-order systems. This we shall do in essentially the same way as we did in section 6.2 for APs. Specifically, we introduce the energy space $H_E(\mathbf{R}^n)$ defined as the completion of $C_0^\infty(\mathbf{R}^n) \times C_0^\infty(\mathbf{R}^n)$ with respect to the norm $\|.\|_E$, where for $\mathbf{f} = \langle f_1, f_2\rangle \in C_0^\infty(\mathbf{R}^n) \times C_0^\infty(\mathbf{R}^n)$,

$$\|\mathbf{f}\|_E^2 := \frac{1}{2}\int_{\mathbf{R}^n} \{|\nabla f_1(x)|^2 + |f_2(x)|^2\}\, dx. \tag{6.128}$$

6.3.1 Propagation Aspects

To fix ideas, we shall consider a nonautonomous version of the IVP (6.1), (6.2). Specifically, we shall consider the NAPs

$$\{\partial_t^2 + A_j(t)\}u_j(x,t) = 0, \qquad (x,t) \in \mathbf{R}^n \times \mathbf{R}, \qquad j = 0, 1, \qquad t \geq 0, \tag{6.129}$$

$$u_j(x,0) = \varphi_j(x), \qquad u_{jt}(x,0) = \psi_j(x), \qquad j = 0, 1, \tag{6.130}$$

where $j = 0$ represents a FP and $j = 1$ a PP. We shall assume that

$$A_j(t) : H(\mathbf{R}^n) \to H(\mathbf{R}^n) \equiv L_2(\mathbf{R}^n), \qquad j = 0, 1, \qquad t \geq 0,$$

where $H(\mathbf{R}^n)$ is a Hilbert space and the $A_j(t)$ are the representations in $H(\mathbf{R}^n)$ of differential expressions $L_j(x,t)$ that characterise the media supporting the wave motions.

We now proceed as in section 6.2 and reduce the problems (6.129), (6.130) to the following IVPs for ordinary differential equations in $H_E(\mathbf{R}^n)$.

$$\{d_t - i\mathbf{G}_j(t)\}\mathbf{u}_j(t) = 0, \qquad \mathbf{u}_j(s) = \mathbf{u}_j^s \quad \text{for all } t > 0, \tag{6.131}$$

where for $j = 0, 1$ and $t \geq 0$,

$$\mathbf{u}_j(t) = \langle u(t), u_t(t)\rangle, \qquad i\mathbf{G}_j(t) = -\begin{bmatrix} 0 & -I \\ A_j(t) & 0 \end{bmatrix}, \qquad \mathbf{u}_j \in D(\mathbf{G}_j(t)),$$

$$\mathbf{G}_j(t) : H_E(\mathbf{R}^n) \supseteq D(\mathbf{G}_j(t)) \to H_E(\mathbf{R}^n),$$

$$D(\mathbf{G}_j(t)) := \{\boldsymbol{\psi} = \langle \psi_1, \psi_2\rangle \in H_E(\mathbf{R}^n) : A_j(t)\psi_1 \in L_2(\mathbf{R}^n),\ \psi_2 \in H_D(\mathbf{R}^n)\}$$

and, symbolically,

$$A_j(t) = L_j(x,t).$$

In contrast to the AP case the matrix operator in (6.131) is now a function of t. Nevertheless, (6.131) can still be solved by an integrating factor technique. In this case we obtain the required solution in the following form.

$$\mathbf{u}_j(t) = \exp\left\{ i \int_s^t \mathbf{G}_j(\eta)\, d\eta \right\} \mathbf{u}_j^s =: \mathbf{U}_j(t, s)\mathbf{u}_j^s, \qquad j = 0, 1, \tag{6.132}$$

where the $\mathbf{U}_j(t, s)$ are seen to satisfy the following (see theorem 5.27).

(i) $\mathbf{U}_j(t, s)$ have values in $B(H_E(\mathbf{R}^n))$, $0 \leq s \leq t \leq T$, for s fixed and some T.
(ii) $\mathbf{U}_j(t, r)\mathbf{U}_j(r, s) = \mathbf{U}_j(t, s)$, $\mathbf{U}_j(s, s) = \mathbf{I}$
(iii) $\partial_t \mathbf{U}_j(t, s) = i\mathbf{G}_j(t)\mathbf{U}_j(t, s)$.
(iv) $\partial_s \mathbf{U}_j(t, s) = -i\mathbf{U}_j(t, s)\mathbf{G}_j(s)$.

As we pointed out in chapter 5, operators with the properties (i)–(iv) are said to be propagators (fundamental solutions) for (6.131). The propagators $\mathbf{U}_j(t, s)$ can also be introduced and studied in abstract form, taking the properties (i)–(iv) as a starting point rather than the IVP (6.131); see, for example [59, 142].

The usefulness of these propagators in the present context is demonstrated by the following straightforward modification of a result in chapter 5.

Theorem 6.10 *There exists at most one propagator $\mathbf{U}_j$ for the family $\{\mathbf{G}_j(t)\}_{0 \leq t \leq T}$. Furthermore, if $\boldsymbol{\psi}_j(t)$ is any solution of the problem*

$$\{d_t - i\mathbf{G}_j(t)\}\boldsymbol{\psi}_j(t) = \mathbf{f}_j(t), \qquad \boldsymbol{\psi}_j(s) = \boldsymbol{\psi}_j^s, \qquad j = 0, 1, \tag{6.133}$$

where $\mathbf{f}_j \in C([s, T], H_E(\mathbf{R}^n))$, then

$$\boldsymbol{\psi}_j(t) = \mathbf{U}_j(t, s)\boldsymbol{\psi}_j^s + \int_s^t \mathbf{U}_j(t, \tau)\mathbf{f}_j(\tau)\, d\tau, \qquad j = 0, 1. \tag{6.134}$$

6.3.2 Scattering Aspects

Scattering theories for NAPs can be developed by paralleling the procedures adopted when developing such theories for APs. This can be illustrated as follows.

For all $t \geq 0$, let $i\mathbf{G}_j(t)$, $j = 0, 1, 2$, be self-adjoint operators in a Hilbert space H that satisfy conditions ensuring that the IVPs

$$\{d_t - i\mathbf{G}_j(t)\}\boldsymbol{\psi}_j(t) = 0, \qquad \boldsymbol{\psi}_j(s) = \boldsymbol{\psi}_j^s, \qquad j = 0, 1,$$

are well posed.

Following along much the same lines as for APs, we can also introduce here the notions of AE and hence of WOs, denoted now by $\mathbf{W}_{\pm}(\mathbf{G}_0(.), \mathbf{G}_1(.))$, according to the strong limits

$$\mathbf{W}_{\pm s}(\mathbf{G}_0(.), \mathbf{G}_1(.)) = s - \lim_{t \to \pm\infty} \mathbf{U}_0(s, t)\mathbf{U}_1(t, s) \tag{6.135}$$

whenever these limits exist.

If

$$\text{range}(\mathbf{W}_{+s}) = \text{range}(\mathbf{W}_{-s}),$$

then the WOs are said to be complete, and a scattering operator, denoted $\mathbf{S}_s(\mathbf{G}_0(.), \mathbf{G}_1(.))$, can be defined according to

$$\mathbf{S}_s(\mathbf{G}_0(.), \mathbf{G}_1(.)) = \mathbf{W}_{+s}(\mathbf{G}_0(.), \mathbf{G}_1(.))^* \mathbf{W}_{-s}(\mathbf{G}_0(.), \mathbf{G}_1(.)). \tag{6.136}$$

For ease of presentation we shall occasionally write

$$\mathbf{W}_{\pm s} \equiv \mathbf{W}_{\pm s}(\mathbf{G}_0(.), \mathbf{G}_1(.)), \qquad \mathbf{S}_s \equiv \mathbf{S}_s(\mathbf{G}_0(.), \mathbf{G}_1(.)).$$

The following results can be established [123, 142].

1. The existence and completeness of $\mathbf{W}_{\pm s}$ for all s follows from the existence and completeness for any particular s.
2. The wave operators $\mathbf{W}_{\pm s}$, when they exist, are partial isometries with H as the initial domain. Also,

$$\mathbf{U}_0(s, r)\mathbf{W}_{\pm r} = \mathbf{W}_{\pm s}\mathbf{U}_1(s, r).$$

3. When the WOs are complete, the SO is unitary and

$$\mathbf{U}_1(r, s)\mathbf{S}_s\mathbf{U}_1(s, r) = \mathbf{S}_r.$$

4. If $\mathbf{W}_{\pm s}(\mathbf{G}_0(.), \mathbf{G}_1(.))$ and $\mathbf{W}_{\pm s}(\mathbf{G}_1(.), \mathbf{G}_2(.))$ exist, then $\mathbf{W}_{\pm s}(\mathbf{G}_0(.), \mathbf{G}_2(.))$ exists and

$$\mathbf{W}_{\pm s}(\mathbf{G}_0(.), \mathbf{G}_2(.)) = \mathbf{W}_{\pm s}(\mathbf{G}_0(.), \mathbf{G}_1(.))\mathbf{W}_{\pm s}(\mathbf{G}_1(.), \mathbf{G}_2(.)).$$

5. If $\mathbf{W}_{\pm s}(\mathbf{G}_0(.), \mathbf{G}_1(.))$ and $\mathbf{W}_{\pm s}(\mathbf{G}_1(.), \mathbf{G}_0(.))$ exist, then $\mathbf{W}_{\pm s}(\mathbf{G}_0(.), \mathbf{G}_1(.))$ are complete and unitary.

We remark that in the majority of applications the operator $\mathbf{G}_0$, for instance, is independent of t.

We notice that even if the above properties can be shown to obtain for a particular problem under consideration, nevertheless, in practical applications they all depend on a detailed knowledge of the propagator.

6.3.3 On the Construction of Propagators and Solutions

We have seen in the previous subsections that the NAPs (6.129), (6.130) can be reduced to consideration of the IVPs

$$\{d_t - i\mathbf{G}_j(t)\}\mathbf{u}_j(t) = 0, \qquad \mathbf{u}_j(s) = \mathbf{u}_j^s, \qquad j = 0, 1. \tag{6.137}$$

An integration factor technique indicated that this problem has a solution of the form

$$\mathbf{u}_j(t) = \exp\left\{ i \int_s^t \mathbf{G}_j(\eta)\, d\eta \right\} \mathbf{u}_j^s =: \mathbf{U}_j(t, s)\mathbf{u}_j^s. \tag{6.138}$$

Relations of the form (6.138) have been discussed in chapter 5. Consequently, slightly modifying the notation used in chapter 5 for our present purposes, we proceed as follows.

Let H_1 be a subspace of the energy Hilbert space H_E such that $H_1 \overset{d}{\hookrightarrow} H_E$ (continuously and densely embedded) and denote their norms by $\|.\|_1$ and $\|.\|_E$, respectively. We shall assume the following.

(i) $\{i\mathbf{G}_j(t)\}_{0\leq t\leq T}$, $j=0, 1$, is a family of closed linear operators in H_E.

(ii) For all $t \in [0, T]$ and $j=0, 1$,

$$[\mathrm{Re}\,\mu \geq 0] := \{\mu \in \mathbf{C} : \mathrm{Re}\,\mu \geq 0\} \subset \rho(-i\mathbf{G}_j(t)),$$

where $\rho(A)$ denotes the resolvent set of an operator A.

(iii) For $j=0, 1$, there exist constants $a_j \geq 1$, independent of $t \in [0, T]$, such that

$$\|(\lambda I - i\mathbf{G}_j(t))^{-1}\| \leq \frac{a_j}{1+|\lambda|}, \qquad j=0, 1,$$

for all $(\lambda, t) \in [\mathrm{Re}\,\mu \geq 0] \times [0, T]$, where $\|.\|$ denotes the usual operator norm.

(iv) For $j=0, 1$, there exist constants $b_j \in (0, 1)$ such that

$$i\mathbf{G}_j(.) \in C^{b_j}([0, T], B(H_1, H_E)).$$

The assumptions (i)–(iii) ensure that for each $t \in [0, T]$ and $j=0, 1$, the operators $i\mathbf{G}_j(t)$ are the infinitesimal generators of strongly continuous semigroups [94, 142].

The following result is a direct consequence of the abstract theory outlined in chapter 5.

THEOREM 6.11 *Let the assumptions (i)–(iv) be satisfied. The propagator $\mathbf{U}_j(t, s)$, $j=0, 1$, satisfies the Volterra integral equation*

$$\mathbf{U}_j(t, s) = \mathbf{U}_j(t-s) + \int_s^t \mathbf{U}_j(t, \tau)\mathbf{R}_{j1}(\tau, s)\, d\tau, \tag{6.139}$$

which has a solution that can be written in the form

$$\mathbf{U}_j(t, s) = \mathbf{U}_j(t-s) + \int_s^t \mathbf{U}_j(t-\tau)\mathbf{R}_j(\tau, s)\, d\tau, \tag{6.140}$$

where

$\mathbf{U}_j(t-s) = \exp\{i(t-s)\mathbf{G}_j(s)\}$

$\mathbf{R}_{j1}(t, s) = i\{\mathbf{G}_j(t) - \mathbf{G}_j(s)\}\mathbf{U}_j(t-s)$

$\mathbf{R}_j(t, s) = \sum_{m=1}^{\infty} \mathbf{R}_{jm}(t, s)$

$\mathbf{R}_{jm}(t, s) = \int_s^t \mathbf{R}_{j(m-1)}(t, \tau)\mathbf{R}_{j1}(\tau, s)\, d\tau,\ m \geq 2,\ 0 \leq s \leq t \leq T.$

Furthermore, the resolvent kernel can be estimated by

$$\|\mathbf{R}_j(t,s)\| \le c(t-s)^{b_j-1}$$

for all (t,s) *such that* $0 \le s \le t \le T$.

Combining these several results, we obtain, as the required solution to (6.137), the expression

$$\mathbf{u}_j(t) = \left\{ \mathbf{U}_j(t-s) + \sum_{m=1}^{\infty} \int_s^t \mathbf{U}_j(t-\tau)\mathbf{R}_{jm}(\tau,s)\,d\tau \right\} \mathbf{u}_j^s. \tag{6.141}$$

This is an exact representation of the solution to (6.137), a problem posed in $H_E(\mathbf{R}^n)$. Following the arguments used in the previous section, we know that the first component of the solution $\mathbf{u}_j(t)$ to (6.137) given by (6.141) yields the solution $u_j(x,t)$ of the NAP (6.129), (6.130). Properties of $u_j(x,t)$ can be obtained with increasing accuracy by taking an increasing number of terms in the summation appearing in (6.141). However, the practical problem of determining the associated spectral families is still present. This difficulty can be eased considerably, just as for APs, by using generalised eigenfunction expansions. In the following chapters we shall indicate how this is done by considering various specific problems.

In practice, instead of using (6.141), a rather more immediate way of generating a sequence of approximate solutions for $\mathbf{U}(t,s)$ and $\mathbf{u}(t)$ can be obtained as follows from (5.12). For ease of presentation we shall for the moment suppress the subscript j. With a slight abuse of notation (5.12) can be written in the form

$$\mathbf{U}(t,s) =: \mathbf{U}(t-s) + \mathbf{N}\mathbf{U}(t,s) =: \mathbf{V}\mathbf{U}(t,s).$$

When $\mathbf{V}$ is a contraction, (5.12) is uniquely solvable for the fixed point $\mathbf{U}(t,s)$. Consequently, this will be the case provided $\|\mathbf{N}\| < 1$. When this latter condition holds, we know that the expansion

$$(\mathbf{I}-\mathbf{N})^{-1} = \sum_{n=0}^{\infty} \mathbf{N}^n$$

exists and is well defined. Hence if for convenience we temporarily set $\mathbf{U}(t-s) =: \mathbf{T}(t,s)$, then we obtain

$$\mathbf{U}(t,s) = (\mathbf{I}-\mathbf{N})^{-1}\mathbf{T}(t,s) = \sum_{n=0}^{\infty} \mathbf{N}^n\mathbf{T}(t,s) =: \sum_{n=0}^{\infty} \mathbf{M}_n(t,s),$$

where

$$\mathbf{M}_n(t,s) = \mathbf{N}^n\mathbf{T}(t,s) = \mathbf{N}(\mathbf{N}^{n-1}\mathbf{T}(t,s)) = \mathbf{N}\mathbf{M}_{n-1}(t,s).$$

We can now construct a sequence of approximations of the following form.

Zero approximation:

$$\mathbf{U}_0(t,s) = \mathbf{M}_0(t,s) = \mathbf{T}(t,s).$$

First approximation:

$$\mathbf{U}_1(t, s) = \sum_{n=0}^{1} \mathbf{M}_n(t, s) = (\mathbf{I}+\mathbf{N})\mathbf{T}(t, s).$$

Second approximation:

$$\mathbf{U}_2(t, s) = \sum_{n=0}^{2} \mathbf{M}_n(t, s) = (\mathbf{I}+\mathbf{N}+\mathbf{N}^2)\mathbf{T}(t, s).$$

Clearly, whenever $\mathbf{V}$ is a contraction, this sequence converges to $\mathbf{U}(t, s)$ as $n \to \infty$.
The solution of the IVP (6.131) can be written as in (6.132) in the form

$$\mathbf{u}(t) = \mathbf{U}(t, s)\mathbf{u}(s). \tag{6.142}$$

Consequently, we can construct a sequence of approximate solutions in the form

$$\mathbf{u}_n(t) = \sum_{k=0}^{n} \mathbf{M}_k(t, s)\mathbf{u}(s). \tag{6.143}$$

The zero approximation is

$$\mathbf{u}_0(t) = \mathbf{U}_0(t, s)\mathbf{u}(s) = \mathbf{M}_0(t, s)\mathbf{u}(s) = \mathbf{T}(t, s)\mathbf{u}(s) = \exp\{i(t-s)\mathbf{G}(s)\}\mathbf{u}(s). \tag{6.144}$$

This is the solution of an autonomous problem in which the target is *fixed* at $\Omega(s)$.
Using the representation (2.72), we obtain the first component of $\mathbf{u}_0(t)$, namely, $u_0(t)$, in the form

$$u_0(t) = \{\cos(t-s)A^{1/2}\}\varphi + \{A^{-1/2}\sin(t-s)A^{1/2}\}\psi. \tag{6.145}$$

Finally, using the generalised eigenfunction expansions introduced in (6.78)–(6.80), we obtain

$$u_0(t) = \int_{\mathbf{R}} w_{\pm}(x, p)\left\{[\cos(t-s)\,|p|]\tilde{\varphi}_{\pm}(p) + \left[\frac{\sin(t-s)\,|p|}{|p|}\right]\tilde{\psi}_{\pm}(p)\right\} dp. \tag{6.146}$$

We see that this zero approximation is the autonomous approximation of the solution to the given nonautonomous IBVP. Subsequent approximations can be obtained as outlined above. Specifically, using (6.143), we can obtain the following first approximation.

$$\mathbf{u}_1(t) = \sum_{k=0}^{1} \mathbf{M}_k(t, s)\mathbf{u}(s) = \{\mathbf{M}_0(t, s) + \mathbf{M}_1(t, s)\}\mathbf{u}(s), \tag{6.147}$$

$$\mathbf{M}_0(t, s) = \mathbf{T}(t, s)\mathbf{u}(s) = \exp\{i(t-s)\mathbf{G}(s)\},$$

$$\mathbf{M}_1(t, s) = \mathbf{N}\mathbf{M}_0(t, s) = i\int_s^t \mathbf{U}(t-\eta)\{\mathbf{G}(\eta) - \mathbf{G}(s)\}\mathbf{U}(\eta - s)\,d\eta. \tag{6.148}$$

The semigroup properties of $\mathbf{U}(t-s)$ and its commuting properties with respect to its generator $\{i\mathbf{G}(s)\}$ yield

$$\mathbf{M}_1(t,s) = i\int_s^t (\mathbf{G}(\eta)\mathbf{U}(t-\eta)\mathbf{U}(\eta-s) - \mathbf{U}(t-\eta)\mathbf{U}(\eta-s)\mathbf{G}(s))\,d\eta. \tag{6.149}$$

Now

$$\begin{aligned}\mathbf{U}(t-\eta)\mathbf{U}(\eta-s) &= \exp\{i(t-\eta)\mathbf{G}(\eta)\}\exp\{i(\eta-s)\mathbf{G}(s)\}\\ &= \exp\{it\mathbf{G}(\eta) - i\eta[\mathbf{G}(\eta)-\mathbf{G}(s)] - is\mathbf{G}(s)\}\\ &= \exp\{i(t-s)\mathbf{G}(s) + i(t-\eta)[\mathbf{G}(\eta)-\mathbf{G}(s)]\}\\ &= \mathbf{U}(t-s)\exp\{i(t-\eta)\mathbf{B}(t,s)\}\\ &= \mathbf{U}(t-s)\sum_{n=0}^{\infty}\frac{[i(t-\eta)\mathbf{B}(\eta,s)]^n}{n!}\\ &= \mathbf{U}(t-s)\{\mathbf{I}+\mathbf{B}(\eta,s)+\cdots\},\end{aligned} \tag{6.150}$$

where

$$\mathbf{B}(\eta,s) = [\mathbf{G}(\eta)-\mathbf{G}(s)].$$

Hence from (6.148)–(6.150),

$$\mathbf{M}_1(t,s) = \mathbf{U}(t-s)\left\{\int_s^t [i\mathbf{G}(\eta) - i\mathbf{G}(s)+\cdots]\,d\eta\right\},$$

$$\mathbf{M}_1(t,s) = \mathbf{U}(t-s)\left\{-i(t-s)\mathbf{G}(s) + \int_s^t i\mathbf{G}(\eta)\,d\eta+\cdots\right\}. \tag{6.151}$$

We then obtain $\mathbf{u}_1(t)$, the first approximation to $\mathbf{u}(t)$, in the form

$$\mathbf{u}_1(t) = \mathbf{U}(t-s)\left\{\mathbf{I} - i(t-s)\mathbf{G}(s) + \int_s^t i\mathbf{G}(\eta)\,d\eta\right\}\mathbf{u}(s). \tag{6.152}$$

We recall that

$$i\mathbf{G}(t) = -\begin{bmatrix} 0 & -I \\ A(t) & 0\end{bmatrix}. \tag{6.153}$$

Furthermore, expressing the integral in (6.152) in terms of Riemann sums, we have

$$\begin{aligned}\int_s^t i\mathbf{G}(\eta)\,d\eta &= \lim_{\pi\to 0}\sum_{j=1}^{n} i\mathbf{G}(\eta')(\eta_j-\eta_{j-1})\\ &=: \sum_{\pi}(\eta_j-\eta_{j-1})\begin{bmatrix} 0 & -I \\ A(\eta') & 0\end{bmatrix},\end{aligned} \tag{6.154}$$

where π denotes a partition of the interval $[s, t]$ according to

$$\pi : s = \eta_0 < \eta_1 < \cdots < \eta_n = t, \tag{6.155}$$

where $\hat{\pi} = \max\{(\eta_j - \eta_{j-1})\}$ and η' denotes a fixed point in $(\eta_{j-1}, \eta_j]$. Now substitute (6.154) into (6.152) and use the matrix representation to obtain, on recalling that $\mathbf{u}(s) = \langle \varphi, \psi \rangle$,

$$\begin{aligned}
\mathbf{u}_1(t) &= \mathbf{U}(t-s)\left\{\begin{bmatrix}\varphi\\ \psi\end{bmatrix} - (t-s)\begin{bmatrix}\psi\\ -A(s)\varphi\end{bmatrix} + \sum_{\pi}(\eta_j - \eta_{j-1})\begin{bmatrix}\psi\\ -A(\eta')\varphi\end{bmatrix}\right\}\\
&= [\cos(t-s)A^{1/2}(s)]\\
&\quad\times\left\{\begin{bmatrix}\varphi\\ \psi\end{bmatrix} - (t-s)\begin{bmatrix}\psi\\ -A(s)\varphi\end{bmatrix} + \sum_{\pi}(\eta_j - \eta_{j-1})\begin{bmatrix}\psi\\ -A(\eta')\varphi\end{bmatrix}\right\}\\
&\quad + A^{-1/2}(s)[\sin(t-s)A^{1/2}(s)]\\
&\quad\times\left\{\begin{bmatrix}\psi\\ -A(s)\varphi\end{bmatrix} - (t-s)\begin{bmatrix}-A(s)\varphi\\ -A(s)\psi\end{bmatrix} + \sum_{\pi}(\eta_j - \eta_{j-1})\begin{bmatrix}-A(\eta')\varphi\\ -A(s)\psi\end{bmatrix}\right\}.
\end{aligned} \tag{6.156}$$

We now take the first component of (6.156) to obtain $u_1(t)$, the first approximation to the required solution. This yields

$$\begin{aligned}
u_1(t) &= [\cos(t-s)A^{1/2}(s)]\left\{\varphi - (t-s)\psi + \sum_{\pi}(\eta_j - \eta_{j-1})\psi\right\}\\
&\quad + [A^{-1/2}(s)\sin(t-s)A^{1/2}(s)]\left\{\psi + (t-s)A(s)\varphi - \sum_{\pi}(\eta_j - \eta_{j-1})A(\eta')\varphi\right\}.
\end{aligned} \tag{6.157}$$

If we now define

$$\begin{aligned}
g &:= \psi + (t-s)A(s),\\
h &:= \sum_{\pi}(\eta_j - \eta_{j-1})A(\eta'),
\end{aligned} \tag{6.158}$$

then (6.157) can be written

$$u_1(t) = [\cos(t-s)A^{1/2}(s)]\varphi + [A^{-1/2}(s)\sin(t-s)A^{1/2}(s)]\{g - h\}. \tag{6.159}$$

In (6.159) we have an expression for the first approximation of the required solution of the given nonautonomous IBVP as a perturbation of an associated autonomous IBVP.

The various terms in (6.159) can be interpreted, as before, in terms of a generalised eigenfunction expansion. Specifically, we obtain

$$u_1(x,t) = \int_{\mathbf{R}} w_{\pm}(p) \left\{ \cos(t-s)\,|p|\,\tilde{\varphi}(p) + \frac{\sin(t-s)\,|p|}{|p|}[\tilde{g}_{\pm}(p) - \tilde{h}_{\pm}(p)] \right\} dp. \tag{6.160}$$

These various results can now be applied to a given scattering problem simply by including the subscript j as indicated in (6.132).

An alternative and perhaps more attractive approach for many practical scattering problems can be developed using the associated wave operators and scattering operator. This is outlined in the next chapter.

Chapter Seven

Echo Analysis

7.1 INTRODUCTION

In the last subsection of chapter 2 we gave an indication of some of the influences of time-dependent perturbations by considering a one-dimensional problem involving a moving bead on a string. This was a simple illustration of a target scattering problem. In this chapter we turn our attention to a more detailed study of this class of problem using the various notations and techniques introduced in chapters 1, 5 and 6.

As mentioned in chapter 1, we consider systems consisting of a medium containing a transmitter and a receiver. The transmitter emits a signal that is eventually detected at the receiver, possibly after it has been perturbed, that is, scattered, by some inhomogeneity in the medium. We are interested in the manner in which the emitted signal evolves through the medium and, in particular, the form that it assumes at the receiver. As an illustration of how such results can be obtained, we shall confine our attention in this chapter to a specific nonautonomous acoustic scattering problem in which a signal is transmitted into a medium that is initially at rest.

7.2 CONCERNING THE MATHEMATICAL MODEL

For convenience we recall the following notations used in chapters 1 and 5.

$$Q \in \left\{(x,t) \in \mathbf{R}^n \times \mathbf{R}\right\},$$

$$\Omega(t) = \{x \in \mathbf{R}^n : (x,t) \in Q\}, \qquad B(t) = \{x \in \mathbf{R}^n : (x,t) \notin Q\}.$$

The region Q is assumed to be open in $\mathbf{R}^n \times \mathbf{R}$, and $\Omega(t)$ denotes the exterior, at time t, of a scattering target $B(t)$. For each value of t, the domain $\Omega(t)$ is open in $(\mathbf{R}^n \times \mathbf{R})$ and is assumed to have a smooth boundary $\partial\Omega(t)$. The lateral surface of Q, denoted ∂Q, is defined by $\partial Q := \cup_{t\in I} \partial\Omega(t)$, where for $T \geqslant 0$ fixed, $I := \{t \in \mathbf{R} : 0 \leq t \leq T\}$. We shall assume throughout that $B(t)$ remains in a fixed ball in $\mathbf{R}^n$ and that ∂Q is timelike in the sense that the boundary speed is less than the speed of propagation.

The scattering problems with which we shall be concerned are centred on the following IBVPs.

For $j=0, 1$, determine a quantity $u_j(x,t)$ satisfying the IBVP

$$\{\partial_t^2 + L_j(x,t)\}u_j(x,t) = f_j(x,t), \qquad (x,t)\in Q, \tag{7.1}$$

$$u_j(x,s) = \varphi_j(x,s), \qquad u_{jt}(x,s) = \psi_j(x,s), \qquad x\in\Omega_j(s),\ s\in\mathbf{R}, \tag{7.2}$$

$$u_j(x,t)\in(bc)_j(t), \qquad (x,t)\in\partial\Omega_j(t)\times\mathbf{R}, \tag{7.3}$$

where

$L_j(x,t) =$ a differential expression characterising the wave field

$f_j, \varphi_j(.,s), \psi_j(.,s) =$ given data functions

$s\in\mathbf{R} =$ a fixed initial time

$(bc)_j(t) =$ boundary conditions to be satisfied by $u_j(.,.)$.

We remark that in (7.1) the inhomogeneous term f_j characterises the transmitter and the signals that it emits.

We shall assume that the case $j=0$ denotes a FP, whilst the case $j=1$ denotes a PP.

In this chapter we will reduce the generality of the problem (7.1)–(7.3) by confining our attention to *acoustic Dirichlet problems* associated with a medium that is initially at rest. In this case we have, for $j=0, 1$,

$$L_j(x,t) = -\Delta, \quad \text{for all } (x,t), \tag{7.4}$$

$$(bc)_j(t) = \{u_j(.,.) : u_j(x,t) = 0,\ (x,t)\in\partial\Omega_j(t)\times\mathbf{R})\}, \tag{7.5}$$

$$u_j(x,s) = 0, \qquad u_{jt}(x,s) = 0. \tag{7.6}$$

Our aim here is to discuss and determine solutions of the acoustic Dirichlet problems defined above and, in particular, investigate the nature of the solution at large distances from both the target and the transmitter. We have already indicated that a very convenient method of tackling such problems is to represent them in some suitable energy space. This approach has been outlined in some detail in chapter 5 for just one IBVP. Nevertheless, the analysis provided in subsection 5.4.2 can be followed through for each of the problems (7.1)–(7.6). The appropriate settings and associated results are distinguished by employing subscripts.

As before, we use the notation

$$\mathbf{g} = \begin{bmatrix} g_1 \\ g_2 \end{bmatrix} = \langle g_1,\ g_2\rangle$$

and introduce the energy norm

$$\|\mathbf{g}\|_0^2 = \|\langle g_1, g_2\rangle\|_0^2 = \frac{1}{2}\int_{\mathbf{R}^n} \{|\nabla g_1(x)|^2 + |g_2(x)|^2\}\,dx, \tag{7.7}$$

where we assume $g_1, g_2 \in C_0^\infty(\mathbf{R}^n)$.

Associated with the norm in (7.7) is the inner product

$$(\mathbf{f},\mathbf{g})_0=(\nabla f_1,\nabla g_1)+(f_2,g_2), \tag{7.8}$$

where on the right-hand side of (7.8) the notation (., .) denotes the usual $L_2(\mathbf{R}^n)$ inner product.

We shall write $H_0=H_0(\mathbf{R}^n)$ to denote the completion of $C_0^\infty(\mathbf{R}^n)\times C_0^\infty(\mathbf{R}^n)$ with respect to the energy norm (7.7) and introduce, for $j=0$, 1,

$$H_j(t)=\{\mathbf{g}\in H_0:\mathbf{g}=0 \text{ on } B_j(t)\}, \tag{7.9}$$

$$H_j^{\text{loc}}(t)=\{\mathbf{g}=\langle g_1,g_2\rangle:\zeta\mathbf{g}\in H_j(t)\forall\zeta\in C_0^\infty(\mathbf{R}^n)\}. \tag{7.10}$$

In most practical cases the FP models a free-space problem. In this case solutions to the FP will indicate the evolution of a signal through the medium in the absence of any scatterers. We shall assume that this is the case here. We notice that the first component of $\mathbf{g}\in H_j^{\text{loc}}(t)$ must vanish on $\partial\Omega_j(t)$ and on $B_j(t)$.

For $j=0$, 1, the function $u_j=u_j(.,.)$ is a solution of *locally finite energy* of the IBVP (7.1)–(7.3) if

(i) $\mathbf{u}_j=\langle u_{1j},u_{2jt}\rangle\in C(\mathbf{R},H_j^{\text{loc}}(t))$,
(ii) $\{\partial_t^2+L_j(x,t)\}u_j(x,t)=f_j(x,t)$, $(x,t)\in Q$
(in the sense of distributions).

We shall say that u_j defines a *solution of finite energy* if $\mathbf{u}_j\in C(\mathbf{R},H_0)$ and $\mathbf{u}_j(t)\in H_j(t)$ for each t, whilst u_j defines a *free solution of finite energy* if $\mathbf{u}_j\in C(\mathbf{R},H_0)$.

If $\mathbf{u}_j\in H_0$, then u_j has finite energy, and we write

$$\|\mathbf{u}_j(t)\|^2=\|u_j(t)\|^2=\frac{1}{2}\int_{\mathbf{R}^n}\{|\nabla u_j(x,t)|^2+|u_{jt}(x,t)|^2\}\,dx \tag{7.11}$$

for the total energy of u_j at time t.

The wave energy in a sphere is obtained from (7.11) by restricting the range of integration appropriately. For example, we could define a quantity $H_E(t)$ to be the completion of $C_0^\infty(\Omega_j(t))\times C_0^\infty(\Omega_j(t))$ with respect to the modified energy norm $\|.\|_{E_j(t)}$ given by

$$\|\mathbf{f}\|_{E_j(t)}^2=\int_{\Omega_j(t)}\{|\nabla f_{1j}(x,t)|^2+|f_{1jt}(x,t)|^2\}\,dx.$$

With this preparation we follow the analysis in subsection 5.4.3 and find that the IBVPs (7.1)–(7.6) lead to IVPs of the following form. For $j=0$, 1,

$$\{\partial_t+\mathbf{N}_j(t)\}\mathbf{u}_j(x,t)=\mathbf{F}_j(x,t), \qquad (x,t)\in Q, \tag{7.12}$$

$$\mathbf{u}_j(x,s)=\langle\varphi_j(.,s),\psi_j(.,s)\rangle(x), \qquad x\in\Omega(s), \tag{7.13}$$

where

$$\mathbf{u}_j(x,t) = \langle u_j, u_{jt}\rangle(x,t), \qquad \mathbf{F}_j(x,t) = \langle 0, f_j\rangle(x,t),$$

$$\mathbf{N}_j(t) := \begin{bmatrix} 0 & -I \\ A_j(t) & 0 \end{bmatrix},$$

and on denoting by $L_2^D(\Omega(t))$ the completion of $C_0^\infty(\Omega(t))$ in $L_2(\Omega(t))$, we have introduced

$$A_j(t) : L_2^D(\Omega(t)) \to L_2^D(\Omega(t)),$$

$$A_j(t)u(.,t) = L_j(.,t)u_j(.,t), \qquad u_j(.,t) \in D(A_j(t)),$$

$$D(A_j(t)) = \{u \in L_2^D(\Omega(t)) : L_j(.,t)u(.,t) \in L_2^D(\Omega(t))\}.$$

Throughout we shall assume that the receiver and the transmitter are in the far field of the $B_j(t)$ and, furthermore, that supp $f_j \subset \{|(x,t)|,\ t_0 \leq t \leq T,\ |x - x_0| \leq \delta_0\}$, where x_0 denotes the position of the transmitter and t_0, T, δ_0 are constants.

If we introduce

$$\mathbf{G}_j(t) : H(t) \to H(t),$$

$$\mathbf{G}_j(t)\mathbf{u}_j(t) = i\mathbf{N}_j(t)\mathbf{u}(t), \qquad \mathbf{u}_j(t) \in D(\mathbf{G}_j(t)),$$

$$D(\mathbf{G}_j(t)) = \{\mathbf{u}_j(t) \in H(t) : \mathbf{N}_j(t)\mathbf{u}_j(t) \in H_j(t)\},$$

then the IVP (7.12) can be realised as a first-order system in $H_j(t)$ in the form

$$\{d_t - i\mathbf{G}_j(t)\}\mathbf{u}_j(t) = \mathbf{F}_j(t), \qquad t \in \mathbf{R},$$
$$\mathbf{u}_j(s) = \mathbf{u}_{js}. \tag{7.14}$$

In chapters 5 and 6 we indicated conditions ensuring that the IVP (7.14) was well posed and, furthermore, had a solution that could be written in the form (see section 5.3, theorem 5.26 and subsection 5.4.3)

$$\mathbf{u}_j(t) = \mathbf{U}_j(t,s)\mathbf{u}_s + \int_s^t \mathbf{U}_j(t,\tau)\mathbf{F}_j(\tau)\,d\tau, \qquad j = 0, 1, \tag{7.15}$$

where $\mathbf{U}_j(t,s)$ is the propagator for (7.14), which we have determined in the form

$$\mathbf{U}_j(t,s) = \exp\left\{i\int_s^t \mathbf{G}_j(\eta)\,d\eta\right\}, \qquad j = 0, 1. \tag{7.16}$$

We notice that when $\mathbf{G}_j(t) = \mathbf{G}_j$, $j = 0, 1$, and for all $t \geqslant 0$, as might have been expected, we recover the evolution operator $\mathbf{U}(t-s)$, which was introduced in (1.85) for APs.

The relation (7.15) is a version of the familiar variation of parameter formula, and the integral involved is an associated Duhamel-type integral.

7.3 SCATTERING ASPECTS AND ECHO ANALYSIS

As we have seen in the previous chapter, scattering theory is concerned with the (asymptotic) comparison of two systems. This is the type of theory we want to have available in practice since experimental measurements are usually made in the far field, that is, far distant from the receiver and the transmitter. In the present case these systems are assumed to be characterised, for some $T > 0$, by the families of operators $\{i\mathbf{G}_j(t)\}_{0\leq t\leq T}$, $j=0, 1$, respectively.

We shall assume the following.

(i) $\{i\mathbf{G}(t)\}_{0\leq t\leq T}$, $j=0, 1$, are self-adjoint operators defined on suitable Hilbert space(s).

(ii) The families $\{i\mathbf{G}(t)\}_{0\leq t\leq T}$, $j=0, 1$, satisfy conditions ensuring that the IVPs

$$\{d_t - i\mathbf{G}_j(t)\}\mathbf{u}_j(t) = \mathbf{F}_j(t), \qquad \mathbf{u}_j(s) = \mathbf{u}_{sj}, \qquad j=0, 1, \tag{7.17}$$

have unique solutions.

(iii) $\mathbf{U}_j(t, s)$, $j=0, 1$, denote the associated propagators.

Of course, when dealing with specific problems, it has to be *proved* that these assumptions are valid and available.

Following the development in subsection 6.3.2, we can now introduce, in much the same manner as for APs (see [123, 142]),

Wave operators:

$$\mathbf{W}_{\pm s}(\mathbf{G}_0, \mathbf{G}_1) := s - \lim_{t\to\pm\infty} \mathbf{U}_0(s, t)\mathbf{U}_1(t, s). \tag{7.18}$$

Scattering operator:

$$\mathbf{S}_s(\mathbf{G}_0, \mathbf{G}_1) = \mathbf{W}_{+s}(\mathbf{G}_0, \mathbf{G}_1)\mathbf{W}_{-s}(\mathbf{G}_0, \mathbf{G}_1)^*. \tag{7.19}$$

We would point out that (ii) and theorem 5.20 (Stone's theorem) ensure that (7.18) is meaningful. The various properties of these two items are much the same as for similar IVPs in the AP case. These are introduced and discussed in detail in ([105, 123, 142]).

We remark that in practice it can often turn out that one or the other of $\mathbf{G}_0$ and $\mathbf{G}_1$ is independent of t.

In developing here an echo analysis for the IBVP (7.1)–(7.5), we shall assume that the medium is initially at rest and concern ourselves with the IVPs

$$\{d_t - i\mathbf{G}_j(t)\}\mathbf{u}_j(t)\} = \mathbf{F}_j(t), \qquad \mathbf{u}_j(s) = 0, \qquad j=0, 1. \tag{7.20}$$

The free (unperturbed) problem obtains when $j=0$ and $\Omega(t) = \Omega = \mathbf{R}^n$. The perturbed problem (PP) obtains when $j=1$ and $\Omega(t) \subset \mathbf{R}^n$.

Using the properties of propagators listed after (6.132), together with theorem 6.10 and the fact that $\mathbf{U}_j(t, s) : H_j(s) \to H_j(t)$, the variation of parameters formula indicates that solutions of (7.20) can be written, for $j=0, 1$, in the form

$$\mathbf{u}_j(t) = \mathbf{U}_j(t, s) \int_s^t \mathbf{U}_j(s, \tau)\mathbf{F}_j(\tau)\, d\tau =: \mathbf{U}_j(t, s)\mathbf{h}_j(s). \tag{7.21}$$

Scattering phenomena involve three fundamental items: the incident field $\mathbf{u}_0$, the total field $\mathbf{u}_1$, and the scattered wave field $\mathbf{u}^s$. In the present case we have

Free wave field = incident wave field $\mathbf{u}_0(t) = \mathbf{U}_0(t, s)\mathbf{h}_0(s)$.

Perturbed wave field = total wave field $\mathbf{u}_1(t) = \mathbf{U}_1(t, s)\mathbf{h}_1(s)$.

Scattered wave field $\mathbf{u}_s(t) = \mathbf{u}_1(t) - \mathbf{u}_0(t)$.

The definition of the WO and the SO enables us to write

$$\begin{aligned}\mathbf{u}_1(t) &= \mathbf{U}_1(t, s)\mathbf{h}_1(s) \\ &= \mathbf{U}_0(t, s)\mathbf{U}_0(s, t)\mathbf{U}_1(t, s)\mathbf{h}_1(s) \\ &= \mathbf{U}_0(t, s)\mathbf{W}_{+s}(\mathbf{G}_0, \mathbf{G}_1)\mathbf{h}_1(s) + \sigma_t(1) \quad \text{as } t \to \infty,\end{aligned}$$

where $\sigma_t(1)$ is an $L_2(\mathbf{R}^n)$ function of t with $\sigma_t(1) \to 0$ as $t \to \infty$.
Similarly, we can obtain

$$\mathbf{u}_s(t) = \mathbf{U}_0(t, s)\{\mathbf{W}_{+s}(\mathbf{G}_0, \mathbf{G}_1)\mathbf{h}_1(s) - \mathbf{h}_0(s)\} + \sigma_t(1). \tag{7.22}$$

For $t \ll 0$, there is no scattered field. Hence $\mathbf{u}_1(t) = \mathbf{u}_0(t)$; that is, $\mathbf{U}_1(t, s)\mathbf{h}_1(s) = \mathbf{U}_0(t, s)\mathbf{h}_0(s)$. Consequently, using the properties (6.132), (6.135) and (6.136),

$$\mathbf{h}_0(s) = \mathbf{W}_{-s}(\mathbf{G}_0, \mathbf{G}_1)\mathbf{h}_1(s) + \sigma_{x_o}(1) \quad \text{as } |x_0| \to \infty,$$

where $\sigma_{x_o}(1)$ is an $L_2(\mathbf{R}^n)$ function of x with $\sigma_{x_o}(1) \to 0$ as $|x_0| \to \infty$. Consequently, operating on both sides of this result with $\mathbf{S}_s$ and recalling (7.19), we obtain

$$\begin{aligned}\mathbf{S}_s(\mathbf{G}_0, \mathbf{G}_1)\mathbf{h}_0(s) &= \mathbf{S}_s(\mathbf{G}_0, \mathbf{G}_1)\mathbf{W}_{-s}(\mathbf{G}_0, \mathbf{G}_1)\mathbf{h}_1(s) + \sigma_{x_0}(1) \\ &= \mathbf{W}_{+s}(\mathbf{G}_0, \mathbf{G}_1)\mathbf{h}_1(s) + \sigma_{x_0}(1).\end{aligned} \tag{7.23}$$

If we now substitute (7.23) in (7.22), then we obtain

$$\mathbf{u}_s(t) = \mathbf{U}_0(t, s)\{\mathbf{S}_s(\mathbf{G}_0, \mathbf{G}_1) - \mathbf{I}\}\mathbf{h}_0(s) + \sigma_t(1) + \sigma_{x_0}(1). \tag{7.24}$$

Thus we see that the scattered (echo) field is determined in the far field by the SO and the FP data.

7.4 ON THE CONSTRUCTION OF THE ECHO FIELD

In this section we use the results of chapters 5 and 6 to indicate how approximations can be derived for $u_s(t)$, the nonautonomous scattered field, as perturbations of an associated free wave field.

Although (7.24) is appropriate for the case when both the FP and the PP are NAPs, nevertheless, in most practical applications the FP is taken to be an AP since it is almost always used to characterise wave propagation in the medium in the absence of any perturbations (scatterers). This will be assumed here to be the case.

From (6.140) and (5.11) we see that the propagators $\mathbf{U}_j(t, s)$, $j=0, 1$, for the FP and the PP are given, respectively, by (see theorem 6.11)

$$\mathbf{U}_j(t, s) = \mathbf{U}_j(t-s) + \int_s^t \mathbf{U}_j(t-\tau)\mathbf{R}_j(\tau, s)\, d\tau, \tag{6.140}$$

where

$$\mathbf{R}_j(t, s) = \sum_{m=1}^{\infty} \mathbf{R}_{jm}(t, s)$$

$$\mathbf{R}_{jm}(t, s) = \textstyle\int_s^t \mathbf{R}_{j(m-1)}(t, \tau)\mathbf{R}_{j1}(\tau, s)\, d\tau, \qquad m \geqslant 2$$

$$\mathbf{R}_{j1}(t, s) = i\{\mathbf{G}_j(t) - \mathbf{G}_j(s)\}\mathbf{U}_j(t-s)$$

$$\mathbf{U}_j(t-s) = \exp\{i(t-s)\mathbf{G}_j(s)\}$$

We shall write (6.140) in the more convenient form

$$\mathbf{U}_j(t, s) = \mathbf{T}_j(t, s) + \mathbf{P}_j(t, s), \tag{7.25}$$

where

$$\mathbf{T}_j(t, s) := \mathbf{U}_j(t-s) = \exp\{i(t-s)\mathbf{G}_j(s)\}$$

and the perturbation $\mathbf{P}_j(t, s)$ is defined by

$$\mathbf{P}_j(t, s) = \int_s^t \mathbf{U}_j(t-\tau)\mathbf{R}_j(\tau, s)\, d\tau = \int_s^t \mathbf{T}_j(t, \tau)\mathbf{R}_j(\tau, s)\, d\tau.$$

In practice it is often sufficient to work with approximations of the $\mathbf{U}_j(t, s)$ rather than to use the full, exact form (7.25). Progressively more refined approximations can be obtained from (7.25) by increasing the number of terms in the summation appearing in the definition of $\mathbf{R}_j(t, s)$. For instance, the nth approximation for $\mathbf{U}_j(t, s)$ will be assumed to have the form

$$\mathbf{U}_{jn}(t, s) = \mathbf{T}_j(t, s) + \mathbf{P}_{jn}(t, s), \tag{7.26}$$

where

$$\mathbf{P}_{jn}(t, s) = \int_s^t \mathbf{T}_j(t, \tau) \sum_{m=1}^{n} \mathbf{R}_{jm}(\tau, s)\, d\tau. \tag{7.27}$$

Corresponding to any approximation we might adopt for $\mathbf{U}_j(t, s)$, associated approximations for the wave operators and scattering operator can be obtained in principle by using (7.18) and (7.19). Hence, using these various results in the exact analytical result (7.24), corresponding approximations for the echo field in the far field can be obtained. Whilst this is a relatively straightforward matter, nevertheless it is quite a lengthy process in practice. However, the situation can be eased considerably by realising that in most practical problems the main interest is centred on the properties of the first component of such vectors as $\mathbf{u}_j(t)$ in (7.21) and $\mathbf{u}^s(t)$ in (7.24). To give an indication of how these various properties can be obtained, we consider in $\mathbf{R}^3$ the acoustic Dirichlet problem (7.1)–(7.6) in the specific case of the scattering of a single pulse of duration T emitted at time t_0 by a transmitter localised near a point x_0. Hence the source functions f_j, $j=0, 1$, that characterise the transmitter will be assumed to have the space-time support

$$\text{supp } f_j \subset \{|(x, t)| : t_0 \leq t \leq t_0 + T \text{ and } |x_0 - x| \leq \delta_0\},$$

where t_0 and δ_0 are constants. We shall also assume, in keeping with the formulation of (7.1)–(7.6), the following.

(i) The scatterer $B(t)$ is contained in a closed bounded set in $\mathbf{R}^3$ with complement $\Omega_1(t) = \mathbf{R}^3 - B(t)$. Hence

$$B(t) \subset \{x : |x| \leq \delta\}, \qquad \text{where } \delta \text{ is fixed for all } t.$$

(ii) The origin of coordinates lies in $B(t)$.
(iii) $\partial\Omega_1(t)$, the boundary of $\Omega_1(t)$, is for all t, a smooth surface.
(iv) The scatterer and transmitter are disjoint, which implies

$$\delta + \delta_0 < |x_0| .$$

(v) The transmitter stops transmitting before the signal reaches the scatterer, which implies

$$T < |x_0| - \delta - \delta_0.$$

With these assumptions in mind we see that (7.21) now assumes the form

$$\mathbf{u}_j(t) = \mathbf{U}_j(t, s) \int_{t_0}^{t_0+T} \mathbf{U}_j(s, \tau)\mathbf{F}_j(\tau)\, d\tau =: \mathbf{U}_j(t, s)\mathbf{h}_j(s), \tag{7.28}$$

which leads to the equivalent representation

$$\mathbf{u}_j(t) = \int_{t_0}^{t_0+T} \mathbf{U}_j(t, \tau)\mathbf{F}_j(\tau)\, d\tau. \tag{7.29}$$

From (6.140) we can obtain

$$\begin{aligned} \mathbf{U}_j(t, \tau) &= \mathbf{U}_j(t - \tau) + \int_{\tau}^{t} \mathbf{U}_j(t - \rho)\mathbf{R}_j(\rho, \tau)\, d\rho \\ &= \mathbf{U}_j(t - \tau) \left\{ \mathbf{I} + \int_{\tau}^{t} \mathbf{U}_j(\tau - \rho)\mathbf{R}_j(\rho, \tau)\, d\rho \right\} \\ &= \mathbf{U}_j(t - \tau)\{\mathbf{I} + \mathbf{P}_j(t, \tau)\}. \end{aligned} \tag{7.30}$$

Combining (7.29) and (7.30), we obtain

$$\mathbf{u}_j(t) = \int_{t_0}^{t_0+T} \mathbf{U}_j(t-\tau)\left\{\mathbf{I}+\mathbf{P}_j(t,\tau)\right\}\mathbf{F}_j(\tau)\,d\tau \tag{7.31}$$

$$= \int_{t_0}^{t_0+T} \mathbf{U}_j(t-\tau)\mathbf{F}_j(\tau)\,d\tau + \int_{t_0}^{t_0+T} \mathbf{U}_j(t-\tau)\mathbf{P}_j(t,\tau)\mathbf{F}_j(\tau)\,d\tau. \tag{7.32}$$

In order to obtain the required first component of the $\mathbf{u}_j(t)$ in (7.32), we simplify the notation by writing

$$\mathbf{P}_j(t,\tau) =: \begin{bmatrix} p_{1j} & p_{2j} \\ p_{3j} & p_{4j} \end{bmatrix}(t,\tau),$$

$$\mathbf{P}_j(t,\tau)\mathbf{F}_j(\tau) = \begin{bmatrix} p_{1j} & p_{2j} \\ p_{3j} & p_{4j} \end{bmatrix}(t,\tau)\begin{bmatrix} 0 \\ f_j \end{bmatrix}(\tau) =: \begin{bmatrix} q_{1j} \\ q_{2j} \end{bmatrix}(t,\tau).$$

Recalling the representation indicated in (6.25), we obtain

$$\mathbf{U}_j(t-\tau)\mathbf{P}_j(t,\tau)\mathbf{F}_j(\tau) = \begin{bmatrix} \cos(t-\tau)A_j^{1/2}(\tau) & A_j^{-1/2}(\tau)\sin(t-\tau)A_j^{1/2}(\tau) \\ -A_j^{1/2}(\tau)\sin(t-\tau)A_j^{1/2}(\tau) & \cos(t-\tau)A_j^{1/2}(\tau) \end{bmatrix} \times \begin{bmatrix} q_{1j} \\ q_{2j} \end{bmatrix}(t,\tau).$$

It then follows that the first component of $\mathbf{u}_j(t)$ in (7.32) is given by

$$\begin{aligned} u_j(t) &= \int_{t_0}^{t_0+T} \{A_j^{-1/2}(\tau)\sin(t-\tau)A_j^{1/2}(\tau)\}f_j(\tau)\,d\tau \\ &\quad + \int_{t_0}^{t_0+T} [\{\cos(t-\tau)A_j^{1/2}(\tau)\}q_{1j}(t,\tau) \\ &\quad + \{A_j^{-1/2}(\tau)\sin(t-\tau)A_j^{1/2}(\tau)\}q_{2j}(t,\tau)]\,d\tau \\ &= \{I_{1j+}I_{2j}\}(t,s), \end{aligned} \tag{7.33}$$

where

$$I_{1j}(t,s) = \mathrm{Re}\{z_j(t,s)\}, \tag{7.34}$$

$$z_j(t,s) = \exp\{-itA_j^{1/2}(s)\}b_j(t,s),$$

$$b_j(t,s) = \int_{t_0}^{t_0+T} A_j^{-1/2}(\tau)\{\exp(i\tau A_j^{1/2}(s))\} \times\{\exp(-i[t-\tau][A_j^{1/2}(\tau)-A_j^{1/2}(s)])\}f_j(\tau)\,d\tau$$

and

$$I_{2j}(t,s) = \text{Re}\{w_j(t,s)\}, \tag{7.35}$$

$$w_j(t,s) = \exp\{-itA_j^{1/2}(s)\}e_j(t,s),$$

$$e_j(t,s) = \int_{t_0}^{t_0+T} \{\exp(i\tau A_j^{1/2}(s))\}$$

$$\times \{\exp(-i[t-\tau][A_j^{1/2}(\tau) - A_j^{1/2}(s)])\}g_j(t,\tau)\,d\tau,$$

$$g_j(t,\tau) = q_1(t,\tau) + iA_j^{-1/2}(\tau)q_2(t,\tau).$$

Hence the wave fields $u_j(t)$, $j=0,1$, are determined in the form

$$u_j(t) = \text{Re}\{[\exp(-itA_j^{1/2}(s))]h_j(t,s)\} = \text{Re}\{v_j(t,s)\}, \tag{7.36}$$

where

$$h_j(t,s) = \{b_j(t,s) + e_j(t,s)\}.$$

The representation (7.36) can be used to develop various levels of approximation of the wave fields $u_j(t)$, $j=0,1$. To illustrate how this can be done, we begin with the observation that underlying the nonautonomous scattering problem we are presently discussing is an associated autonomous scattering problem. We shall say that this autonomous scattering problem provides the zero approximation to the nonautonomous scattering problem. Subsequent (higher-level) approximations of the nonautonomous wave fields are then obtained as perturbations of the associated autonomous wave fields.

7.4.1 Zero Approximation for the Echo Field

In this subsection we discuss the autonomous scattering problem associated with the nonautonomous problem introduced initially. In this case we have for $j=0,1$ and for all $t \in \mathbf{R}$,

$$\Omega_j(t) = \Omega_j, \qquad A_j(t) = A_j, \qquad \mathbf{G}_j(t) = \mathbf{G}_j. \tag{7.37}$$

Consequently, bearing in mind (6.140) and the notation used there together with (7.33), (7.36) and (7.37), we obtain wave fields of the form

$$u_j(t) = \text{Re}\{v_j(t,s)\} = \text{Re}\{[\exp(-itA_j^{1/2})]h_j\}, \qquad j=0,1, \tag{7.38}$$

$$h_j = \int_{t_0}^{t_0+T} A_j^{-1/2}\{\exp(i\tau A_j^{1/2}\}f_j(\tau)\,d\tau.$$

For this scattering problem the FP is taken to be the special case when there are no scatterers present, that is, when $\Omega_0 = \mathbf{R}^3$. Thus $A_0 : L_2(\mathbf{R}^3) \to L_2(\mathbf{R}^3)$

defined by

$$A_0 u_0 = -\Delta u_0 \quad \text{for all } u_0 \in D(A_0), \tag{7.39}$$
$$D(A_0) = \{u \in L_2(\mathbf{R}^3) : -\Delta u \in L_2(\mathbf{R}^3)\},$$

is self-adjoint in $L_2(\mathbf{R}^3)$ [154, 105].

Our main aim is to calculate the scattered (echo) field $u_s(x, t)$ produced by the signal (incident) field $u_0(x, t)$, where

$$u_s(x, t) = u_1(x, t) - u_0(x, t). \tag{7.40}$$

Now u_1 is defined on Ω_1 and u_0 on $\Omega_0 = \mathbf{R}^3$. To be able to compare these two quantities, we introduce the operator

$$J : L_2(\Omega_1) \to L_2(\mathbf{R}^3), \tag{7.41}$$
$$Jg(x) = \begin{cases} j(x)g(x) & \text{for } x \in \Omega_1, \\ 0 & \text{for } x \in \mathbf{R}^3 - \Omega_1, \end{cases}$$

where $j \in C^\infty(\mathbf{R}^3)$ is such that $0 \leq j(x) \leq 1$ with $j(x) = 1$ for $|x| \geqslant \delta$ and $j(x) = 0$ in a neighbourhood of B_1. It will be convenient to extend the definition of u_s in (7.40) to the complex plane by defining

$$u_s(x, t) = \operatorname{Re}\{v_s(x, t)\}, \tag{7.42}$$

where

$$v_s(x, t) = Jv_1(x, t) - v_0(x, t)$$

with v_j, $j = 0, 1$, defined as in (7.36). Clearly, the far-field form of the echo u_s can be obtained from that of v_s.

Calculation of the far-field form of u_s and v_s can be based on the theory of wave operators introduced in chapter 6 and in [105] and developed fully in [154].

For the present scattering problem the wave operators are defined by (see subsection 6.2.4)

$$W_\pm = s - \lim_{t \to \pm\infty} \{\exp(itA_0^{1/2})\} J \{\exp(-itA_1^{1/2})\}. \tag{7.43}$$

It is proved in [154] that these limits exist and that they define unitary operators $W_\pm : L_2(\Omega_1) \to L_2(\mathbf{R}^3)$.

It now follows that for each $h_1 \in L_2(\Omega_1)$,

$$\begin{aligned} Jv_1(t) &= J\{\exp(-itA_1^{1/2})\}h_1(x) \\ &= \{\exp(-itA_0^{1/2})\}W_+ h_1(x) + \sigma_t(1) \quad \text{as } t \to +\infty, \end{aligned} \tag{7.44}$$

where $\sigma_t(1)$ is an $L_2(\mathbf{R}^3)$-valued function of t that tends to zero in $L_2(\mathbf{R}^3)$ as $t \to +\infty$.

The equations (7.38), (7.42) and (7.44) combine to give

$$v_s(x,t) = \{\exp(-itA_0^{1/2})\}\{W_+h_1(x) - h_0(x)\} + \sigma_t(1) \quad \text{as } t \to +\infty. \tag{7.45}$$

With the assumptions (i)–(v) in mind we see that $v_s(x,t) = 0$ for $t_0 + T \leq t \leq t_0 + |x| - \delta - \delta_0$ and $x \in \mathbf{R}^3$. Furthermore, if we choose

$$t_0 = -|x| + \delta + \delta_0,$$

then the arrival time of the signal at the scatterer B_1 will be non-negative. With this understanding we see that

$$J\{\exp(-itA_1^{1/2})\}h_1(x) = \{\exp(-itA_0^{1/2})\}h_0(x) \quad \text{for } t_1 \leq t \leq 0, \tag{7.46}$$

where

$$t_1 = t_0 + T = -|x| + \delta + \delta_0 + T. \tag{7.47}$$

Setting $t = 0$ in (7.46) yields

$$Jh_1 = h_0,$$

whilst for $t = t_1$,

$$\{\exp(it_1A_0^{1/2})\}J\{\exp(-it_1A_1^{1/2})\}h_1(x) = h_0(x). \tag{7.48}$$

The scatterer B_1 will be in the far field of the transmitter if either $|x_0| \gg 1$ or, by (7.47), $t_1 \ll -1$. Combining this observation with (7.48) and (7.43) indicates that

$$h_0(x) = W_-h_1(x) + \sigma_{x_0}(1) \quad \text{as } |x_0| \to \infty, \tag{7.49}$$

where $\sigma_{x_0}(1)$ is an $L_2(\mathbf{R}^3)$-valued function of x_0 such that $\sigma_{x_0}(1)$ tends to zero in $L_2(\mathbf{R}^3)$ when $|x_0| \to \infty$.

We now introduce, as in subsection 6.2.4, the scattering operator S defined by

$$S = W_+W_-^*, \tag{7.50}$$

where W_-^* denotes the adjoint of W_-.

Multiplying (7.49) by S gives

$$W_+h_1(x) = Sh_0(x) + \sigma_{x_0}(1) \quad \text{as } |x_0| \to \infty. \tag{7.51}$$

This follows since by the unitarity of W_- we have $W_-W_-^* = I$ [105, 154].

We now combine (7.45) and (7.51) to obtain

$$v_s(x,t) = \{\exp(-itA_0^{1/2})\}\{(S-I)h_0(x)\} + \sigma_t(1) + \sigma_{x_0}(1). \tag{7.52}$$

The result (7.52) shows that in the far field an approximation is given in terms of the scattering operator S and the FP data $h_0(x)$. (Notice the symbolic similarity to (7.24).)

The construction of the scattering operator for B_1 can be achieved in terms of an associated generalised eigenfunction expansion. The expansions were introduced in chapter 6. For convenience we recall some of the salient features here. For details see [105, 154] and chapter 6.

The operator A_0 is a self-adjoint operator on $L_2(\mathbf{R}^3)$ and has a purely continuous spectrum.

The plane waves

$$w_0(x, p) := (2\pi)^{1/2}\{\exp(ix.p)\}, \qquad x, p \in \mathbf{R}^3, \tag{7.53}$$

form a complete family of generalised eigenfunctions for A_0.

For scattering by bounded objects the generalised eigenfunctions are distorted plane waves [105, 126, 154]

$$w_\pm(x, p) = w_0(x, p) + w_\pm^s(x, p), \qquad x \in \Omega_1 \quad p \in \mathbf{R}^3. \tag{7.54}$$

We have seen in chapter 6 that these distorted plane waves satisfy

$$(\Delta + |p|)w_\pm(x, p) = 0, \qquad x \in \Omega_1, \tag{7.55}$$

$$\left\{\frac{\partial w_\pm^s}{\partial |x|} \mp i\,|p|\,w_\pm^s\right\}(x, p) = \mathrm{O}\left\{\frac{1}{|x|^2}\right\}, \qquad |x| \to \infty. \tag{7.56}$$

The existence and uniqueness properties of these distorted plane waves can be found in [126, 154] and in chapter 6. Physically, $w_\pm^s(x, p)$ represents the steady-state scattered (echo) field when the plane wave (7.53) is scattered by B_1. The far-field form of $w_\pm^s(x, p)$ can be shown to be [154]

$$w_\pm^s(x, p) = \frac{\exp(\pm|p|\,|x|)}{4\pi\,|x|} T_\pm(|p|\,\theta, p) + \mathrm{O}\left\{\frac{1}{|x|^2}\right\} \quad \text{as } |x| \to \infty, \tag{7.57}$$

where $\theta = x/|x|$ and $T(p, p')$, the *scattering amplitude* or *differential cross section* of B_1, is defined for all $p, p' \in \mathbf{R}^3$ such that $|p| = |p'|$. See also definition 1.1.

Following the development in chapter 6, the plane and distorted plane waves w_0 and $w_\pm$ generate the generalised eigenfunction expansions (6.50)–(6.52) and (6.81)–(6.83), respectively. As a consequence, we saw in subsection 6.2.4 that the wave operators $W_\pm$ had the representation

$$W_+ = F_0^* F_- \quad \text{and} \quad W_- = F_0^* F_+. \tag{7.58}$$

Combining (7.50) and (7.58), we obtain

$$S = W_+ W_-^* = F_0^* \hat{S} F_0, \tag{7.59}$$

where

$$\hat{S} = F_- F_+^*, \tag{7.60}$$

is called the S-matrix for the scatterer B_1.

In order to be able to make use of (7.52), we need an interpretation of the term $(S-I)h_0$. This can be achieved by first noticing that

$$(S-I)h_0 = (F_0^* \hat{S} F_0 - I)h_0 = F_0^*(\hat{S} - I)\overset{\wedge}{h_0}. \tag{7.61}$$

For acoustic scattering problems in $\mathbf{R}^3$ it has been shown that

$$(\hat{S} - I)\overset{\wedge}{h_0} = \frac{i\,|p|}{2(2\pi)^{1/2}} \int_{S^2} T_+(p, |p|\,\theta)\,\overset{\wedge}{h_0}(|p|\,\theta)\,d\theta. \tag{7.62}$$

The integration in (7.62) is over points θ of the unit sphere S^2 in $\mathbf{R}^3$. The first proof of the integral representation was given by Shenk [126].

Finally in this subsection, we investigate the nature of the signal and of the echo in the far field.

The complex wave function $v_0(x,t)$ defined by (7.38) has the Fourier representation

$$v_0(x,t) = (2\pi)^{-1/2} \int_{\mathbf{R}^3} \{\exp(i(x.p - t\,|p|)\}\,\overset{\wedge}{h_0}(p)\,dp, \tag{7.63}$$

where

$$\overset{\wedge}{h_0}(p) = (2\pi)^{1/2} i\,|p|^{-1}\,\overset{\wedge}{f_0}(-|p|\,,\,p) \tag{7.64}$$

and

$$\overset{\wedge}{f_0}(\omega, p) = (2\pi)^{-2} \int_{\mathbf{R}^4} \{\exp(-i(x.p + \omega t))\}\,f_0(x,t)\,dx\,dt \tag{7.65}$$

denotes the four-dimensional Fourier transform of f_0.

The notion of an asymptotic wave function was introduced in definition 1.1. For this particular problem the asymptotic wave function u_0^∞ associated with the signal wave field $u_0(x,t) = \mathrm{Re}\{v_0(x,t)\}$ is defined to be (see [154](chap. 2)]

$$u_0^\infty(x,t) = \frac{s(|x| - t, \theta)}{|x|}, \qquad x = |x|\,\theta, \tag{7.66}$$

where $s \in L_2(\mathbf{R} \times S^2)$ is defined by

$$\begin{aligned} s(\tau,\theta) &= \mathrm{Re}\left\{(2\pi)^{-1/2} \int_0^\infty \{\exp(i\tau\omega)\}\{-i\omega\}\,\overset{\wedge}{h_0}(\omega\theta)\,d\omega\right\} \\ &= \mathrm{Re}\left\{\int_0^\infty \{\exp(i\tau\omega)\}\,\overset{\wedge}{f_0}(-\omega, \omega\theta)\,d\omega\right\}. \end{aligned} \tag{7.67}$$

It is proved in [154] that u_0^∞ describes the asymptotic behaviour of u_0 in $L_2(\mathbf{R}^3)$ as $t \to \infty$ in the sense that

$$u_0(.,t) = u_0^\infty(.,t) + \sigma_t(1) \quad \text{as } t \to \infty. \tag{7.68}$$

From (7.52) we see that the echo wave field can be represented in the form

$$u_s(x,t) = \mathrm{Re}\{[\exp(itA_0^{1/2})](S-I)h_0(x,t)\} + \sigma_t(1) + \sigma_{x_0}(1). \tag{7.69}$$

The first term on the right-hand side is the same as that for the signal u_0 as defined in (7.38) but with h_0 replaced by $(S-I)h_0$. Consequently, with this in mind it follows from our treatment of u_0 that we can write

$$u_s(x,t) = u_s^\infty(x,t) + \sigma_t(1) + \sigma_{x_0}(1) \quad \text{as } t \to \infty, \tag{7.70}$$

where

$$u_s^\infty(x,t) := \frac{e(|x|-t,\theta)}{|x|}, \qquad x = |x|\,\theta, \tag{7.71}$$

and

$$e(\tau,\theta) = \operatorname{Re}\left\{(2\pi)^{-1/2}\int_0^\infty [\exp(i\tau\omega)][(S-I)h_0]^\wedge(\omega,\theta)\,d\omega\right\}, \tag{7.72}$$

where $[\ldots]^\wedge$ denotes the Fourier transform of $[\ldots]$.

From (7.61) and (7.62) it follows that

$$e(\tau,\theta) = \frac{1}{4\pi}\operatorname{Re}\left\{\int_0^\infty [\exp(i\tau\omega)]\omega^2 \int_{S^2} T_+(\omega\theta,\omega\theta')\,\hat{h}_0(\omega\theta')\,d\theta'\,d\omega\right\} \tag{7.73}$$

and, by using (7.64),

$$e(\tau,\theta) = \frac{1}{2(2\pi)^{1/2}}\operatorname{Re}\left\{i\int_0^\infty [\exp(i\tau\omega)]\omega \int_{S^2} T_+(\omega\theta,\omega\theta')\,\hat{f}_0(-\omega,\omega\theta')\,d\theta'\,d\omega\right\}. \tag{7.74}$$

Thus (7.67) and (7.74) provide a representation of the asymptotic signal wave form and the asymptotic echo waveform, respectively.

7.4.2 Concerning Higher-Order Approximations

The influence of time-dependent perturbations on the various wave fields is governed analytically by the representation (7.36). The term h_j in (7.36) fully characterises the nature of the perturbation being considered. We see that the representations (7.36) are in such a form that they allow a similar development to that given in the previous subsection provided that the h_j in (7.38) is replaced by the h_j given in (7.36). When this is done, we develop a hierarchy of approximations as follows.

- Decide on the number of terms to be taken in the approximation to be made for the P_j. This sets the matrix $[p_{jk}]$, and as a consequence the approximation being made for h_j in (7.36). (See (7.26), (7.27).)
- Introduce Riemann sums as in (6.154). The effects of doing this are threefold:
  1. The number of points taken in the partitioning of the interval (t_0, t_0+T) will contribute further to the level of approximation being obtained.
  2. All the operators appearing in the definition of h_j in (7.36) will, as a consequence of the partitioning, appear as being at a fixed time $t=\eta'$, where η' denotes a fixed point in the partitioning (see (6.154)).

3. At the fixed time $t = \eta'$ we will be concerned with a *fixed* scatterer $B_1(\eta')$ and a *fixed* exterior region $\Omega_1(\eta')$. It then follows that we can establish, for each fixed $t = \eta'$ in the partition, generalised eigenfunction expansion theorems as in the previous subsection in terms of plane wave and distorted plane waves associated with $\Omega_1(\eta')$. This can be done for each fixed η' in the partition and the associated approximation obtained by forming the appropriate Riemann sum.

7.5 A REMARK ABOUT ENERGY IN THE SYSTEM

In chapter 2 we discussed a one-dimensional NAP and showed that in such a system energy was not conserved (see (2.144)). Indeed, it is quite possible that the energy in the system could increase without bound. The same feature is present in problems, such as those discussed in this chapter, that are posed in higher dimensions. This situation can be handled by first recalling that in our treatment of NAP we have obtained solutions with the symbolic form

$$\mathbf{u}(t) = \mathbf{U}(t, s)\mathbf{f}(s),$$

where $\mathbf{u}(t)$ is the required solution, $\mathbf{f}(s)$ with $s \in \mathbf{R}$ fixed characterises the given data and $\mathbf{U}(t, s)$ is the propagator for the problem. Consequently, if appropriate bounds can be obtained for the propagator and if the given data is suitably bounded, then we can ensure that bounded solutions can be obtained despite the fact that the energy in the system is not conserved. This aspect was not considered explicitly in this chapter. We shall discuss it in some detail in the next chapter.

Chapter Eight

Wave Scattering from Time-Periodic Perturbations

8.1 INTRODUCTION

There are many cases of interest in the applied sciences that involve vibrating or pulsating or rotating media. A powerful diagnostic for investigating the properties of such systems is provided by the effect that such systems have on waves that are incident on them. Typical examples include the ultrasonic investigation of the heart and the reflection of radio, television and radar signals from moving objects.

In [61] and [148] many engineering applications of scattering of electromagnetic waves by rotating bodies can be found. However, one of the main difficulties in the analytical study of such problems lies in being unable to use the usual separation of variables technique. A way of overcoming this difficulty is to use the quasi-stationary approximation introduced in chapter 2. Indeed, the majority of contributions in this general area seem to use the QSA. As we have seen, in the QSA a given NAP is replaced by an associated AP. The AP is then solved, and the solutions obtained are used as a basis from which to develop approximations to the various properties of the given NAP. Here and throughout this monograph we concentrate on obtaining an exact, abstract solution form for the given NAP and using that as a basis from which to develop approximations. Throughout, our approach has always been very much directed towards developing constructive methods. Consequently, we shall be concerned here with generalised eigenfunction expansion techniques developed by Wilcox [154] rather than with the approach adopted by Lax and Phillips [68].

8.2 CONCERNING THE MATHEMATICAL MODEL

We shall consider IBVPs of the form (7.1)–(7.8). Proceeding as in chapter 7, we represent these IBVPs in the form of an equivalent first-order system posed in an appropriate energy space. Specifically, we obtain the IVPs (7.14). In the first instance we shall suppress the subscript j in (7.14) and investigate the abstract IVP

$$\{d_t - i\mathbf{G}(t)\}\mathbf{u}(t) = \mathbf{f}(t), \qquad t \in \mathbf{R}, \qquad \mathbf{u}(s) = \boldsymbol{\theta}(s), \tag{8.1}$$

where $s \in \mathbf{R}$ is fixed. This is essentially a propagation problem. We shall return to scattering aspects in later sections.

In this chapter we investigate whether or not the IVP (8.1) has solutions in the presence of periodic perturbations, that is, when the operator $i\mathbf{G}(t)$ is time-periodic, and when solutions exist, what their basic properties are. We shall also be interested in the asymptotic properties of any solutions and in particular under what conditions the solutions can be said to exhibit scattering phenomena.

8.3 BASIC ASSUMPTIONS, DEFINITIONS AND RESULTS

For the purposes of this chapter we introduce two Banach spaces X_0 and X_1 such that $X_1 \overset{d}{\hookrightarrow} X_0$ (that is, continuously and densely embedded). In applications, X_0 and X_1 will be energy spaces endowed with a Hilbert space structure.

Let $T > 0$ and consider the family $\{i\mathbf{G}(t)\}_{t\in\mathbf{R}}$ of closed linear operators on X_0. We shall make the following assumptions regarding this family of operators.

(i) $D(i\mathbf{G}(t)) = X_1$ for all $t \in [0, T]$.

(ii) For all $t \in [0, T]$, we have

$$[\operatorname{Re}\mu \geq 0] \subset \rho\{i\mathbf{G}(t)\},$$

where we have introduced the notation

$$[\operatorname{Re}\mu \geq 0] = \{\mu \in \mathbf{C} : \operatorname{Re}\mu \geq 0\}.$$

Furthermore, there exists a constant $M \geq 1$, independent of $t \in [0, T]$, such that

$$\left\| (\lambda \mathbf{I} - i\mathbf{G}(t))^{-1} \right\| \leq \frac{M}{1 + |\lambda|}$$

for all $(\lambda, t) \in [\operatorname{Re}\mu \geq 0] \times [0, T]$.

(iii) There exists a constant $\alpha \in (0, 1)$ such that

$$\mathbf{G}(.) \in C^{\alpha}([0, T],\ B(X_1, X_0)).$$

(iv) $\mathbf{G}(t)$ is periodic of period $T > 0$. That is,

$$\mathbf{G}(t + T) = \mathbf{G}(t) \quad \text{for all } t \in \mathbf{R}.$$

We notice that the assumptions (i) and (ii) imply that for each $t \in [0, T]$ the operator $(i\mathbf{G}(t))$ is the infinitesimal generator of a strongly continuous analytic semigroup on X_0. (See chapters 5 and 11.)

Remark 8.1 *When we come to study wave problems, we will have to remember that $(-i\mathbf{G}(t))$ is a matrix operator defined on a suitably chosen energy space.*

For the periodic family $\{iG(t)\}_{t\in\mathbf{R}}$ we introduce, more precisely than previously, the notion of a propagator as follows [142].

DEFINITION 8.2 *Let*

$$\Delta_T = \{(t, s) : 0 \leq s \leq t \leq T\}, \qquad \overset{\bullet}{\Delta}_T = \{(t, s) : 0 \leq s < t \leq T\}.$$

The propagator for the family $\{G(t)\}_{t \in \mathbf{R}}$ *is a mapping*

$$U : \Delta_T \to B(X_0)$$

with the following properties.

(i) $U \in C(\Delta_T, B(X_0)) \cap C(\Delta_T, B_s(X_1)) \cap C(\overset{\bullet}{\Delta}_T, B(X_0, X_1))$,
where $C(Z, B_s(X, Y))$ *denotes the space of all functions* $Z \to B(X, Y)$ *that are strongly continuous at all points.*

(ii)

$$U(t, t) = I(X_0) = \textit{identity on } X_0,$$

$$U(t, s) = U(t, \tau)U(\tau, s) \quad \textit{for all } 0 \leq s \leq t \leq T.$$

(iii) The mapping $(t, s) \to G(t)U(t, s)$ *is in* $C(\overset{\bullet}{\Delta}_T, B(X_0))$, *and*

$$\sup_{(t,s) \in \overset{\bullet}{\Delta}_T} (t - s) \, \|G(t)U(t, s)\| < \infty.$$

(iv) $U(., s) \in C^1((s, T], B(X_0))$ *for each* $s \in [0, T)$, *and for all* $t \in (s, T]$,

$$\partial_1 U(t, s) = -G(t)U(t, s).$$

$U(t, .) \in C^1([0, t), B_s(X_1, X_0))$ *for each* $t \in (0, T]$, *and for all* $s \in [0, t)$,

$$\partial_2 U(t, s) f = U(t, s)G(s) f \quad \textit{for all } f \in X_1.$$

It follows from this definition that there is a unique mapping

$$U : \Delta_\infty \to B(X_0)$$

satisfying properties (i)–(iv) on Δ_{nT} for each $n \in \mathbb{N}$. This function (mapping) we will call the propagator for the T-periodic family $\{G(t)\}_{t \in \mathbf{R}}$.

For the propagator just introduced, we have the following results.

THEOREM 8.3 *For any* $(t, s) \in \Delta_\infty$, *the following identities hold.*

(i) $U(t + T, s + T) = U(t, s)$,

(ii) $U(t + nT, s) = U(t + T, t)^n U(t, s) = U(t, s)U(s + T, s)^n$ *for each* $n \in \mathbb{N}$.

Proof. (i) Let $0 \leq s < nT$ for some $n \in \mathbb{N}$. Take $f \in X_0$ and define

$$v(t) := U(t + T, s + T) f - U(t, s) f \quad \text{for } t \in [s, nT].$$

Then by property (iv) in definition 8.2 and the T-periodicity of $G(.)$,

$$\{\partial_t + G(t)\} v(t) = 0, \qquad v(s) = 0, \tag{8.2}$$

holds for every $t \in (s, nT)$. In the light of assumptions (i)–(iv) and the earlier discussions of the abstract Cauchy problem, the IVP has the unique solution $v(t) = 0$ for all $t \in [s, nT]$. This proves part (i).

(ii) By property (ii) and part (i), we have for any $s \geq 0$ and $n \in \mathbb{N}$,

$$\begin{aligned} U(s+T, s)^n &= \prod_{k=0}^{n-1} U(s+(n-k)T, s+(n-k-1)T) \\ &= U(s+nT, s), \end{aligned} \tag{8.3}$$

where the product must be taken in the order of the index as the operators involved do not necessarily commute.

Using property (ii) and part (i) again, we obtain for any $(t, s) \in \Delta_\infty$,

$$\begin{aligned} U(t+nT, t)U(t, s) &= U(t+nT, s) \\ &= U(t+nT, s+nT)U(s+nT, s) \\ &= U(t, s)U(s+nT, s). \end{aligned} \tag{8.4}$$

The results (8.3) and (8.4) establish part (ii). □

We now introduce

$$\Delta := \{(t, s) \in \mathbf{R}^2 : s \leq t\}, \qquad \dot{\Delta} = \{(t, s) \in \mathbf{R}^2 : s < t\}. \tag{8.5}$$

It follows from theorem 8.3(i) that we can define $U(t, s)$ for any $(t, s) \in \Delta$ by

$$U(t, s) := U(t+nT, s+nT) \tag{8.6}$$

for any $n \in \mathbb{N}$ such that $(t+nT, s+nT) \in \Delta_\infty$.

Furthermore, the mapping $U : \Delta \to B(X_0)$ satisfies properties (i)–(iv) in definition 8.2 on $\Delta \cap [-nT, nT]^2$ for any $n \in \mathbb{N}$. It also follows that theorem 8.3 holds on Δ.

In later sections we shall be interested in T-periodic nonhomogeneous abstract Cauchy problems of the general form

$$\begin{aligned} \{\partial_t + G(t)\}u(t) &= f(t), \qquad t > s, \\ u(s) &= \varphi, \end{aligned} \tag{8.7}$$

where $(s, \varphi) \in \mathbf{R} \times X_0$ and $f \in C(\mathbf{R}_+, X_0)$. We would emphasise that the operator $G(t)$ in (8.7) assumes various forms depending on the problem being considered. For example, when dealing with acoustic wave problems, $G(t)$ has to be replaced by a matrix operator of the form $(-iG(t))$ defined on an appropriate energy space.

Following our discussions in chapters 5–7, we introduce the following.

Definition 8.4 *A solution of (8.7) is a function u with the properties*
(i) $u \in C([s, \infty), X_0) \cap C^1((s, \infty), X_0)$,
(ii) $u(t) \in X_1$ for $t > s$,
(iii) u satisfies (8.7).

The variation of parameters formula indicates that u is given for all $t > s$ by

$$u(t) = U(t, s)\varphi(s) + \int_s^t U(t, \tau) f(\tau)\, d\tau. \tag{8.8}$$

An important notion in the study of T-periodic problems is $V(s)$, the *shift operator* or *period map* associated with problem (8.7). It is defined according to

$$V(s) := U(s+T, s) \in B(X_0). \tag{8.9}$$

Some of the more important properties of V are contained in the following.

THEOREM 8.5

(i) $V(s+t) = V(s)$ for all $s \geq 0$.
(ii) $U(t+nT, s) = V(t)^n U(t, s) = U(t, s)V(s)^n$ for all $(t, s) \in \Delta$ and $n \in \mathbb{N}$.
(iii) $\sigma_p(V(s))\setminus\{0\}$ is independent of $s \in \mathbf{R}$.
(iv) $\sigma(V(s))\setminus\{0\}$ is independent of $s \in \mathbf{R}$.

Proof. Parts (i) and (ii) are restatements of theorem 8.3 allowing for definition (8.9).

(iii) let μ be an eigenvalue of $V(s)$ and let $t \in [s, s+T]$. If $f \in X_0\setminus\{0\}$ is the corresponding eigenfunction, then $V(s)f = \mu f$. Now set $g = U(t, s)f$ and use part (ii) to obtain

$$V(t)g = U(t, s)V(s)f = \mu U(t, s)f = \mu g.$$

Hence μ is an eigenvalue of $V(t)$.

Furthermore, we have

$$U(s+T, t)g = U(s+T, t)U(t, s)f = V(s)f \neq 0,$$

which implies that $\mu \in \sigma_p(V(t))$.

(iv) let $\lambda \in \varrho(V(s))\setminus\{0\}$ and let $t \in [s, s+T]$. For any $g \in X_0$, set

$$h := \lambda^{-1}(g + U(t, s)(\lambda - V(s))^{-1}U(s+T, t)g).$$

Then

$$V(t)h = \lambda^{-1}(V(t)g + V(t)U(t, s)(\lambda - V(s))^{-1}U(s+T, t)g).$$

Now

$$\begin{aligned} & V(t)U(t, s)(\lambda - V(s))^{-1}U(s+T, t)g \\ &= -U(t, s)\{-\lambda + (\lambda - V(s))\}(\lambda - V(s))^{-1}U(s+T, t)g \\ &= \lambda U(t, s)(\lambda - V(s))^{-1}U(s+T, t)g - V(t)g, \end{aligned}$$

where, by theorem 8.3, we have used

$$\begin{aligned} U(t, s)U(s+T, t) &= U(t+T, s+T)U(s+T, t) \\ &= U(t+T, t) = V(t). \end{aligned}$$

Hence

$$V(t)h = \lambda h - g;$$

that is,

$$(\lambda - V(t))h = g,$$

which establishes the surjectivity (that is, one-to-one and onto) of $(\lambda - V(t))$. It then follows that $(\lambda - V(t))$ is continuously invertible, which together with the periodicity of $V(t)$ establishes part (ii). □

The next result indicates how the nature of the nonhomogeneous term in the equation influences the solutions to equation (8.7).

THEOREM 8.6 *If*
(i) $f : \mathbf{R} \to X_0$ *is* T*-periodic,*
(ii) $1 \in \varrho(V(0))$,
then equation (8.7) has a unique T*-periodic solution*

$$u(t) = U(t,0)\varphi + \int_0^t U(t,\tau) f(\tau)\, d\tau, \qquad t \geq 0, \tag{8.10}$$

where

$$\varphi = (1 - V(0))^{-1} \int_0^T U(T,\tau) f(\tau)\, d\tau. \tag{8.11}$$

Proof. Assume that u is a T-periodic solution of equation (8.7) and set $\varphi = u(0)$. The variation of parameters formula then indicates

$$u(t) = U(t,0)\varphi + \int_0^t U(t,\tau) f(\tau)\, d\tau. \tag{8.12}$$

Hence

$$u(T) = U(T,0)\varphi + \int_0^T U(T,\tau) f(\tau)\, d\tau. \tag{8.13}$$

However, by the assumed periodicity of u, we have $u(T) = u(0)$. Therefore (8.13) can be written in the form

$$(1 - V(0))\varphi = \int_0^T U(T,\tau) f(\tau)\, d\tau,$$

and we recover (8.11).

Conversely, if φ is defined by (8.11), then we see that $u(T) = \varphi$. Using theorem 8.5(i), (ii), we obtain

$$\begin{aligned}
u(t+T) &= U(t+T,0)\varphi + \int_0^{t+T} U(t+T,\tau) f(\tau)\, d\tau \\
&= U(t+T,T)\left\{U(T,0)\varphi + \int_0^T U(T,\tau) f(\tau)\, d\tau \right. \\
&\quad \left. + \int_T^{t+T} U(t+T,\tau) f(\tau)\, d\tau \right\} \\
&= U(t,0)\varphi + \int_0^t U(t,\tau) f(\tau)\, d\tau = u(t) \quad \text{for all } t \geq 0.
\end{aligned}$$

Hence u is T-periodic. □

The variation of parameters formula (5.9) together with theorem 5.26 establishes uniqueness.

8.4 SOME REMARKS ON ESTIMATES FOR PROPAGATORS

In the previous chapters and sections we have seen that the solutions u of NAPs can be represented in the typical form

$$u(t) = U(t, s)\varphi(s),$$

where $U(t, s)$ is the propagator associated with the problem of interest and $\varphi(s)$ characterises the given data. Consequently, the "size" of the required solution u is largely governed by the size of the propagator U. In practical problems this situation can be controlled to a large extent by working in a suitable energy space and investigating the existence of solutions with finite energy. However, when working with NAPs, there is an additional difficulty that must be taken into account. Specifically, energy is not conserved (see chapter 2). This, coupled with the fact that estimates for U play a crucial role in the general theory of evolution equations [5, 142] suggests that we should always have an awareness of the need to be able to estimate the size of the propagator U. This is a study in itself [5, 142]. For our purposes in this monograph it is often sufficient to notice that our various assumptions of properties (i)–(iv) in definition 8.2 implicitly provide the required control on $U(t, s)$. For example, with property (iii) in mind we see that

$$\begin{aligned}(t-s)\,\|G(t)U(t,s)\| &\leq (t-s)\,\|G(t)\|\,\|U(t,s)\| \\ &= (t-s)M_1\,\|U(t,s)\|,\end{aligned}$$

where M_1 is a constant. Therefore property (iii) indicates that we must have

$$\|U(t,s)\| \leq M_2(t-s)^{-1}, \tag{8.14}$$

where M_2 is a constant for all $(t, s) \in \Delta_\infty$ and, of course, whilst more refined estimates than (8.14) can be obtained [5, 142], the result (8.14) is adequate for the moment.

8.5 SCATTERING ASPECTS

In this section we investigate the scattering of an acoustic wave in the presence of a T-periodic in time perturbation. Although the approach will be abstract, nevertheless, it will provide a framework for studying practical problems involving potential scattering problems or target scattering problems or both.

Proceeding as in previous chapters, we see that scattering problems can be expressed in terms of IVPs of the form (8.1). In terms of the notations introduced in this chapter and bearing in mind remark 8.1, we shall understand the following.

Definition 8.7 *A solution of (8.1) is a function*

$$\mathbf{u} \in C([s, \infty], X_0) \cap C^1((s, \infty), X_0)$$

such that $\mathbf{u}(t) \in X_1$ *for* $t > s$, *where*

$$\{d_t - i\mathbf{G}(t)\}\mathbf{u}(t) = \mathbf{f}(t), \qquad t \in \mathbf{R}, \qquad \mathbf{u}(s) = \boldsymbol{\varphi}(s)$$

in the sense of distributions $(s, \varphi) \in (\mathbf{R}_+, X_0)$ *and* $\mathbf{f} \in C(\mathbf{R}_+, \mathbf{X_0})$.

The solution in definition 8.7 has, by the variation of parameters formula, the unique representation

$$\mathbf{u}(t) = \mathbf{U}(t, s)\varphi(s) + \int_s^t \mathbf{U}(t, \tau)\mathbf{f}(\tau)\, d\tau.$$

We will assume that **f** is rather better than being just continuous; we will take it to be at least locally Hölder-continuous.

When dealing with time-periodic problems, it might be expected that something like a Floquet theory would be available. An indication that this can indeed be the case can be given as follows.

Definition 8.8 *A complex number* σ *such that* $\exp(i\sigma T)$ *is an eigenvalue of* $\mathbf{V}(0) = \mathbf{U}(T, 0)$ *is a scattering frequency associated with the IVP (8.1). The associated eigenfunctions are referred to as scattering eigensolutions.*

In the above definition we have used the matrix form of (8.9).

In applications, our next result provides a particularly useful characterisation of scattering frequencies.

Theorem 8.9 *The following statements are equivalent.*

(i) $\sigma \in \mathbf{C}$ *is a scattering frequency.*

(ii) $\exp(i\sigma T)$ *is an eigenvalue of* $\mathbf{V}(s)$ *for all* $s \in \mathbf{R}$.

(iii) There exists a solution $\mathbf{u}(t)$ *of (8.1), not identically zero, that is such that* $\exp(-i\sigma t)\mathbf{u}(t)$ *has period* T *in* t.

Remark 8.10 *When (8.1) is obtained from a wave problem, then* **u** *is understood to be of the form* $\mathbf{u} = \langle u, u_t \rangle$, *where* u *is a solution of the given wave problem. Consequently, properties of* **u** *are reflected into its components and in particular into the required solution* u.

Proof. (ii) $\Rightarrow$ (i): This follows immediately from (8.9) and theorem 8.5.

(iii) $\Rightarrow$ (ii): By assumption, $\exp(-i\sigma t)\mathbf{u}(t)$ has period T in t.
Define

$$\mathbf{g}(t) = \exp(-i\sigma t)\mathbf{u}(t),$$

which implies $\mathbf{g}(0) = \mathbf{u}(0)$.

By assumption, we have $\mathbf{g}(t + T) = \mathbf{g}(t)$, which implies

$$\exp(-i\sigma(t + T))\mathbf{u}(t + T) = \exp(-i\sigma t)\mathbf{u}(t).$$

Hence

$$\exp(-i\sigma T)\mathbf{u}(t + T) = \mathbf{u}(t),$$

and in particular,

$$\exp(-i\sigma T)\mathbf{u}(T) = \mathbf{u}(0). \tag{8.15}$$

Also, since for the IVP we are dealing with we have written $\mathbf{u}(0) = \boldsymbol{\varphi}$, we have

$$\mathbf{V}(0)\boldsymbol{\varphi} = \mathbf{U}(T, 0)\varphi = \mathbf{u}(T). \tag{8.16}$$

Combining (8.15) and (8.16), we obtain

$$\mathbf{V}(0)\boldsymbol{\varphi} = \mathbf{u}(T) = \exp(i\sigma T)\mathbf{u}(0) = \exp(i\sigma T)\boldsymbol{\varphi}. \tag{8.17}$$

Thus we can conclude that $\mathbf{V}(0)$ has an eigenvalue $\exp(i\sigma T)$ and associated eigenfunction $\boldsymbol{\varphi} = \mathbf{u}(0)$.

Recalling (8.9) and theorems 8.3 and 8.5, we can write

$$\begin{aligned}
\mathbf{U}(s+T, s)\mathbf{U}(s, 0)\boldsymbol{\varphi} &= \mathbf{U}(s+T, T)\mathbf{U}(T, s)\mathbf{U}(s, 0)\boldsymbol{\varphi} \\
&= \mathbf{U}(s+T, T)\mathbf{U}(T, 0)\boldsymbol{\varphi} \\
&= \mathbf{U}(s+T, T)\mathbf{u}(T) \\
&= \mathbf{U}(s+T, T)\exp(i\sigma T)\boldsymbol{\varphi} \\
&= \exp(i\sigma T)\mathbf{U}(s, 0)\boldsymbol{\varphi},
\end{aligned}$$

where this last equality has been obtained by using theorem 8.3(i).

Hence we can conclude that $\mathbf{V}(s)$ has an eigenvalue $\exp(i\sigma T)$ with associated eigenfunction

$$\mathbf{U}(s, 0)\boldsymbol{\varphi} = \mathbf{V}(0)\boldsymbol{\varphi}.$$

Therefore we have shown that (iii) implies (ii).

(i) $\Rightarrow$ (iii) The proof of this implication is rather lengthy and technical. More immediate proofs become available when specific problems are considered. Several works of this nature can be found in the references cited in chapter 11. □

8.5.1 Some Results for Potential Scattering

For the purpose of illustration we consider the acoustic wave equation perturbed by a real potential $q(x, t)$ that has compact support in $\mathbf{R}^n$ and is periodic in time. We shall be concerned with an equation of the form

$$\{\partial_t^2 - \Delta + q(x, t)\}u(x, t) = 0. \tag{8.18}$$

We shall be interested in solutions of (8.18) when the potential q has the following properties.

(i) q is defined on $\mathbf{R}^n \times \mathbf{R}$ with n odd,
(ii) $q = 0$ for $|x| \geq \rho$ and for all t,

(iii) there is a real number $T > 0$ such that

$$q(x,t+T) = q(x,t) \quad \text{for all } (x,t) \in \mathbf{R}^n \times \mathbf{R},$$

(iv) the mapping $t \to q(.,t)$ is continuous on $\mathbf{R}$ and takes values in $L_p(\mathbf{R}^n)$ for $p > n$.

We shall denote the set of all such potentials by $\mathcal{P}$.

With theorem 8.9 in mind we shall say here that $\sigma \in \mathbf{C}$ is a scattering frequency for $q \in \mathcal{P}$ if there exists a quantity $u(x,t)$ of locally finite energy for each t, which we shall call a scattering eigensolution associated with q, that has the following properties.

(i) $\{\partial_t^2 - \Delta + q(x,t)\}u(x,t) = 0$ for $(x,t) \in \mathbf{R}^{n+1} = \mathbf{R}^n \times \mathbf{R}$.
(ii) $u(x,t)$ is an outgoing solution.
(iii) $\exp(-i\sigma t)u(x,t)$ has period T.

By saying that $u(x,t)$ is an outgoing solution, we mean that if $u(x,t)$ is continued for $t > T$ as though it were a free solution $u_0(x,t)$, then $u_0(x,t) = 0$ in the cone $\{|x| \leq t - T - \rho\}$. This is consistent with the similar notion mentioned in chapters 6 and 7.

We would emphasise two things at this stage. First, we assume that the well- posedness of IBVPs and IVPs associated with the equation in property (i) has been settled by investigating them by means of their equivalent representation as a first-order system. Second, T in the above might well depend on q.

An elegant extension of the Lax-Phillips theory of scattering [68] for such problems as we have mention here has been developed by Cooper and Strauss [27]. Their abstract theory has yielded the following results.

Theorem 8.11 *If $q \in \mathcal{P}$, then its set of scattering frequencies is a discrete set in $\mathbf{C}$. Furthermore, the set of scattering eigensolutions corresponding to each scattering frequency is finite-dimensional.*

Theorem 8.12 *Let*
(i) $\{q_m : \mathbf{R} \to \mathcal{P}\}$ be a continuous family of potentials,
(ii) $\sigma_0(m_0)$ be a scattering frequency for q_{m_0}.

Then there are scattering frequencies $\sigma(m)$ for q_m that form a continuous curve in $\mathbf{C}$ passing through $\sigma_0(m_0)$. The scattering frequencies can disappear only when $|\sigma_0(m)| \to \infty$.

In particular, the theory of Cooper and Strauss has shown that the asymptotic expansion

$$u(x,t) \sim \sum_m c_m \exp(i\sigma(m)t).p_m(x,t)$$

is valid as $t \to \infty$, where p_m is periodic in time.

The proofs of these results are long and very technical. However, for most practical purposes the statement of these results is often sufficient. Nevertheless, interested readers will find the proofs fully presented in the references cited here and in chapter 11.

Chapter Nine

Concerning Inverse Problems

9.1 INTRODUCTION

One of the more intriguing and, indeed, difficult problems in mathematical physics and the applied sciences is the determination of an impurity in an otherwise homogeneous region from the measurements available of a field scattered by the inhomogeneity. This is the standard inverse scattering problem. Such problems frequently arise when analysing, for example, various ultrasonic diagnostic techniques and other nondestructive testing processes. Typical areas include remote sensing problems associated with radar, sonar, geophysics and medical diagnosis.

We remark that in the majority of practical problems measurements of the scattered field can only be made in the far field of the transmitter and scatterer. Consequently, we can expect that the scattering theory developed in earlier chapters can be used to provide a satisfactory means of investigating the sensing problem mentioned above. To see this, return for a moment to the development of scattering theories in an AP setting. In this case we recall that we work with self-adjoint operators A_0, A_1 on a Hilbert space H. Associated with these operators are the following.

Evolution operators:

$$U_0(t) := \exp(-itA_0), \qquad U_1(t) := \exp(-itA_1).$$

Wave operators:

$$W_\pm := s - \lim_{t\to\pm\infty} U_0^*(t)U_1(t).$$

Scattering operator:

$$S := W_+^* W_-.$$

If the operators A_0, A_1 characterise initial value problems with solutions u_0, u_1, respectively, which are required to satisfy given initial data f_0, f_1, respectively, then we have seen that the solutions can be written in the form

$$u_0(t) = U_0(t)f_0, \qquad u_1(t) = U_1(t)f_1.$$

In practice, f_1 is given from the outset and u_1 is regarded as the solution of a PP. When the intention is to use scattering theory techniques to approach solutions to this problem, the aim will be to determine initial data for an associated FP, having a reasonably easily obtainable solution u_0, that will ensure that u_0 and u_1 are AE as $t \to \pm\infty$. To this end, we introduce initial data $f_0^{\pm}$ for the FP that generates FP solutions $u_{\pm}$ such that $(u_1 - u_{\pm}) \to 0$ in a suitable energy norm as $t \to \pm\infty$, respectively. The initial data $f_0^{\pm}$ are related by means of the scattering operator S in the form

$$Sf_0^- = f_0^+.$$

With the above understanding we can describe two types of scattering problems as follows.

Direct scattering problems: Knowing U_0, U_1 and f_1, determine S and $f_0^{\pm}$.
Inverse scattering problems: Knowing S and $U_0(t)$, determine $U_1(t)$.

For NAPs the same situations obtain, but now an evolution operator of the form $U(t-s)$, where $s \in \mathbf{R}$ is a fixed initial time, must be replaced by a propagator of the form $U(t, s)$ (see chapter 5). In this chapter, with practical problems such as those mentioned above in mind, we shall reduce considerably the generality of these abstract problems by restricting our attention to a study of a perturbed wave equation of the form

$$\{\partial_t^2 - \Delta + q(x,t)\}u(x,t) = 0, \qquad (x,t) \in \mathbf{R}^n \times \mathbf{R}, \tag{9.1}$$

where $1 \leq n \leq \infty$. This equation is also referred to as the *plasma wave equation* and as such is intimately connected with potential scattering problems. It arises in the modelling of many physical systems ranging, for example, from the study of the electron density in the atmosphere of the earth to the vibrations of an elastically braced string. We concentrate here on indicating methods used to determine the potential term q when u, the state of the system, is known (in the far field). As before, the main difficulty associated with any analysis of (9.1) when q is time-dependent centres on the fact that the familiar separation of variables technique is no longer immediately available.

9.2 PRELIMINARIES

We are interested in the situation when solutions of (9.1) have finite energy. With this in mind we introduce the following notations.

Let X_j, $j = 0, 1$, denote the completion of the linear space of pairs

$$\mathbf{f} := \begin{bmatrix} f_1 \\ f_2 \end{bmatrix} =: \langle f_1, f_2 \rangle, \qquad f_i \in C_0^{\infty}(\mathbf{R}^n), \qquad i = 1, 2,$$

with respect to the energy norm

$$\|\mathbf{f}\|_{X_j} := \frac{1}{2}\int_{\mathbf{R}^n} \{|\nabla f_1(x)|^2 + |f_2(x)|^2 + jq(x,t)\,|f_1|^2\}\,dx, \qquad j = 0, 1. \tag{9.2}$$

The spaces X_j, $j=0,1$, will be referred to as energy spaces associated with (9.1). We notice that X_0 is the familiar energy space associated with the free or unperturbed wave equation in which $q\equiv 0$. For the perturbed problem, $q\neq 0$ and has properties for which the perturbed problem has solutions with finite energy. Specifically, we shall assume the following conditions hold.

Definition 9.1 *The potential q will be said to satisfy condition (H) if*
(i) $q\in L_\infty(\mathbf{R}^{n+1})$,
(ii) q is non-negative,
(iii) for each $t\in\mathbf{R}$, the function $q(.,t)$ has compact support in x and

$$\operatorname{supp} q(.,t)\subset B_M(0)=\{x\in\mathbf{R}^n : |x|<M\}, \tag{9.3}$$

where $B_M(0)$ is a fixed ball.

Although the norm associated with X_1 is time-dependent, nevertheless, the following result can be established [40].

Theorem 9.2 *The norms $\|.\|_{X_j}$, $j=0,1$, are equivalent.*

Proof. Since q is assumed to be non-negative, the proof will follow if we can show that there exists a positive constant c such that

$$\int_{\mathbf{R}^n} q(x,t)\psi^2(x)\,dx\le c\int_{\mathbf{R}^n}|\nabla\psi(x)|^2\,dx$$

for all $\psi\in C_0^\infty(\mathbf{R}^n)$.

Using Hölder's inequality and then Sobolev's inequality (see [101] and chapter 11) we obtain, in the case $n=3$, for example,

$$\begin{aligned}\int_{\mathbf{R}^3} q(x,t)\psi^2(x)\,dx &= \int_{|x|\le M} q(x,t)\psi^2(x)\,dx\\ &\le \|q\|_{L_\infty(\mathbf{R}^4)}\int_{|x|\le M}\psi^2(x)\,dx, \qquad \mathbf{R}^4=\mathbf{R}^3\times\mathbf{R}\\ &\le \left(\frac{4}{3}\pi M^3\right)^{2/3}\|q\|_{L_\infty(\mathbf{R}^4)}\|\psi\|^2_{L^6(|x|\le M)}\\ &\le \text{const}\,\|q\|_{L_\infty(\mathbf{R}^4)}\|\nabla\psi\|^2_{L^2(|x|\le M)},\end{aligned}$$

which establishes the proof. □

It will be convenient to state the following result due to Tamura [140], which we shall make use of in later sections.

Theorem 9.3 *Let the potential term q in (9.1) satisfy the condition (H) and assume*
(i) $q\in C^1$ with bounded derivatives,
(ii) for some $0<\alpha\le 1$,

$$q_t(x,t)=O(|t|^{-\alpha})\quad \text{as } t\to\infty \tag{9.4}$$

uniformly in x.

If u is a solution of (9.1) with initial data at $t=0$ with compact support and finite energy, then for $0<\rho<\infty$ there exists a constant $c=c(\rho)$ and a $\theta>0$ such that

$$E_\rho(t) = c\ \exp\{-\theta\ |t|\} E_\infty(0), \tag{9.5}$$

where

$$E_\rho(t) := \frac{1}{2}\int_{|x|\le\rho} \{|u_t(x,t)|^2 + |\nabla u(x,t)|^2 + q(x,t)\ |u(x,t)|^2\}\, dx,$$

$$E_\infty(t) := \frac{1}{2}\int_{\mathbf{R}^n} \{|u_t(x,t)|^2 + |\nabla u(x,t)|^2 + q(x,t)\ |u(x,t)|^2\}\, dx.$$

It can be shown (using this theorem) [96] that with q defined as above there exists a scattering operator $\mathbf{S}: X_0 \to X_0$ associated with (9.1) in the following sense. For each $\mathbf{f}_- := \langle f_{1^-}, f_{2^-}\rangle \in X_0$, there exists a unique solution u of (9.1) and a unique element $\mathbf{f}_+ := \langle f_{1^+}, f_{2^+}\rangle \in X_0$ such that

$$\lim_{t\to\pm\infty} \|\mathbf{U}_0(t)\mathbf{f}_\pm - \mathbf{u}(.,t)\|_{X_0} = 0,$$

where $\mathbf{u}(.,t) := \langle u(.,t), u_t(.,t)\rangle$ and $\mathbf{U}_0(t)$ denotes the unitary group of operators on X_0 associated with the free-wave equation. In this case the scattering operator $\mathbf{S}$ is defined in the familiar form as $\mathbf{S}\mathbf{f}_- = \mathbf{f}_+$.

We remark that some authors define $\mathbf{S}$ according to

$$\mathbf{S}(\mathbf{U_0(t)f_-}) = \mathbf{U}_0(t)\mathbf{f}_+$$

and that the first component of $\mathbf{U}_0(t)\mathbf{f}_\pm$ is denoted by $\mathbf{u}_\pm$.

With regard to the inverse problem associated with (9.1), it can be shown [40] that if q satisfies condition (H) and is small in a suitable sense, then $\mathbf{S}$ determines q uniquely.

With this in mind we make the following remarks and observations.

When the potential term is time-independent, a number of methods are available for tackling inverse scattering problems. In particular, taking the Fourier transform with respect to time of the plasma wave equation yields a stationary Schrödinger equation. The potential is then recovered by means of the celebrated Marchenko equations in one or another of its many forms in both the frequency and time domains (see chapter 11). A comprehensive and self-contained account of the derivation of the two basic types of Marchenko equations is given in [93]. Essentially, the required potential term is obtained in terms of a certain functional of a reflection coefficient.

When the potential is time-dependent, the situation changes dramatically. This is principally because, as we have already mentioned a number of times, separation of variable techniques are no longer available. However, a number of authors have studied a plasma wave equation that has a time-dependent potential term. In this connection the studies described in [40, 96, 97] have established a sound analytical basis for the study of such problems by providing existence and uniqueness results and contributions to the development of associated scattering theories. In this chapter we shall simply state these various results as we need them; their proofs are quite long and technically demanding. We shall assume that they hold and concentrate on indicating how they might be used when developing constructive methods. The account will be almost entirely formal.

9.3 REDUCTION OF THE PLASMA WAVE EQUATION TO A FIRST-ORDER SYSTEM

We consider the initial value problem

$$\{\partial_t^2 - \Delta + q(x,t)\}u(x,t) = 0, \qquad (x,t) \in \mathbf{R}^n \times (0,T), \tag{9.6}$$

$$u(x,s) = \varphi(x,s), \qquad u_t(x,s) = \psi(x,s), \qquad 0 \le s \le t \le T, \tag{9.7}$$

where $s \in \mathbf{R}$ is a fixed initial time.

Proceeding as in previous chapters (see equations (7.11)–(7.14)), we introduce the energy space H_0 that is the completion of $C_0^\infty(\mathbf{R}^n) \times C_0^\infty(\mathbf{R}^n)$ with respect to the norm generated by the inner product

$$(\mathbf{f}, \mathbf{g})_0 := (\nabla f_1, \nabla g_1) + (f_2, g_2), \tag{9.8}$$

where $\mathbf{f} = \langle f_1, f_2\rangle$, $\mathbf{g} = \langle g_1, g_2\rangle$ and $(.,.)$ denotes the usual $L_2(\mathbf{R}^n)$ inner product.

We now set

$$\mathbf{u}(.,t) = \langle u(.,t), u_t(.,t)\rangle, \qquad \boldsymbol{\varphi} = \langle \varphi, \psi\rangle, \tag{9.9}$$

$$\mathbf{Q}(.,t) = \begin{bmatrix} 0 & 0 \\ q(.,t) & 0 \end{bmatrix}, \qquad \mathbf{G}_0 = -i\begin{bmatrix} 0 & I \\ A_0 & 0 \end{bmatrix}, \tag{9.10}$$

where A_0 is an $L_2(\mathbf{R}^n)$ realisation of the Laplacian and where, for convenience, we have suppressed the parameter s. Hence the initial value problem (9.6), (9.7) can be realised as a first-order system in H_0 in the form

$$\{\partial_t - i\mathbf{G}_0\}\mathbf{u}(t) + \mathbf{Q}(t)\mathbf{u}(t) = 0, \qquad t \in (0,T), \tag{9.11}$$

$$\mathbf{u}(s) = \boldsymbol{\varphi}(s). \tag{9.12}$$

In chapters 5 and 6 we indicated conditions that ensure that the initial value problem (9.11), (9.12) is well posed. Consequently, when these conditions hold, the variation of parameters formula applied to (9.11), (9.12) yields

$$\mathbf{u}(t) = \mathbf{U}(t-s)\boldsymbol{\varphi}(s) - \int_s^t \mathbf{U}(t-\tau)\mathbf{Q}(\tau)\mathbf{u}(\tau)\,d\tau, \tag{9.13}$$

where

$$\mathbf{U}(t) = \exp(i\mathbf{G}_0) \tag{9.14}$$

is a semigroup that is well-defined since it can be shown [154] (chap. 5)] that $\mathbf{G}_0$ is self-adjoint on H_0.

When dealing with the direct problem, the potential term $\mathbf{Q}$ is known and (9.13) is a Volterra integral equation of the second kind for $\mathbf{u}$. This equation can be solved using the methods outlined in section 5.6.

For the inverse problem the *state vector* $\mathbf{u}$ is known and (9.13) is now a Volterra integral equation of first kind for the potential term $\mathbf{Q}$. We shall investigate this first-kind equation in the following sections and will see that it can be used quite successfully as a base from which to develop constructive methods for the determination of $\mathbf{Q}$ to various degrees of approximation. This is not by any means an easy equation to deal with, especially when matrix quantities are involved. From a practical point of view, it appears that a rather more attractive approach to the inverse scattering problem can be developed by working directly with the IVPs associated with the plasma wave equation (9.1) rather than with their representation as first-order systems. We outline in the next section a particularly promising approach.

9.4 A HIGH-ENERGY METHOD

The inverse scattering problem is a nonlinear, ill-posed problem. When dealing with the plasma wave equation, there are three basic questions that have to be addressed.

1. Does a potential term q exist that is compatible with measured data?
2. Is the potential term unique?
3. How can the potential term be construed from measured data?

Newton [90] addressed these questions in an AP setting by investigating an extension of the Marchenko equations in $\mathbf{R}^1$. However, the potential term was required to satisfy quite strict conditions. Furthermore, either the scattering data or the associated spectral measure was required on the whole real line and for all energy numbers $k \geq 0$.

A high-energy method for investigating inverse problems associated with the plasma wave equation in the AP case has been introduced by Saito [116, 119]. This method, when combined with suitable extensions of Wilcox's spectral method outlined in previous chapters, offers good prospects for developing approximation methods for reconstructing the potential q when the scattering data are only required to be known for a sufficiently large value of k. We shall also see that it provides a promising means of studying NAPs that will allow the use of well-known methods involving distorted plane waves and generalised eigenfunction expansion theorems.

In the next two subsections we give some results for related APs before dealing with the inverse problem for the plasma wave equation.

9.4.1 Some Asymptotic Formulae for the Plasma Wave Equation

We first consider the plasma wave equation (9.1) in an AP setting for which the potential $q(x)$ satisfies the following.

ASSUMPTION 9.4 *The potential function q is defined in $\mathbf{R}^3$ and is a non-negative bounded function with compact support; that is, q is a measurable function on $\mathbf{R}^3$ satisfying*

$$0 < q(x) \leq c_0, \qquad x \in \mathbf{R}^3,$$

$$q(x) = 0, \qquad |x| \geq R_0,$$

where c_0 and R_0 are positive constants.

We introduce the operator

$$A : L_2(\mathbf{R}^3) \to L_2(\mathbf{R}^3) \equiv H, \tag{9.15}$$

$$Au = -\Delta u + q(x), \qquad u \in D(A),$$

$$D(A) = \{u \in H : (-\Delta u + qu) \in H\},$$

which is the self-adjoint realisation of $(-\Delta u + q(x)u)$ in H.

We also introduce a real-valued function $s \in C^2$ defined on $\mathbf{R}$ such that

$$\text{supp}(s) \subset [a, b],$$

with $-\infty < a < b < \infty$. We denote by θ a unit vector in $\mathbf{R}^3$ and set

$$u_0(x, t) \equiv u_0(x, t, \theta, s) = s(x.\theta - t), \tag{9.16}$$

where $x.\theta$ denotes the scalar product in $\mathbf{R}^3$. The form of the argument of s indicates that the right-hand side of (9.16) represents a plane wave and that consequently u_0 satisfies the free equation (see [105], chap. 7).

$$(\partial_t^2 - \Delta)u_0(x, t) = 0. \tag{9.17}$$

Furthermore, since

$$q(x)u_0(x, t, \theta) = 0,$$

with $(x, t) \in \mathbf{R}^3 \times (-\infty, t_0)$ and $(x, t) \in \mathbf{R}^3 \times (t_1, \infty)$, where

$$t_0 := -b - R_0 \quad \text{and} \quad t_1 := -a + R_0,$$

u_0 satisfies the (free) plasma wave equation for $t \leq t_0$ and $t \geq t_1$.

Let $u(x, t) \equiv u(x, t, \theta)$ denote the solution of (9.1), the perturbed wave equation, that satisfies the initial conditions

$$u(x, t_0, \theta) = u_0(x, t, \theta) \quad \text{and} \quad \partial_t u(x, t_0, \theta) = \partial_t u_0(x, t, \theta). \tag{9.18}$$

We shall refer to $u(x, t)$ as the *total field*. We now denote the scattered wave by $u_{sc}(x, t) = u_{sc}(x, t, \theta)$, where

$$u_{sc}(x, t, \theta) = u(x, t, \theta) - u_0(x, t, \theta).$$

Then the scattered field satisfies the IVP

$$\{\partial_t^2 - \Delta + q(x)\}u_{sc}(x,t) = -q(x)u_0(x,t,\theta), \tag{9.19}$$

$$u_{sc}(x,t,\theta) = \partial_t u_{sc}(x,t,\theta) = 0, \qquad t \le t_0, \qquad x \in \mathbf{R}^3, \tag{9.20}$$

where we have recognised (9.18).

In practical cases it is usually u_{sc} that is measured and, moreover, measured in the far field of the transmitter and receiver. Consequently, we need to study the asymptotic behaviour of $u_{sc}(x,t,\theta)$ as $t \to \infty$. In order to do this we need some preparation. With this in mind we introduce the notion of a *scattering amplitude* denoted by $F(k,\omega,\omega')$, where

$$F(k,\omega,\omega') = \frac{1}{4\pi}\int_{\mathbf{R}^3} \varphi(y,-k\omega)q(y)\{\exp(iky\omega')\}\,dy, \tag{9.21}$$

with $k > 0$ and $\omega, \omega' \in S^2$ the unit sphere in three dimensions. Here $\varphi(x,\xi)$, with $x, \xi \in \mathbf{R}^3$, is the unique solution of the celebrated Lippmann-Schwinger equation [6, 54, 88, 101]

$$\varphi(x,\xi) = \exp(ix.\xi) - \frac{1}{4\pi}\int_{\mathbf{R}^3} \frac{\exp(i\,|\xi|\,|x-y|)}{|x-y|} q(y)\varphi(y,\xi)\,dy. \tag{9.22}$$

The solutions φ are the *distorted plane waves* associated with the plasma wave equation (9.1).

We now introduce the *far-field solution* u_{sc}^{∞} defined by

$$u_{sc}^{\infty}(x,t) \equiv u_{sc}^{\infty}(x,t,\theta,s) = |x|^{-1}\,K(|x|-t,\omega_x,\theta,s), \tag{9.23}$$

where $\omega_x = x/\,|x|$,

$$K(\upsilon,\omega,\theta,s) = \frac{1}{(2\pi)^{1/2}}\int_{-\infty}^{\infty} \{\exp(i\upsilon\rho)\}\hat{s}(\rho)F(\rho,\omega,\theta)\,d\rho \tag{9.24}$$

is the *asymptotic wave function* (*profile*) and $\hat{s}(\rho)$ is the usual one-dimensional Fourier transform of $s(\tau)$ given by

$$\hat{s}(\rho) = \frac{1}{(2\pi)^{1/2}}\int_{-\infty}^{\infty} \{\exp(-i\rho\tau)\}s(\tau)\,d\tau.$$

The following result can be obtained [119, 154].

THEOREM 9.5 *If the requirements of assumption 9.4 hold, then*

$$u_{sc}(x,t,\theta,s) = u_{sc}^{\infty}(x,t,\theta,s) \quad as\, t \to \infty$$

in the sense that

$$\lim_{t\to\infty} \left\| u_{sc}(.,t,\theta,s) - u_{sc}^{\infty}(.,t,\theta,s) \right\|_H = 0.$$

9.4.2 On the Autonomous Inverse Scattering Problem

In the previous subsection we described a possible strategy for attacking the inverse problem associated with the plasma wave equation. Specifically, we can take the following steps.

- Launch the incident pulses $u_0(x,t,\theta,s)$ characterised by (9.16).
- Evaluate $u_{sc}(x,t,\theta,s)$ from experimental data in practice and recover the asymptotic wave function $K(\upsilon,\omega,\theta,s)$ via (9.23). Here we acknowledge the results of theorem 9.5.
- Obtain the scattering amplitude using (9.24).
- Construct the potential function q by solving the first kind of Volterra integral equation (9.21).

The above set of steps can involve a great deal of hard, technical work [6, 90]. Another line of attack is to use the *high-energy method* introduced by Saito [116, 119]. This is an elegant method that offers good prospects for developing constructive methods using some of the generalised eigenfunction techniques introduced earlier.

The high-energy method of Saito begins by introducing

$$\varphi_{k,x}(\omega) = \exp(-ikx.\omega), \qquad \omega \in S^2,$$

and regarding $\varphi_{k,x}$ as a function on S^2 with parameters $k > 0$ and $x \in \mathbf{R}^3$. We then define

$$g(x,k) := k^2 \int_{S^2} \int_{S^2} F(k,\omega,\omega')\varphi_{k,x}(\omega)\overline{\varphi_{k,x}(\omega')}\, d\omega\, d\omega', \tag{9.25}$$

where $F(k,\omega,\omega')$ is the scattering amplitude defined in (9.21). The main result obtained by Saito [116, 117] is as follows.

Theorem 9.6 *Let the potential term $q(x)$ satisfy assumption 9.4 and let $\varphi_{k,x}(\omega)$ and $g(x,k)$ be as defined above. Then the following limit exists.*

$$g(x,\infty) = \lim_{k\to\infty} g(x,k) = -2\pi \int_{\mathbf{R}^3} \frac{q(y)}{|x-y|^2}\, dy. \tag{9.26}$$

Furthermore, the potential $q(x)$ is recovered in the form

$$q(x) = -\frac{1}{4\pi} F_0^*(|\xi|\, F_0 g(.,\infty))(x), \tag{9.27}$$

where F_0 is the usual Fourier transform

$$(F_0 f)(\xi) = s - \lim_{R\to\infty} \frac{1}{(2\pi)^{1/2}} \int_{|x|<R} \{\exp(-i\xi.y)\} f(y)\, dy$$

and F_0^* denotes its adjoint (section 3.5).

We remark that in proving theorem 9.6, a generalised Fourier transform associated with $A = -\Delta + q(x)$ given by

$$(F_+h)(\xi) = s - \lim_{R\to\infty} \frac{1}{(2\pi)^{1/2}} \int_{|x|<R} \varphi(y, -\xi) h(y)\, dy$$

is used, where φ is the distorted plane wave introduced in (9.22).

Let

$$q(x, k) := -\frac{1}{4\pi} F_0^*(|\xi|\, F_0 g(., k))(x). \tag{9.28}$$

Then the following approximation result is available [118].

Theorem 9.7 *If*
(i) $q(x) \le c(1+|x|)^{-2}$, $c > 0$,
(ii) $q \in C^2(\mathbf{R}^3)$ *satisfying*

$$|D^\alpha q(x)| \le c_1(1+|x|)^{-\beta}, \qquad x \in \mathbf{R}^3,\ |\alpha| = 1, 2,$$

with constants $c_1 > 0$ *and* $\beta > 5/2$, *where* $\alpha = (\alpha_1, \alpha_2, \alpha_3)$ *is a multi-index of non-negative integers such that*

$$|\alpha| = \alpha_1 + \alpha_2 + \alpha_3.$$

Furthermore, we have written

$$D^\alpha = D_1^{\alpha_1} D_2^{\alpha_2} D_3^{\alpha_3}, \qquad D_j = \frac{\partial}{\partial x_j}, \quad j = 1, 2, 3.$$

Then

$$\|q(.) - q(., k)\|_H \le c_2(k)^{-1} \quad \textit{as } k \to \infty,$$

where c_2 *depends on* c_1, β *and* $\max_{x\in\mathbf{R}^3} |q(x)|$ *but does not depend on* k.

This last result indicates that we need only work with measurements taken at sufficiently high values of k.

9.4.3 Extension to Nonautonomous Inverse Scattering Problems

For the NAP, we are required to determine the time-dependent potential $q(x, t)$ that appears in the plasma wave equation (9.1). In practical experiments, measurements of the scattered field are made at discrete times $t = t_j$, $j = 1, 2, \ldots$. At these instants the potential $q(x, t)$, if it were known, would assume the values $q(x, t_j)$, $j = 1, 2, \ldots$. Furthermore, at each of these instants $q(x, t_j)$ is a function of x that involves a parameter t_j. Therefore, at each of these instants, we are effectively faced with an autonomous inverse scattering problem. That is, determine $q(x, t_j)$ from scattering data measured at $t = t_j$, and the high-energy method outlined above can be used at each of the instants when measurements are made. Consequently, for a given nonautonomous inverse problem, we partition the time interval of interest as indicated above and assume that for all values of j the potentials $q(x, t_j)$ satisfy the conditions given in assumption 9.4. Therefore, paralleling the development in subsection 9.4.2, we introduce the quantity

$$g(x, t_j, k) := k^2 \int_{S^2} \int_{S^2} F(t_j, k, \omega, \omega') \varphi_{k,x}(\omega) \overline{\varphi_{k,x}(\omega')} \, d\omega \, d\omega', \tag{9.29}$$

where $F(t_j, k, \omega, \omega')$ is the scattering amplitude associated with the potential $q(x, t_j)$. Proceeding as before, we recover the potential $q(x, t_j)$ by means of the relation

$$q(x, t_j) := -\frac{1}{4\pi} F_0^*(|\xi| \, F_0 g(., t_j, \infty))(x). \tag{9.30}$$

Furthermore, paralleling (9.28), we have the approximation formula

$$q(x, t_j, k) := -\frac{1}{4\pi} F_0^*(|\xi| \, F_0 g(., t_j, k))(x), \tag{9.31}$$

which has the same implications as in the previous subsection. In particular, theorem 9.7 is available for use at each of the instants t_j, $j = 1, 2, \ldots$.

Once the $q(x, t_j)$ has been computed in this way, it is used to determine the associated scattering data for such a potential using (9.21), (9.23) and (9.24). These scattering data are then compared with the measured data and their quality assessed. If the quality is not adequate, then the above procedure is repeated, with an adjusted partitioning of the time interval, until a satisfactory agreement is obtained. With $q(x, t_j)$, $j = 1, 2, \ldots$, satisfactorily determined, $\|q(, , t_j)\|_H$ is computed and plotted against t_j. A simple curve-fitting process applied to the points obtained will provide a means of assessing the t-dependence of the potential q.

The approach outlined here is felt to offer good prospects for obtaining the salient properties of the (time-dependent) potential q. Stability and convergence results have been obtained [93] to support this view. Admittedly, the process is a rather lengthy one. Unfortunately, there does not appear to be as yet any particularly short way of tackling the inverse scattering problem in either the AP or the NAP case. Nevertheless, detailed investigations of specific practical problems, with particular interest being directed towards the influence of inherent resonances, are being made and are referred to in the references cited here and/or in chapter 11.

Finally in this chapter, we remark that here we have confined our attention to potential scattering problems. There is of course a similar set of problems dealing with target scattering issues. These problems are at least equally challenging but will not be discussed here.

Chapter Ten

Some Remarks on Scattering in Other Wave Systems

10.1 INTRODUCTION

In the previous chapters we have been concerned with acoustic wave scattering problems. In dealing with such problems we adopted the Wilcox theory of acoustic scattering introduced in [154]. The main reason for adopting this particular approach was that the Wilcox theory uses quite elementary results from functional analysis, the spectral theory of self-adjoint operators on Hilbert spaces and semi-group theory and leads quite readily to the development of constructive methods based on generalised eigenfunction expansion theorems. Furthermore, unlike the Lax-Phillips theory [68], the Wilcox theory applies to scattering problems in both even and odd space dimensions.

In this monograph we have been largely concerned with the scattering of acoustic waves by time-dependent perturbations, the NAPs. We have seen that in order to investigate such problems it was first necessary to understand and to deal with the associated APs. In the following sections we indicate how scattering problems associated with electromagnetic waves and with elastic waves can be placed in a similar framework to that used when studying acoustic wave scattering problems. Indeed, we will see that electromagnetic and elastic wave problems can be given the same symbolic form as acoustic wave problems. Consequently, the constructive methods developed for acoustic wave problems can in principle become available for electromagnetic and for elastic wave problems.

10.2 SCATTERING OF ELECTROMAGNETIC WAVES

The analysis of these problems closely parallels the analysis of acoustic wave scattering problems. Each step in the analysis of acoustic wave scattering problems has an analogue in the analysis of electromagnetic wave scattering problems. In particular, the scalar d'Alembert equation that arises in studying the acoustic field is replaced by a vector wave equation when investigating the electromagnetic field. The vector nature of electromagnetic problems requires more demanding algebraic manipulations and calculations. Despite this, we shall see that each of the main steps in the analysis of acoustic problems has an analogue in electromagnetic problems.

We consider here electromagnetic problems in $\mathbf{R}^3$ and understand $x=(x_1,x_2,x_3)$. Furthermore, we write $(x,t)=(x_1,x_2,x_3,t)$ to denote the space-time coordinates. When studying target scattering, we let $B\subset\mathbf{R}^3$ denote a closed bounded set such that $\Omega = \mathbf{R}^3 - B$ is connected. The set B represents the scattering body. For potential-type scattering problems there is no need to introduce the set B. In this case the scattering arises as a result of perturbations of coefficients in the governing field equations.

We assume that the medium filling Ω is characterised by a dielectric constant ε and a magnetic permeability μ.

Electromagnetic phenomena are governed by the celebrated *Maxwell's equations*, which we write in the form

$$\{\varepsilon \mathbf{E}_t - \nabla \times \mathbf{H}\}(x, t) = \mathbf{J}(x, t), \tag{10.1}$$

$$\{\mu \mathbf{H}_t + \nabla \times \mathbf{E}\}(x, t) = \mathbf{K}(x, t), \tag{10.2}$$

where $\mathbf{E}$ and $\mathbf{H}$ represent electric and magnetic field vectors, respectively, and $\mathbf{J}$ and $\mathbf{K}$ denote electric and magnetic currents, respectively. Furthermore, we have introduced the symbol ∇, referred to as nabla, to denote the vector differential expression

$$\nabla := \mathbf{i}\frac{\partial}{\partial x_1} + \mathbf{j}\frac{\partial}{\partial x_2} + \mathbf{k}\frac{\partial}{\partial x_3},$$

where $\mathbf{i}, \mathbf{j}, \mathbf{k}$ are a triad of unit vectors used to characterise $\mathbf{R}^3$. In terms of this symbol we define

$$\operatorname{grad} \varphi = \nabla \varphi,$$

$$\operatorname{div} \mathbf{v} = \nabla \cdot \mathbf{v},$$

$$\operatorname{curl} \mathbf{v} = \nabla \times \mathbf{v},$$

where the dot denotes the usual scalar product and $\times$ the usual vector product [87].

Maxwell's equations (10.1), (10.2) can be written more conveniently in the form

$$\begin{bmatrix} \mathbf{E} \\ \mathbf{H} \end{bmatrix}_t (x, t) + \begin{bmatrix} 0 & -\frac{1}{\varepsilon}\nabla\times \\ \frac{1}{\mu}\nabla\times & 0 \end{bmatrix} \begin{bmatrix} \mathbf{E} \\ \mathbf{H} \end{bmatrix} (x, t) = \begin{bmatrix} \frac{1}{\varepsilon}\mathbf{J} \\ \frac{1}{\mu}\mathbf{K} \end{bmatrix} (x, t). \tag{10.3}$$

This in turn can be written

$$(\mathbf{U}_t - i\mathbf{G}\mathbf{U})(x, t) = \mathbf{f}(x, t), \tag{10.4}$$

where

$$\mathbf{U}(x, t) = \langle \mathbf{E}, \mathbf{H} \rangle (x, t), \qquad \mathbf{f}(x, t) = \left\langle \frac{1}{\varepsilon}\mathbf{J}, \frac{1}{\mu}\mathbf{K} \right\rangle (x, t),$$

$$-i\mathbf{G} = \mathbf{M}^{-1} \begin{bmatrix} 0 & -\nabla\times \\ \nabla\times & 0 \end{bmatrix}, \qquad \mathbf{M} = \begin{bmatrix} \varepsilon & 0 \\ 0 & \mu \end{bmatrix}$$

and, as usual $\langle ., . \rangle$ denotes the transpose of the vector $\begin{bmatrix} \cdot \\ \cdot \end{bmatrix}$.

Equation (10.4) has the same symbolic form as that discussed in chapter 7 (see (7.17)). However, here the unknown $\mathbf{U}$ has a more complicated structure since it has the general form $\mathbf{U}=\langle \mathbf{u}_1, \mathbf{u}_2\rangle$, where in our particular case $\mathbf{u}_1=\mathbf{E}$ and $\mathbf{u}_2=\mathbf{H}$.

We shall use the notation

$$\mathbf{v}(x,t)=\langle v_1, v_2, v_3\rangle(x,t), \qquad v_j \in \mathbf{C}, \quad j=1,2,3,$$

to denote a vector in $\mathbf{R}^3$ with complex-valued components. Throughout we shall be interested in vectors $\mathbf{v}$ having the following properties.

$$\mathbf{v}\in (L_2(\Omega))^3 := \{\mathbf{v}=\langle v_1, v_2, v_3\rangle : v_j \in L_2(\Omega),\ j=1,2,3\}. \tag{10.5}$$

Occasionally, we will emphasise matters by using the notation

$$L_2(\Omega, \mathbf{C}^3)=(L_2(\Omega))^3 \cap \{v_j : \|v_j\|^2 < \infty,\ j=1,2,3\},$$

where $\|.\|$ denotes the usual $L_2(\Omega)$ norm.

The collection $L_2(\Omega, \mathbf{C}^3)$ is a Hilbert space, which we shall denote by H, with respect to the inner product and norm structure

$$(\mathbf{u}, \mathbf{v})_H=(u_1, v_1)+(u_2, v_2)+(u_3, v_3), \tag{10.6}$$

$$\|\mathbf{u}\|_H^2=\|u_1\|^2+\|u_2\|^2+\|u_3\|^2, \tag{10.7}$$

where $(.,.)$ and $\|.\|$ denote the usual $L_2(\Omega)$ inner product and norm.

We also introduce

$$\mathbb{D}(\Omega) := \{\mathbf{u}\in (L_2(\Omega))^3 : \operatorname{div}\mathbf{u} \in L_2(\Omega)\}, \tag{10.8}$$

$$\mathbb{R}(\Omega) := \{\mathbf{u}\in (L_2(\Omega))^3 : \operatorname{curl}\mathbf{u} \in (L_2(\Omega))^3\}, \tag{10.9}$$

with structure

$$\|\mathbf{u}\|_{\mathbb{D}}^2 := \|\mathbf{u}\|_H^2+\|\operatorname{div}\mathbf{u}\|^2, \tag{10.10}$$

$$\|\mathbf{u}\|_{\mathbb{R}}^2 := \|\mathbf{u}\|_H^2+\|\operatorname{curl}\mathbf{u}\|_H^2.$$

Furthermore, we define

$$\mathbb{R}_0(\Omega) := \{\mathbf{u}\in \mathbb{R}(\Omega) : (\nabla\times\mathbf{u}, \mathbf{v})=(\mathbf{u}, \nabla\times\mathbf{v})\}. \tag{10.11}$$

With this preparation it therefore seems natural to look for solutions $\mathbf{U}$ of (10.4) that belong to the class

$$\mathbf{L} := (L_2(\Omega))^3 \times (L_2(\Omega))^3. \tag{10.12}$$

On this class we will find it convenient to define a product of the two elements $\mathbf{W}=\langle \mathbf{w}_1, \mathbf{w}_2\rangle$ and $\mathbf{V}=\langle \mathbf{v}_1, \mathbf{v}_2\rangle$ to be

$$\begin{aligned}(\mathbf{W}, \mathbf{V})_{\mathbf{L}} &:= \left(\begin{bmatrix}\mathbf{w}_1\\ \mathbf{w}_2\end{bmatrix}, \begin{bmatrix}\mathbf{v}_1\\ \mathbf{v}_2\end{bmatrix}\right)\\ &= (\mathbf{w}_1, \mathbf{v}_1)_H+(\mathbf{w}_2, \mathbf{v}_2)_H.\end{aligned} \tag{10.13}$$

When dealing with Maxwell's equations (10.1), (10.2), we find it convenient to use, instead of $\mathbf{L}$ as defined above, the class

$$\mathcal{H} = (L_2(\Omega))^3 \times (L_2(\Omega))^3,$$

which is a Hilbert space with respect to a weighted inner product defined by

$$(\mathbf{W}, \mathbf{V})_{\mathcal{H}} = (\mathbf{W}, \mathbf{MV})_{\mathbf{L}}, \qquad (10.14)$$

where the weight matrix $\mathbf{M}$ is defined in (10.4).

If we now introduce the operator

$$\mathbf{G} : \mathcal{H} \to \mathcal{H}, \qquad (10.15)$$

$$-i\mathbf{GU} = \mathbf{M}^{-1} \begin{bmatrix} 0 & -\nabla\times \\ \nabla\times & 0 \end{bmatrix} \begin{bmatrix} \mathbf{u}_1 \\ \mathbf{u}_2 \end{bmatrix},$$

where

$$\mathbf{U} = \langle \mathbf{u}_1, \mathbf{u}_2 \rangle \in D(\mathbf{G}),$$

$$D(\mathbf{G}) = \mathbb{R}_0(\Omega) \times \mathbb{R}(\Omega) \subset \mathcal{H},$$

then we see that (10.1), (10.2) have the following representation in $\mathcal{H}$.

$$\mathbf{U}_t(t) - i\mathbf{GU}(t) = \mathbf{f}(t), \qquad \mathbf{U}(0) = \mathbf{U}_0 \in \mathcal{H}. \qquad (10.16)$$

As in previous chapters, we understand

$$\mathbf{U}(.,.) : t \to \mathbf{U}(.,t) =: \mathbf{U}(t).$$

Implicit in this development are the assumptions of the following.

Initial conditions:

$$\mathbf{E}(0) = \mathbf{E}_0, \qquad \mathbf{H}(0) = \mathbf{H}_0.$$

Boundary conditions: Total reflection at the boundary $\partial\Omega$ of Ω, that is,

$$(\mathbf{n} \times \mathbf{E}) = 0 \quad \text{on } \partial\Omega,$$

where $\mathbf{n}$ is the outward drawn unit vector normal to $\partial\Omega$. Other boundary conditions can be imposed as required [71].

We require solutions of (10.5) that have finite energy. With the above structure we have that the energy $\mathcal{E}$ of the electromagnetic field can be expressed in the form

$$\mathcal{E} := \|\mathbf{U}\|^2_{\mathcal{H}} = (\mathbf{u}_1, \varepsilon\mathbf{u}_1)_H + (\mathbf{u}_2, \mu\mathbf{u}_2)_H. \qquad (10.17)$$

Once we have a Hilbert space representation of (10.1), (10.2), the matter of existence and uniqueness of solutions to (10.16) arises. These results can be obtained using the methods outlined in chapter 5. To this end, the following results can be obtained in much the same manner as for acoustic equations, but now they have to be specialised bearing in mind the structure of $\mathcal{H}$. Specifically, the following theorem can be established [71].

THEOREM 10.1 *If* $\mathbf{G}$ *is defined as in (10.15), then*
(i) $\mathbf{G}$ *is a self-adjoint operator,*
(ii) $\mathbf{C}\setminus\mathbf{R}\subset\rho(\mathbf{G})$,
(iii) $\|(\mathbf{G}+i\lambda)^{-1}\|\leq 1/|\lambda|$.

The results in theorem 8.1 when combined with Stone's theorem yield, as outlined in chapter 5, the following important result.

THEOREM 10.2 *The initial boundary value problem*

$$\mathbf{U}_t(t)-i\mathbf{G}\mathbf{U}(t)=0, \qquad \mathbf{U}(0)=\mathbf{U}_0, \tag{10.18}$$

is uniquely solvable in $\mathcal{H}$. *Moreover, the solution* $\mathbf{U}$ *has finite energy.*

Once questions of existence and uniqueness have been settled, as in previous chapters, we can take the first component (say) of the solution $\mathbf{U}$ to provide details solely of the electric field $\mathbf{E}$. However, once it has been established that theorem 10.2 holds, that is, that (10.1), (10.2) is a well-posed system, an attractive alternative approach is to deal with equations (10.1), (10.2) directly, and this we shall do.

If we recall the vector identity

$$\text{curl curl } \mathbf{v} = \text{grad div } \mathbf{v} - \Delta\mathbf{v}, \tag{10.19}$$

then we can eliminate $\mathbf{H}$ in (10.1), (10.2) and obtain the second-order vector equation

$$\{\partial_t^2+\text{curl curl}\}\mathbf{E}(x,t)=-\partial_t\mathbf{J}(x,t). \tag{10.20}$$

Whilst we shall use (10.20) in this section, we would point out that a similar vector wave equation to (10.20) can be obtained for $\mathbf{H}$ by eliminating $\mathbf{E}$ in (10.1), (10.2).

The equation (10.20) is equivalent to a system of three second-order partial differential equations for the components of the electric field. Once these are solved, the magnetic field can be found from (10.2).

In the following it will be convenient to use some of the more detailed notations and conventions of matrix algebra.

If the electric field vector $\mathbf{E}$ is written in the form

$$\mathbf{E}=E_1\mathbf{i}_1+E_2\mathbf{i}_2+E_3\mathbf{i}_3, \tag{10.21}$$

where $(\mathbf{i}_1,\mathbf{i}_2,\mathbf{i}_3)$ denote the orthonormal basis associated with the coordinate system (x_1,x_2,x_3), then (10.21) defines a one-to-one correspondence of the form

$$\mathbf{E}\leftrightarrow u=(E_1,E_2,E_3)^T, \tag{10.22}$$

where T denotes the transpose of a matrix. Consequently, for two vectors $\mathbf{a}$ and $\mathbf{b}$ with associated correspondences,

$$\mathbf{a}=a_1\mathbf{i}_1+a_2\mathbf{i}_2+a_3\mathbf{i}_3\leftrightarrow a=(a_1,a_2,a_3)^T, \tag{10.23}$$

$$\mathbf{b}=b_1\mathbf{i}_1+b_2\mathbf{i}_2+b_3\mathbf{i}_3\leftrightarrow a=(b_1,b_2,b_3)^T, \tag{10.24}$$

the associated scalar and vector products of vector algebra correspond to the matrix operations

$$\mathbf{b}.\mathbf{a} \leftrightarrow b^T a = b.a = b_1 a_1 + b_2 a_2 + b_3 a_3, \tag{10.25}$$

$$\mathbf{b} \times \mathbf{a} \leftrightarrow b \times a = \mathcal{M}(b)a, \tag{10.26}$$

where the matrix $\mathcal{M}(b)$ is defined by

$$\mathcal{M}(b) = \begin{bmatrix} 0 & -b_3 & b_2 \\ b_3 & 0 & -b_1 \\ -b_2 & b_1 & 0 \end{bmatrix}. \tag{10.27}$$

Similarly, the vector differential operator

$$\text{curl} = \nabla \times = \left(\mathbf{i}_1 \frac{\partial}{\partial x_1} + \mathbf{i}_2 \frac{\partial}{\partial x_2} + \mathbf{i}_3 \frac{\partial}{\partial x_3} \right) \tag{10.28}$$

has a matrix representation that has the typical form

$$\text{curl}\, \mathbf{E} = \nabla \times \mathbf{E} \leftrightarrow \nabla \times u = \mathcal{M}(\partial) u, \tag{10.29}$$

where

$$\partial = (\partial_1, \partial_2, \partial_3) = \left(\frac{\partial}{\partial x_1}, \frac{\partial}{\partial x_2}, \frac{\partial}{\partial x_3} \right).$$

Using (10.29) repeatedly, we can obtain

$$\text{curl curl}\, \mathbf{E} = \nabla \times \nabla \times \mathbf{E} \leftrightarrow \nabla \times \nabla \times u = \mathcal{A}(\partial) u, \tag{10.30}$$

where

$$\mathcal{A}(b) = \mathcal{M}(b)^2 = bb - |b|^2 \mathcal{I}. \tag{10.31}$$

The term bb denotes a (3×3) matrix whose jkth element is $b_j b_k$, and $\mathcal{I}$ denotes the (3×3) identity matrix. Hence the identity (10.30) can be written

$$\nabla \times \nabla \times u = (\nabla \nabla . - \Delta \mathcal{I}) u, \tag{10.32}$$

where Δ denotes the three-dimensional Laplacian

$$\Delta = \partial_1^2 + \partial_2^2 + \partial_3^2. \tag{10.33}$$

We remark that (10.19) and (10.32) are corresponding expressions. However, care must be exercised when dealing with the vector identity (10.19) in coordinate systems other than Cartesian. This is particularly the case when interpreting the vector Laplacian $\mathbf{\Delta}$; see [87].

With these various notations in mind we see that the electromagnetic field generated in Ω is characterised by a function u of the form

$$u(x,t) = (u_1(x,t), u_2(x,t), u_3(x,t))^T, \qquad x \in \Omega, \qquad t \in \mathbf{R}, \tag{10.34}$$

where $u_j = E_j$, $j = 1, 2, 3$, are the components of the electric field. The quantity $u(x, t)$ is a solution of the inhomogeneous vector equation

$$\{\partial_t^2 + \nabla \times \nabla \times\} u(x, t) = -\partial_t J(x, t) =: f(x, t), \qquad (x, t) \in \Omega \times \mathbf{R}, \qquad (10.35)$$

and $J(x, t)$ is the electric current density that generates the field. We notice that if we are dealing with divergence-free fields, then the first term on the right-hand side of (10.32) vanishes and (10.35) reduces to a vector form of the familiar scalar wave equation.

The FP associated with (10.35) has a field that we shall denote by $u_0(x, t)$. This is the field generated by the sources (transmitters) in the medium when no scatterers are present. It can be represented, as in the scalar case, in terms of a retarded potential (see appendix A5). To see this, we apply the divergence operator to both sides of (10.35) and use the well-known vector identity $\nabla . \nabla \times u = 0$ to obtain

$$\partial_t^2 \nabla . u(x, t) = -\partial_t J(x, t). \qquad (10.36)$$

We now integrate (10.36) twice over the interval $t_0 \leq \tau \leq t$ and use the initial condition

$$u(x, t) = 0 \quad \text{for } t < t_0, \ x \in \Omega$$

to obtain

$$\nabla . u(x, t) = -\int_{t_0}^{t} \nabla . J(x, \tau)\, d\tau \qquad x \in \Omega, \qquad t \in \mathbf{R}. \qquad (10.37)$$

Using (10.35), (10.37) and the identity (10.32), we see that u satisfies

$$\{\partial_t^2 - \Delta\} u(x, t) = \nabla \nabla . \int_{t_0}^{t} J(x, \tau)\, d\tau - \partial_t J(x, t). \qquad (10.38)$$

Since

$$\partial_t^2 \int_{t_0}^{t} J(x, \tau)\, d\tau = \partial_t J(x, t),$$

we see, on replacing u by u_0 in (10.38), that the free field is determined by the inhomogeneous wave equation

$$\{\partial_t^2 - \Delta\} u_0(x, t) = \{\nabla \nabla - \partial_t^2 \mathbf{I}\} . \int_{t_0}^{t} J(x, \tau)\, d\tau \qquad (10.39)$$

for $(x, t) \in \Omega \times \mathbf{R}_+$ and the initial condition $u_0(x, t) = 0$ for all $t \leq t_0$.

Equation (10.39) is equivalent to three scalar wave equations and, as in the acoustic case, they can be integrated by the retarded potential formula. For ease of presentation, we define

$$\mathcal{I} J(x, t) = \int_{t_0}^{t} J(x, \tau)\, d\tau.$$

Consequently, the function V defined by

$$V(x,t) = \frac{1}{4\pi}\int_{|x'-x_0|\leq\delta_0} \frac{\mathcal{I}J(x', t-|x-x'|)}{|x-x'|}\,dx' \tag{10.40}$$

satisfies

$$\{\partial_t^2 - \Delta\}V(x,t) = \mathcal{I}J(x,t), \qquad (x,t)\in\mathbf{R}^3\times\mathbf{R}, \tag{10.41}$$

and the initial condition $V(x,t)=0$ for $t\leq t_0$. It then follows, by the linearity of the wave equation, that the free field is given by

$$u_0(x,t) = \{\nabla\nabla - \partial_t^2\mathbf{I}\}.V(x,t). \tag{10.42}$$

We have now reached the stage when we can take the same steps for analysing electromagnetic scattering problems as we did when dealing with acoustic scattering problems in chapters 6 and 7. However, working through these steps in detail is a lengthy process. Consequently, since applications are really our main interest, we shall simply state here the salient features and refer the reader to chapter 11 where references to the full working can be found.

For convenience, we gather together in the next two sections the salient features of the strategic steps to be taken when studying APs, first when dealing with the scattering of acoustic waves and then when dealing with the scattering of electromagnetic waves.

10.3 STRATEGY FOR AUTONOMOUS ACOUSTICS PROBLEMS IN $\mathbf{R}^3$

Very briefly and purely for convenience at this stage, we gather together the various aspects that have to be addressed when investigating autonomous scattering problems. In this section we consider the acoustic case, and in the next section the electromagnetic case.

The governing equation in this case is

$$\{\partial_t^2 - \Delta\}u(x,t) = f(x,t), \qquad (x,t)\in\mathbf{R}^3\times\mathbf{R}, \tag{10.43}$$

where u characterises the acoustic field.

It is assumed that the transmitter is localised near the point x_0. It will also be assumed that the transmitter emits a signal at time t_0 that in the first instance is in the form of a pulse of time duration T. Consequently, we will have

$$\text{supp } f\subset\{(x,t): t_0\leq t\leq t_0+T\}, \qquad |x-x_0|\leq\delta_0,$$

where δ_0 is a constant.

If the pulse is scattered by a body B, then we will assume that

$$B\subset\{x: |x|\leq\delta,\ \delta=\text{constant}\}.$$

It will be assumed that the transmitter and the scattering body are very far apart and disjoint; this introduces the far-field approximation

$$|x_0| >> \delta_0 + \delta.$$

The primary, or free problem, acoustic field, that is, the acoustic field that obtains when there are no scatterers present in the medium, is given in terms of a retarded potential in the form (see appendix (A5))

$$u_0(x,t) = \frac{1}{4\pi} \int_{|x'-x_0|\leq\delta_0} \frac{f(x', t - |x - x'|)}{|x - x'|}\, dx' \tag{10.44}$$

for $(x,t) \in \mathbf{R}^3 \times \mathbf{R}$.

If θ_0 denotes a unit vector defined by

$$x_0 = -\,|x_0|\,\theta_0, \tag{10.45}$$

then expanding $|x - x'|$ in powers of $|x_0|$, we find

$$u_0(x,t) = \frac{s(\theta_0, x.\theta_0 - t + |x_0|)}{|x_0|} + O\left(\frac{1}{|x_0|^2}\right)$$

uniformly for $t \in \mathbf{R}$, $|x| \leq \delta$, where

$$s(\theta_0, \tau) = \frac{1}{4\pi} \int_{|x'-x_0|\leq\delta_0} f(x', \theta_0.(x' - x_0) - \tau)\, dx', \qquad \tau \in \mathbf{R},$$

is the signal waveform.

When the error term is dropped in the above, the primary field is a plane wave propagating in the direction of unit vector θ_0.

When the primary acoustic field is a plane wave

$$u_0(x,t) = s(\theta_0, x.\theta_0 - t), \qquad \text{supp}\, s(\theta_0, .) \subset [a, b], \tag{10.46}$$

which is scattered by B, then the resulting acoustic field $u(x,t)$ is the solution of the IBVP

$$\{\partial_t^2 - \Delta\} u(x,t) = 0, \qquad (x,t) \in \Omega \times \mathbf{R}, \tag{10.47}$$

$$u(x,t) \in (bc), \tag{10.48}$$

$$u(x,t) \equiv u_0(x,t), \qquad x \in \Omega, \quad t + b + \delta < 0. \tag{10.49}$$

The echo or scattered field is then defined to be

$$u_{sc}(x,t) = u(x,t) - u_0(x,t), \qquad (x, t \in \Omega \times \mathbf{R}). \tag{10.50}$$

It can then be shown [154] that in the far field,

$$u_{sc}(x,t) \approx \frac{e(\theta_0, \theta, |x| - t)}{|x|}, \qquad x = |x|\,\theta. \tag{10.51}$$

The quantity $e(\theta_0, \theta, \tau)$ is the echo waveform.

A main aim in practical scattering theory is to calculate the relationship between the signal and echo waveforms. A way of doing this is as follows.

Assume the source function f has the form

$$f(x,t) = g_1(x)\cos\omega t + g_2(x)\sin\omega t = \mathrm{Re}\{g(x)\exp(-i\omega t)\}, \tag{10.52}$$

where $g(x) = g_1(x) + ig_2(x)$.

The associated wave field $u(x,t)$ will have the same time dependence and have the typical form

$$u(x,t) = w_1(x)\cos\omega t + w_2(x)\sin\omega t = \mathrm{Re}\{w(x)\exp(-i\omega t)\}, \tag{10.53}$$

where $w(x) = w_1(x) + iw_2(x)$.

The wave field $u(x,t)$ must satisfy the d'Alembert equation (10.43) and the imposed boundary conditions when the source field has the form (10.46).

The boundary value problem for the complex wave function w is

$$\{\partial_t^2 - \Delta\}w(x) = -g(x), \qquad x \in \Omega, \tag{10.54}$$

$$w(x) \in (bc), \tag{10.55}$$

$$\left(\frac{\partial w}{\partial |x|} - i\omega w\right)(x) = O\left(\frac{1}{|x|^2}\right) \quad \text{as } |x| \to \infty. \tag{10.56}$$

The radiation condition (10.56) ensures that the wave field characterised by u will be outgoing at infinity. Furthermore, it guarantees the uniqueness of the wave field u [154].

The field $u_0(x,t)$ generated by $f(x,t)$ in the absence of scatterers is characterised by the complex wave function w_0, which is given by

$$w_0(x) = \frac{1}{4\pi}\int_{|x'-x_0|\le\delta_0} \frac{\exp(i\omega\,|x-x'|)}{|x-x'|} g(x')\,dx', \qquad x \in \mathbf{R}^3. \tag{10.57}$$

Expanding $|x-x'|$ in powers $|x_0|$, as before, we obtain

$$w_0(x) = \frac{T(\omega\theta_0)}{|x_0|}\exp(i\omega\theta_0.x) + O\left(\frac{1}{|x_0|^2}\right) \quad \text{as } |x_0| \to \infty \tag{10.58}$$

uniformly for $|x| \le \delta$, where

$$T(\omega\theta_0) = \frac{1}{4\pi}\int_{|x'-x_0|\le\delta_0} \{\exp -(i\omega\theta_0.x')\}g(x')\,dx'. \tag{10.59}$$

With (10.53) in mind, together with the familiar forms of solutions of the d'Alembert equation, it will be convenient to express the primary wave function w_0 in the form

$$w_0(x, \omega\theta_0) = (2\pi)^{-3/2}\exp(i\omega\theta_0.x).$$

The wave field produced when $w_0(x, \omega\theta_0)$ is scattered by B is denoted $w^+(x, \omega\theta_0)$. It is taken to have the form

$$w^+(x, \omega\theta_0) = w_0(x, \omega\theta_0) + w_{sc}^+(x, \omega\theta_0)$$

and is expected to satisfy the boundary value problem

$$(\Delta + \omega^2) w^+(x, \omega\theta_0) = 0,$$

$$w^+(x, \omega\theta_0) \in (bc),$$

$$\left(\frac{\partial}{\partial |x|} - i\omega\right) w_{sc}^+(x, \omega\theta_0) = O\left(\frac{1}{|x|^2}\right) \quad \text{as } |x| \to \infty.$$

It can then be shown [154] that in the far field of B the scattered, or echo, field is a diverging spherical wave with asymptotic form

$$w_{sc}^+(x, \omega\theta_0) \approx \frac{\exp(i\omega |x|)}{4\pi |x|} T_+(\omega\theta, \omega\theta_0), \qquad x = |x|\,\theta. \tag{10.60}$$

The coefficient $T_+(\omega\theta, \omega\theta_0)$ is the scattering amplitude of B. It determines the amplitude and phase of the scattered field in the direction θ due to a primary field in the direction θ_0.

It can be shown that the echo wave profile e can be expressed in terms of known quantities in the form [154]

$$e(\theta, \theta_0, \tau) = \text{Re}\left\{ \int_0^\infty \{\exp(i\omega\tau)\} T_+(\omega\theta, \omega\theta_0) \hat{s}(\omega, \theta_0)\, d\omega \right\}, \tag{10.61}$$

where $\hat{s}$ denotes the Fourier transform of the signal waveform; that is,

$$\hat{s}(\omega, \theta_0) := \frac{1}{(2\pi)^{1/2}} \int_{-\infty}^{\infty} \{\exp(-i\omega\tau)\} s(\tau, \theta_0)\, d\tau. \tag{10.62}$$

10.4 STRATEGIES FOR ELECTROMAGNETIC SCATTERING PROBLEMS

10.4.1 Concerning Autonomous Problems

Quite simply, we follow the various steps outlined in the previous section, but now we must be careful to recognise that we will be dealing with matrix-valued coefficients, vectors and tensors [3, 115, 131].

The defining equation we must now consider is the inhomogeneous vector wave equation

$$\{\partial_t^2 + \nabla \times \nabla \times\} u(x, t) = f(x, t), \qquad (x, t) \in \Omega \times \mathbf{R}. \tag{10.63}$$

The far-field approximations can be obtained in a similar manner as in the acoustic case. The first result we can obtain is a representation of $s(\tau, \theta_0)$, the signal wave profile, in the form

$$s(\tau, \theta_0) = Q(\theta_0)\frac{1}{4\pi}\int_{|x'-x_0|\le\delta_0} f(x', \theta_0.(x'-x_0)-\tau)\,dx', \tag{10.64}$$

where Q is a tensor defined by

$$Q(\theta) = I - \theta\theta \tag{10.65}$$

and is the projection onto the plane through the origin with normal in the direction θ.

We assume that an electric current density of the form

$$J(x,t) = \text{Re}\{J(x)\exp(-i\omega t)\} \tag{10.66}$$

generates wave fields

$$u(x,t) = \text{Re}\{w(x)\exp(-i\omega t)\}, \tag{10.67}$$

where $w(x) = \langle w_1, w_2, w_3\rangle(x)$ has complex-valued components. The wave $u(x,t)$ must satisfy (10.63) with f defined by (10.66) and (10.35) and the imposed boundary and initial conditions.

In the electromagnetic case the Sommerfeld radiation conditions used in connection with acoustic problems are not adequate; they have to be replaced by the Silver-Muller radiation conditions [71]. Consequently, the boundary value problem for the vector quantity in (10.67) is

$$\begin{aligned} \{\nabla\times\nabla\times-(\omega^2)\}w(x) &= f(x), && x\in\Omega, && (10.68)\\ w(x) &\in (bc), && x\in\partial\Omega, \\ \{\theta\times\nabla\times(+i\omega)\}w(x) &= O\left(\frac{1}{|x|^2}\right), && |x|\to\infty, \end{aligned}$$

where $x = |x|\,\theta$ and

$$f(x) = i\omega J(x).$$

An application of the far-field assumptions yields the following estimate for $w_0(x)$, the incident or primary field,

$$w_0(x) = Q(\theta_0)\left\{\frac{T(\omega\theta_0)}{|x_0|}\exp(-i\omega\theta_0.x)\right\} + O\left(\frac{1}{|x_0|^2}\right), \tag{10.69}$$

where

$$T(\omega\theta_0) = \frac{1}{4\pi}\int_{|x'-x_0|\le\delta_0}\{\exp(-i\omega\theta_0.x')\}f(x')\,dx'. \tag{10.70}$$

When the error term in (10.69) is dropped, w_0 represents a plane wave represented in the form

$$w_0(x, \omega\theta_0) = \{\exp(-i\omega\theta_0.x)\}Q(\theta_0).a, \tag{10.71}$$

where a is a constant vector.

It will be convenient to introduce the matrix plane wave

$$\Psi_0(x, \omega\theta_0) = (2\pi)^{-3/2}\{\exp(-i\omega\theta_0.x)\}Q(\theta_0), \tag{10.72}$$

where the columns of Ψ_0 are plane waves of the form given in (10.71).

The electric field produced when the primary wave $\Psi_0(x, \omega\theta_0)$ is scattered by B will be denoted by $\Psi^+(x, \omega\theta_0)$. It will be assumed to have the form

$$\Psi^+(x, \omega\theta_0) = \Psi_0(x, \omega\theta_0) + \Psi^+_{sc}(x, \omega\theta), \qquad x \in \Omega,$$

and to be a solution of the vector boundary value problem

$$\{\nabla \times \nabla \times (-\omega^2)\}\Psi^+(x, \omega\theta_0) = 0, \qquad x \in \Omega,$$

$$\Psi^+(x, \omega\theta_0) \in (bc), \qquad x \in \partial\Omega,$$

$$\{\theta \times \nabla \times (+i\omega)\}\Psi^+_{sc}(x, \omega\theta_0) = O\left(\frac{1}{|x|^2}\right), \qquad |x| \to \infty,$$

where, as in the acoustic case, the plus sign indicates "outgoing wave."

It can be shown [154] that $\Psi^+_{sc}(x, \omega\theta_0)$ characterises a diverging spherical wave with asymptotic form

$$\Psi^+_{sc}(x, \omega\theta_0) \approx \frac{\exp(i\omega\,|x|)}{4\pi\,|x|} T_+(\omega\theta, \omega\theta_0), \qquad x = |x|\,\theta.$$

We would emphasise that here $T_+(\omega\theta, \omega\theta_0)$ is a matrix-valued coefficient.

With these various modifications of the acoustic case and using the same notations, it can be shown that

$$e(\theta, \theta_0, \tau) = \mathrm{Re}\left\{\int_0^\infty \{\exp(i\omega\tau)\}T_+(\omega\theta, \omega\theta_0)\,\hat{s}\,(\omega, \theta_0)\,d\omega\right\}, \tag{10.73}$$

where, as in the acoustic case, $\hat{s}$ denotes the Fourier transform of the signal profile, namely,

$$\hat{s}\,(\omega, \theta_0) := \frac{1}{(2\pi)^{1/2}} \int_{-\infty}^{\infty} \{\exp(-i\omega\tau)\}s(\tau, \theta_0)\,d\tau. \tag{10.74}$$

Although (10.73) and (10.74) have much the same symbolic form as for the acoustic case, it must be emphasised that $T_+(\omega\theta, \omega\theta_0)$ is matrix-valued, that

$\hat{s}(\omega, \theta_0)$ is vector-valued and that the order of the factors in (10.73) must be maintained.

10.4.2 Concerning Nonautonomous Problems

All the discussion so far in this chapter has been confined to APs. For NAPs we cannot follow the above route. Instead, we must return to the IVP (10.16) and investigate this IVP when $\mathbf{G}$ is replaced by $\mathbf{G}(t)$. This we do for the electromagnetic case by following the method indicated in chapter 7, but paying proper attention to the more complicated nature of the unknown $\mathbf{U}$. The treatment we have just given for the associated AP will stand as the zero approximation in the sense of chapter 7. Higher approximations are then obtained as in chapter 7. The details, whilst being straightforward enough, are nevertheless very lengthy.

10.5 SCATTERING OF ELASTIC WAVES

The strategy outlined in this monograph for investigating wave scattering problems indicates that our first task is to represent the given physical problem as an operator equation problem in a suitable Hilbert space where ideally the operator involved is self-adjoint. Bearing in mind that our intention is to investigate an NAP in a ready and constructive manner, we will require that the Hilbert space representation of the given physical problem be in the form of an IVP for a first-order differential equation.

10.5.1 Strategy for Autonomous Elastic Wave Scattering Problems

In $\mathbf{R}^3$, elastic wave phenomena are governed by IBVPs of the typical form

$$\{\partial_t^2 - \Delta^*\}\mathbf{u}(x,t) = \mathbf{f}(x,t), \qquad x \in \Omega \subseteq \mathbf{R}^3, \qquad t \in \mathbf{R}^+, \tag{10.75}$$

$$\mathbf{u}(x,0) = \boldsymbol{\varphi}(x), \qquad \mathbf{u}_t(x,0) = \boldsymbol{\psi}(x), \qquad x \in \mathbf{R}^3, \tag{10.76}$$

$$\mathbf{u}(x,t) \in (bc), \qquad (x,t) \in \partial\Omega \times \mathbf{R}^+, \tag{10.77}$$

where
$x = (x_1, x_2, x_3)$
$\boldsymbol{\varphi}, \boldsymbol{\psi}$ = given vector functions characterising initial conditions
$\Delta^* = -\mu \operatorname{curl}\operatorname{curl} + \lambda \operatorname{grad}\operatorname{div}$ is the Lamé operator
λ, μ = Lamé constants
$\Omega \subset \mathbf{R}^3$ = connected open region exterior to the scattering target B
$\partial\Omega$ = closed smooth boundary of Ω
$\mathbf{u}(x,t)$ = vector quantity characterising the elastic wave field $= \langle u_1, u_2, u_3 \rangle(x,t)$
$\mathbf{f}(x,t)$ = vector quantity characterising the signal emitted by the transmitter.

Other quantities of interest are

$$
\begin{aligned}
c_1 &= \sqrt{\lambda+2\mu} = \text{longitudinal wave speed}\\
c_2 &= \sqrt{\mu} = \text{shear wave speed}\\
\mathbf{T}(\mathbf{u},\mathbf{n}) &= 2\mu(\partial\mathbf{u}/\partial\mathbf{n}) + \lambda\mathbf{n}\,\operatorname{div}\mathbf{u} + \mathbf{n}\times\operatorname{curl}\mathbf{u} \qquad (10.78)\\
&= \text{vector traction at a boundary point where the normal is } \mathbf{n}.
\end{aligned}
$$

For $\mathbf{u}(x,t) = \langle u_1, u_2, u_3\rangle(x,t)$, we shall write

$$\mathbf{u} \in (L_2(\mathbf{R}^3))^3 \quad \text{whenever } u_j \in L_2(\mathbf{R}^3),\ j=1,2,3.$$

As in the acoustic and electromagnetic cases, we begin an investigation of elastic waves by examining the FP. To this end, we introduce the following notations and function spaces.

$X(\Omega)$ = space of scalar functions defined on a region Ω.

$\mathbf{X}(\Omega)$ = space of vector functions defined on a region Ω.

For example, if $\Omega \subseteq \mathbf{R}^3$ and

$$
\begin{aligned}
X(\Omega) &= L_2(\Omega) := \left\{u : \int_\Omega |u(x)|^2\,dx < \infty\right\}\\
\mathbf{X}(\Omega) &= \mathbf{L}_2(\Omega) := \{\mathbf{u} = \langle u_1, u_2, u_3\rangle : u_j \in L_2(\Omega),\ j=1,2,3\}\\
&= \{\mathbf{u} : \mathbf{u} \in (L_2(\Omega))^3\}\\
&= (X(\Omega))^3,
\end{aligned}
$$

then $L_2(\Omega)$ is a Hilbert space with inner product

$$(u,v) = \int_\Omega u(x)\overline{v(x)}\,dx \qquad (10.79)$$

and $\mathbf{L}_2(\Omega)$ is a Hilbert space with inner product

$$(\mathbf{u},\mathbf{v}) = (u_1,v_1) + (u_2,v_2) + (u_3,v_3). \qquad (10.80)$$

We also introduce

$$\mathbf{L}_2(\Delta^*,\Omega) := \mathbf{L}_2(\Omega) \cap \{\mathbf{u} : \Delta^*\mathbf{u} \in \mathbf{L}_2(\Omega)\}. \qquad (10.81)$$

$\mathbf{L}_2(\Delta^*,\Omega)$ is a Hilbert space with inner product

$$(\mathbf{u},\mathbf{v})_{\Delta^*} := (\mathbf{u},\mathbf{v}) + (\Delta^*\mathbf{u}, \Delta^*\mathbf{v}). \qquad (10.82)$$

Following the approach adopted in earlier chapters, the FP associated with the IBVP, that is, the problem (10.75), (10.76), can now be interpreted as an IVP for a second-order ordinary differential equation in $\mathbf{L}_2(\mathbf{R}^3)$. Specifically, we introduce the operator

$$A_0 : \mathbf{L}_2(\mathbf{R}^3) \to \mathbf{L}_2(\mathbf{R}^3), \tag{10.83}$$
$$A_0\mathbf{u} = -\Delta^*\mathbf{u}, \qquad \mathbf{u} \in D(A_0),$$
$$D(A_0) = \{\mathbf{u} \in \mathbf{L}_2(\mathbf{R}^3) : \Delta^*\mathbf{u} \in \mathbf{L}_2(\mathbf{R}^3)\} \equiv \mathbf{L}_2(\Delta^*, \mathbf{R}^3).$$

This then yields the IVP

$$\left\{\frac{d^2}{dt^2} + A_0\right\}\mathbf{u}(t) = \mathbf{f}(t), \tag{10.84}$$

$$\mathbf{u}(0) = \boldsymbol{\varphi}, \qquad \mathbf{u}_t(0) = \boldsymbol{\psi}, \tag{10.85}$$

where we understand

$$\mathbf{u} \equiv \mathbf{u}(.,.) : t \to \mathbf{u}(.,t) =: \mathbf{u}(t).$$

It is readily seen that A_0 is positive self-adjoint on $\mathbf{L}_2(\Delta^*, \mathbf{R}^3)$ provided λ and μ are strictly positive. Questions about the existence and uniqueness of the solution to the IVP (10.84), (10.85) can be settled by using the limiting absorption principle (see chapter 11). The required solution $\mathbf{u}$ can then be obtained in the form

$$\mathbf{u}(x,t) = \mathrm{Re}\{\mathbf{v}(x,t)\},$$
$$\mathbf{v}(x,t) = \{\exp(-itA^{1/2})\}\mathbf{h}(x),$$

where

$$\mathbf{h}(x) = \boldsymbol{\varphi}(x) + iA^{-1/2}\boldsymbol{\psi}(x)$$

when $\mathbf{f} \equiv 0$; otherwise, $\mathbf{h}$ has to have an additional integral term in its definition.

Since A is self-adjoint, the spectral theorem can be used to interpret terms such as $A^{1/2}$. Generalised eigenfunction expansions can then be established to provide a basis for constructive methods.

However, this approach is not suitable when we come to deal with NAPs. As we have seen in the acoustic and electromagnetic cases, this difficulty can be overcome by reducing the problem (10.84), (10.85) to an equivalent first-order system. The required reduction is obtained in the now familiar manner and yields the IVP

$$\boldsymbol{\Psi}_t(t) + i\mathbf{G}\boldsymbol{\Psi}(t) = \mathbf{F}(t), \qquad \boldsymbol{\Psi}(0) = \boldsymbol{\Psi}_0, \tag{10.86}$$

where

$\boldsymbol{\Psi}(t) = \langle \mathbf{u}, \mathbf{u}_t\rangle(t), \qquad \boldsymbol{\Psi}(0) = \boldsymbol{\Psi}_0 = \langle \boldsymbol{\varphi}, \boldsymbol{\psi}\rangle,$

$\mathbf{F}(t) = \langle 0, \mathbf{f}\rangle(t),$

$$i\mathbf{G} = \begin{bmatrix} 0 & -I \\ A & 0 \end{bmatrix}.$$

Proceeding formally, a simple integrating factor technique indicates that

$$\boldsymbol{\Psi}(t) = \{\exp(-it\mathbf{G})\}\boldsymbol{\Psi}_0 + \int_0^t \{\exp(-i(t-\eta)\mathbf{G}\}\mathbf{F}(\eta)\,d\eta. \tag{10.87}$$

For this approach to be meaningful, we must be able to show that the problem (10.86) is well posed and that the relation (10.87) is well defined. As we have seen in the previous chapters, all this can be settled by showing that $i\mathbf{G}$ is self-adjoint; Stone's theorem will then indicate that $\exp\{-it\mathbf{G}\}$ is a well-defined semigroup, and the result of chapter 5 will indicate when (10.86) is a well-posed problem.

To show that $i\mathbf{G}$ is self-adjoint, we consider $i\mathbf{G}$ to be defined on an "energy space" $\mathbf{H}_E$ that is a Hilbert space with inner product

$$(\boldsymbol{\Psi}, \boldsymbol{\Phi})_E := (\nabla \times \boldsymbol{\psi}_1, \nabla \times \boldsymbol{\varphi}_1)_{\mathbf{L}_2} + (\boldsymbol{\psi}_2, \boldsymbol{\varphi}_2)_{\Delta^*},$$

where

$$\boldsymbol{\Psi} = \langle \boldsymbol{\psi}_1, \boldsymbol{\psi}_2 \rangle, \qquad \boldsymbol{\Phi} = \langle \boldsymbol{\varphi}_1, \boldsymbol{\varphi}_2 \rangle.$$

A straightforward calculation establishes that $i\mathbf{G}$ is symmetric. The full proof that $i\mathbf{G}$ is self-adjoint then follows as in chapters 5 and 6 and in [105].

The required solution $\mathbf{u}(x,t)$ of the IVP (10.75), (10.76) is then meaningfully defined by the first component of the solution $\boldsymbol{\Psi}$ of the IVP (10.86). For example, in the case when $\mathbf{f} \equiv \mathbf{0}$, we obtain the solution form

$$\mathbf{u}(x,t) = (\cos tA^{1/2})\boldsymbol{\varphi}(x) + A^{-1/2}(\sin tA^{1/2})\boldsymbol{\psi}(x). \tag{10.88}$$

The self-adjointness of A ensures that the spectral theorem is available for the interpretation of such terms as $A^{1/2}$. Consequently, (10.88) is well defined and computable. As in the acoustic and electromagnetic cases, the computation of the elastic wave field $\mathbf{u}(x,t)$ is carried out using results of generalised eigenfunction expansion theorems. These theorems, once established, are essentially of two types, one for longitudinal wave phenomena, the other to accommodate shear wave phenomena.

When we deal with perturbed problems, for example, with target scattering problems, the IBVP (10.75)–(10.77) has to be investigated. We remark that the accommodation of the effects of boundary conditions can possibly cause technical difficulties when the self-adjointness of associated operators has to be established. Nevertheless, we have now arrived at the stage when we have managed to give elastic wave scattering problems the same symbolic form that we investigated when dealing with acoustic and electromagnetic wave scattering problems. Consequently, we are now in the position of being able to follow step by step the procedures outlined in chapters 5–7 for acoustic and electromagnetic problems. However, although this is a straightforward matter, it is an extremely lengthy process. It has been worked through in detail by a number of authors for APs, and the details are to be found in the references cited in chapter 11. The manner in which the associated NAPs can be studied is as outlined in subsection 10.4.2 for NA electromagnetic problems.

Chapter Eleven

Commentaries and Appendices

11.1 REMARKS ON PREVIOUS CHAPTERS

From the outset it has been emphasised that this book is an introductory text intended for the use of those wishing to begin studying the scattering of waves by time-dependent perturbations. For this reason we offer in this chapter some additional remarks on the material that has been presented in previous chapters. The main intentions are, on the one hand, to give some indications of the work that either has been or is being done for more general situations than those considered here and, on the other hand, to suggest further reading directions. Whilst it is recognised that it is impossible to give credit to all sources, nevertheless, those cited in the extended bibliography provided here will in turn give additional references. For the sake of completeness, the commentary on each chapter begins with a listing of the basic references cited for it earlier.

Chapter 1: [44, 54, 62, 68, 71, 76, 78, 105, 144, 153, 154].

As its title suggests, this chapter is purely introductory. The various aspects of scattering theory that will be needed in later chapters are illustrated in an entirely formal manner. Most of the notions introduced seem to have appeared initially in the theory of quantum mechanical scattering theory. In this connection see [6, 30, 68, 90, 95, 101]. In a series of papers Wilcox and his collaborators showed how many of the techniques used in the study of quantum mechanical scattering could be extended to deal with wave problems in classical physics. The foundations for this work are fully discussed in [154]. The Wilcox approach to wave scattering problems relies on the availability of suitable generalised eigenfunction expansion theorems [54, 126] and offers good prospects for developing constructive methods of solution. This approach is distinct from that adopted by Lax and Phillips [68]; a reconciliation of the two approaches is given in [67, 153]. Although mainly concerned with quantum mechanical scattering aspects, the texts [12, 17] are worth bearing in mind when developing detailed analyses of wave scattering problems.

Chapter 2: [11, 61, 65, 101, 105, 138, 144, 149, 154].

Wave motion on strings is developed in many texts. We would particularly mention [11, 138, 149]. The approach to solutions of the wave equation by considering an equivalent first-order system is discussed from the standpoint of semigroup theory in [49, 83]. The investigation of solutions to the wave equation represented in the form (2.58) is detailed in [71, 105, 154]. The method of comparing solutions that have the typical form given in (2.58) parallels that used so successfully in quantum scattering theory. In this connection we would particularly mention the texts [6, 12, 17]. A comprehensive account of waves on strings can be found in [69].

The discussion of a scattering problem on a semi-infinite string follows the treatment found in [61].

Chapter 3: [4, 7, 31, 53, 63, 102, 104, 105, 114, 129, 144].

In this chapter a number of mathematical facts used frequently in this monograph are gathered together. The material is included mainly for the newcomer to the area of scattering theory who might possibly not have had the training in mathematical analysis that present-day mathematics students receive. There are a number of fine texts available that provide a thorough development of the various topics introduced in this chapter; see [53, 57, 59, 63, 75, 156]. This being said, particular attention should be paid to the following topics. The notion of completeness is crucial for many of the arguments used in developing scattering theories. A good account of this concept can be found in [63], whilst fine illustrations of its use in practical situations can be found in [87, 138]. Distribution theory is comprehensively developed in [43, 59, 74, 75, 125]. The newcomer to this area should be encouraged to become familiar with distribution theory developed in $\mathbf{R}^n$, $n > 1$, and especially with the notion of the n-dimensional Dirac delta [104]. The theory of linear operators on Hilbert spaces is comprehensively treated in [4, 53].

Chapter 4: [4, 53, 59, 95, 102, 104, 105].

Spectral theory is a highly developed topic that has many far-reaching applications. In this chapter we have outlined the principal notions required when developing a scattering theory either in this or in more advanced texts. Comprehensive accounts of spectral theory from a number of different standpoints are available. In this connection we would particularly mention [4, 12, 32, 34, 41, 53, 59, 63, 77, 95, 101, 102, 110, 122, 156], amongst which will be found something to suit most tastes. The material in this chapter is quite standard, but, nevertheless, we would make the following remarks. In the presentation given here the development has been mainly concerned with bounded linear operators simply for ease of presentation. Similar results can be obtained for unbounded operators, but in this case more care is required when dealing with domains of operators. This aspect is discussed in detail in [53] and [102], the latter giving a fine account of reducing subspaces.

A comprehensive account of measure theory with applications to scattering theory very much in mind can be found in [95]. The account of spectral decomposition of Hilbert spaces given in [95] can usefully be augmented by the associated material given in [59] and [156].

Chapter 5: [5, 49, 50, 59, 83, 86, 94, 102, 104, 105, 130, 141, 142, 145].

In this chapter it has been shown how a number of the techniques used successfully when dealing with quantum mechanical scattering problems can be adjusted to deal with wave scattering problems. Essentially, this amounted to replacing the partial differential equation associated with an IBVP for the wave equation by an equivalent IVP for an ordinary differential equation defined in a suitable energy space H (say) that will have solutions that are H-valued functions of t. Indications were given of how existence and uniqueness results for solutions of the IVP could be obtained using results from semigroup theory. Suggestions were also given as to how constructive methods could be developed using results from the abstract theory of integral equations.

One of the earliest accounts of semigroup theory can be found in [137]. A fine introduction to the subject is given in [83]. More advanced modern texts that will be useful are [13, 14, 49, 94, 142].

Chapter 6: [1, 6, 11, 35, 36, 49, 54, 59, 68, 71, 94, 95, 101, 105, 123, 142, 153, 154].

The modelling of acoustic wave phenomena leading to the IBVPs considered in this chapter is introduced and discussed in [55] and [127]. The material presented in the first part of this chapter is based very much on the work described in [154]. Some of the more important results and concepts are gathered together here in order that the newcomer to the area can gain familiarity, as quickly as possible, with the various strategies involved when developing scattering theories. Many of the results are simply stated, as their proof is usually quite lengthy. However, as the various proofs are given in the literature cited here, it is felt that they can be read more profitably once the overall strategy of the subject has been appreciated. Furthermore, in many cases it will then be readily seen how these proofs should be modified to accommodate other problems than the relatively simple acoustic problems considered here; in this connection see, for instance, [71, 72, 108, 109, 110]. A considerable amount of work has been done, in a quantum mechanics setting, on the existence, uniqueness and completeness of wave operators; see [6, 12, 17, 90, 101]. A comprehensive and unified account is given in [95]. A detailed account of how these various notions can be adapted for a target scattering problem is to be found in [154] and the references cited there. A full account of radiation conditions and incoming and outgoing waves can be found in [35, 36, 147], whilst a treatment of generalised eigenfunction theorems and completeness in this context can be found in [6, 11, 87]. The various types of solutions that can be obtained for the partial differential equations that arise when analysing scattering processes are introduced and discussed in texts dealing with modern theories of partial differential equations; see, for example, [34, 43, 71, 155]. The notion of solutions with finite energy has been used extensively [154]. Methods for determining and discussing such solutions can be readily developed by reducing IBVPs for wave equations to equivalent IVPs for an ordinary differential equation and then using semigroup theory [49, 83, 105, 142].

Central to development of the generalised eigenfunction expansion theorems required when dealing with acoustic scattering problems is an appreciation of the Helmholtz equation and its properties. A comprehensive treatment of boundary value problems for the Helmholtz equation using integral equation methods is given in [62].

The limiting absorption principle was introduced in [35]. It has been applied to problems involving a variety of different differential expressions, boundary conditions and domains. Recently, in conjunction with the related limiting amplitude principle [36], it has been successfully used to analyse scattering problems in domains involving unbounded interfaces [108, 109, 110, 111].

The second part of this chapter indicates how the various notions and strategies introduced when dealing with APs can be modified to deal with NAPs. In this connection see [49, 107, 123, 142] and the references cited there.

Chapter 7: [5, 59, 105, 107, 123, 126, 130, 142, 154].

An understanding of how a given incident wave evolves throughout a medium is the central problem when developing a scattering theory. In this chapter explicit attention has been paid to the nature of the signal transmitted [105] and the forms that it can adopt at large distances from the transmitter and scatterers. This then enables the notions of signal and echo waveforms to be introduced [154]. An alternative method to that adopted in this chapter for obtaining such quantities relies on the properties of retarded potentials. This is outlined in appendix A5 of this chapter.

The final two sections of the chapter indicate how the results obtained and the strategies used when dealing with APs can be used as a basis from which to generate approximations to solutions of related NAPs.

Chapter 8: [5, 27, 61, 68, 142, 148, 154].

An alternative method for discussing similar problems to those appearing in this chapter has been developed in [26, 99]. In this approach a local evolution is defined as a product of the propagator of the problem and of projection operators onto the incoming and outgoing subspaces used in the Lax-Phillips scattering theory [68]. A detailed analysis from this standpoint of Neumann and Robin boundary value problems can be found in [28]. The abstract theory in this connection is developed in [27].

Chapter 9: [6, 40, 54, 88, 90, 93, 96, 97, 101, 105, 116, 117, 118, 119, 123, 140, 154].

The inverse scattering problem for APs has been considered by many authors. In this connection particular reference can be made to [20, 21, 70, 81, 82]. As might be expected, there are also many papers in journals devoted to this area. Typical examples of works that cover many aspects of this general area are [9, 10, 39, 91, 92, 112, 113].

For NAPs an alternative method to that outlined in this chapter for recovering a time-dependent potential term in the plasma wave equation is given in [132, 133, 134]. This work relies on the notion of a generalised scattering kernel introduced in [26].

Chapter 10: [3, 68, 71, 87, 105, 115, 131, 154].

As has been pointed out, working through the details for APs in the electromagnetic and the elastic cases, which potentially might be considered a straightforward matter, is nevertheless an extremely lengthy process. This fact is well illustrated in [8, 55, 71, 73] and in [79, 80], respectively. However, since the methods outlined in this monograph for tackling NAPs rely on a sound knowledge of the corresponding AP, the details given in these references will provide in principle a starting point for studying the associated NAPs in a similar manner to that outlined here for acoustic wave problems.

11.2 APPENDICES

This section is included to provide an easy and convenient reference to some of the more technical concepts that have been referred to but not developed in this monograph. The presentation is brief and is made almost entirely in spaces of one dimension. More details can be found in standard books on mathematical analysis (e.g., [63, 114]).

A1: Limits and Continuity

If B is an interval in $\mathbf{R}$, then we write $B \subseteq \mathbf{R}$.
If $B = (a, b) = \{x \in \mathbf{R} : a < x < b\}$, then B is said to be an *open* interval.
If $B = [a, b] = \{x \in \mathbf{R} : a \leq x \leq b\}$, then B is said to be a *closed* interval, which we denote by $\overline{B}$. It follows that

$$\overline{B} = B \cup \{a, b\},$$

where ∂B denotes the boundary of B.

The notions of half-closed intervals such as $[a, b)$ and $(a, b]$ can be introduced similarly.

A real-valued function f defined on $\mathbf{R}$, which symbolically is characterised by writing $f : \mathbf{R} \to \mathbf{R}$, is said to have a *limit* L as $x \to a \in \mathbf{R}$ if for any real number $\varepsilon > 0$, no matter how small, there is a $\delta > 0$ such that

$$|f(x) - L| < \varepsilon \quad \text{whenever } 0 < |x - a| < \delta,$$

and we write in this case

$$\lim_{x \to a} f(x) = L.$$

An equivalent definition in terms of sequences is available. Specifically, if $\{x_n\}$ is a sequence such that $x_n \to a$ as $n \to \infty$, then it follows that $f(x_n) \to L$.

In order that a function $f : \mathbf{R} \to \mathbf{R}$ have a limit at $x = a$, the function must be defined on each side of $x = a$ but not necessarily *at* $x = a$. For example, the function f defined by

$$f(x) = \frac{\sin x}{x}$$

is not defined at $x = 0$ since there it assumes the form $0/0$. However, on expanding the numerator in series form, we see that it has the value unity at $x = 0$.

It is also possible to introduce the notion of one-sided limits. A function $f : \mathbf{R} \to \mathbf{R}$ is said to have a *limit* L *from the right* at $x = a$ as $x \to a$ if for any number $\varepsilon > 0$ there exists a number $\delta > 0$ such that

$$|f(x) - L| < \varepsilon \quad \text{whenever } 0 < x - a < \delta.$$

When this is the case, we replace L by $f(x^+)$.

Similarly, a function $f : \mathbf{R} \to \mathbf{R}$ is said to have a *limit* L *from the left* at $x = a$ as $x \to a$ if for any number $\varepsilon > 0$ there exists a number $\delta > 0$ such that

$$|f(x) - L| < \varepsilon \quad \text{whenever } 0 < a - x < \delta.$$

When this is the case, we replace L by $f(x^-)$.

Clearly, if the limits from both right and left exist *and* if $f(x^+) = f(x^-)$, then $\lim_{x\to a} f(x)$ exists and equals $f(x^+) = f(x^-)$.

A function $f : \mathbf{R} \to \mathbf{R}$ is *continuous* at a point $x = a \in \mathbf{R}$ if $\lim_{x\to a} f(x)$ exists and equals $f(a)$. In particular, the function has to be defined at $x = a$. For example, the function f defined by $f(x) = (\sin x)/x$ and $f(0) = 1$ is continuous at every point. A function is said to be *continuous in an interval* $a \leq x \leq b$ if it is continuous at each point in the interval.

The following are important results in applications.

Theorem A1: ***Intermediate Value Theorem*** *If $f : \mathbf{R} \to \mathbf{R}$ is continuous in a finite (i.e., bounded) interval $[a, b]$ and if $f(a) < p < f(b)$, there exists at least one point $x = c \in [a, b]$ such that $f(c) = p$.*

Theorem A2: ***Concerning Maxima*** *If $f : \mathbf{R} \to \mathbf{R}$ is continuous in a finite closed interval $[a, b]$, then it has a maximum in that interval. That is, there exists a point $m \in [a, b]$ such that $f(x) \leq f(m)$ for all $x \in [a, b]$.*

Applying this result to the function $(-f)$ indicates that f also has a minimum.

Theorem A3: *Let $f : \mathbf{R} \to \mathbf{R}$ and assume*
(i) f is continuous on a finite closed interval $[a, b]$,
(ii) $f(x) \geq 0,\ x \in [a, b]$,
(iii) $\int_a^b f(\eta)\, d\eta = 0$.
Then $f(x) \equiv 0, x \in [a, b]$.

A function $f : \mathbf{R} \to \mathbf{R}$ has a jump discontinuity if both one-sided limits $f(x^+)$ and $f(x^-)$ exist. The number $(f(x^+) - f(x^-))$ is the value of the jump.

A function $f : \mathbf{R} \to \mathbf{R}$ is piecewise-continuous on a finite closed interval $[a, b]$ if there is a finite number of points

$$a = a_0 \leq a_1 \leq a_2 \leq \cdots \leq a_n = b$$

such that f is continuous on each subinterval (a_{j-1}, a_j), $j = 1, 2, \ldots, n$, and all the one-sided limits $f(x^-)$ for $1 \leq j \leq n$ and $f(x^+)$ for $0 \leq j \leq n$ exist. Such a function has jumps at a finite number of points but otherwise is continuous.

It can be shown [114] that every piecewise-continuous function is integrable. Similar results can be shown to hold in $\mathbf{R}^n$ [102, 114].

A2: Differentiability

A function $f : \mathbf{R} \to \mathbf{R}$ is *differentiable* at a point $x \in [a, b]$ if

$$\lim_{x\to a} \frac{|f(x) - f(a)|}{(x - a)}$$

exists. The value of the limit is denoted by either $f'(a)$ or $(df/dx)(a)$.

A function $f : \mathbf{R} \to \mathbf{R}$ is differentiable in the open interval (b, c) if it is differentiable at each point of the interval.

When we work in $n > 1$ dimensions, some rather more sensitive notation is required. We illustrate this in the following paragraphs.

A3: The Function Classes $C^m(B)$, $C^m(\overline{B})$

Let $B \subset \mathbf{R}^n$, $n \geq 1$, and let $f : \mathbf{R}^n \to \mathbf{R}^n$. We shall assume

(i) $\alpha = (\alpha_1, \alpha_2, \ldots, \alpha_n)$ is a vector with non-negative integer components α_j, $j = 1, 2, \ldots, n$,
(ii) $|\alpha| = \sum_{j=1}^n \alpha_j$,
(iii) $x^\alpha = x_1^{\alpha_1} x_2^{\alpha_2} \ldots x_n^{\alpha_n}$.

We shall denote by $D^\alpha f(x)$ the derivative of order $|\alpha|$ of the function f by

$$D^\alpha f(x) = \frac{\partial^{|\alpha|} f(x_1, x_2, \ldots, x_n)}{\partial_{x_1}^{\alpha_1} \partial_{x_2}^{\alpha_2} \ldots \partial_{x_n}^{\alpha_n}}$$
$$= D_1^{\alpha_1} D_2^{\alpha_2} \ldots D_n^{\alpha_n}, \quad D_j^{\alpha_m} = \frac{\partial^{\alpha_m}}{\partial x_j^{\alpha_m}}, \quad D_j = \frac{\partial}{\partial x_j}, \quad j = 1, 2, \ldots, n,$$

with

$$D^0 f(x) = f(x).$$

A set of (complex-valued) functions f on $B \subset \mathbf{R}^n$ that are continuous together with their derivatives $D^\alpha f(x)$, $|\alpha| \leq p$, where $1 \leq p < \infty$, form a *class of functions* denoted $C^p(B)$.

Functions $f \in C^p(B)$ for which all the derivatives $D^\alpha f(x)$ with $|\alpha| \leq p$ allow continuous extension into the closure $\overline{B}$ form a class of functions $C^p(\overline{B})$. We shall also write

$$C^\infty(B) = \bigcap_{p \geq 0} C^p(B), \qquad C^\infty(\overline{B}) = \bigcap_{p \geq 0} C^p(\overline{B}).$$

These classes of functions are linear sets. Furthermore, if, for example, we endow the class $C(B)$ with a norm by setting

$$\|f\| := \max_{x \in B} |f(x)|,$$

then $C(B)$ is converted into a normed linear space. Similarly, we can convert the class $C(\overline{B})$.

A set of functions $f \in M \subset C(B)$ is said to be *equicontinuous* on B if for any $\varepsilon > 0$ there exists a number $\delta(\varepsilon)$ such that for all $f \in M$ the inequality $|f(x_1) - f(x_2)| \leq \varepsilon$ holds whenever $|x_1 - x_2| < \delta(\varepsilon)$, where $x_1, x_2 \in B$.

A function $f \in C(B)$ is said to be *Hölder-continuous* on B if there are numbers $c > 0$ and $0 < \alpha \leq 1$ such that for all $x_1, x_2 \in B$ the inequality

$$|f(x_1) - f(x_2)| \leq c\,|x_1 - x_2|^\alpha$$

holds. In the case when $\alpha = 1$, the function $f \in C(B)$ is said to be Lipschitz-continuous on B.

A4: Sobolev Spaces

Let $\Omega \subseteq \mathbf{R}^n$ be an open set. The Sobolev space $W_p^m(\Omega)$, $m \in \mathbf{N}$ and $0 \leq p \leq \alpha$, is the space of all distributions that together with their distributional derivatives of order less than or equal to m are associated with functions that are pth-power-integrable on Ω; that is, they are elements of $L_p(\Omega)$. Such a collection is a Banach space with respect to the norm $\|.\|_{m,p}$ defined by

$$\|f\|_{m,p} := \left\{ \int_\Omega \sum_{0 \leq |m| \leq m} \left| \frac{\partial^{|m|} f}{\partial x_1^{m_1} \partial x_2^{m_2} \ldots \partial x_n^{m_n}} \right|^p \right\}^{1/p},$$

where $|m| = m_1 + m_2 + \cdots + m_n$ and m_j, $j = 1, 2, \ldots, n$, are positive integers.

If $p = 2$, then $W_2^m(\Omega)$ is a Hilbert space usually denoted $H^2(\Omega)$.

Of particular interest in this monograph are the Hilbert spaces $H^1(\Omega)$ and $H^2(\Omega)$. These are defined as follows.

$$H^2(\Omega) := \left\{ f \in L_2(\Omega) : \frac{\partial f(x)}{\partial x_j} \in L_2(\Omega),\ j = 1, 2, \ldots, n \right\}$$

with inner product

$$(f, g)_{1,2} = (f, g)_1 = (f, g) + (\nabla f, \nabla g)$$

and norm

$$\|f\|_1^2 = \|f\|^2 + \|\nabla f\|^2,$$

where $(.,.)$ and $\|.\|^2$ denote the usual $L_2(\Omega)$ inner product and norm, respectively.

Similarly,

$$H_2(\Omega) = \left\{ f \in L_2(\Omega) : \frac{\partial f(x)}{\partial x_j} \in L_2(\Omega),\ \frac{\partial^2 f(x)}{\partial x_j \partial x_k} \in L_2(\Omega),\ j, k = 1, 2, \ldots, n \right\}$$

with inner product and norm

$$(f, g)_{2,2} = (f, g) + (\nabla f, \nabla g) + (D^\alpha f, D^\alpha g),$$

$$\|f\|_2^2 = \|f\|^2 + \|\nabla f\|^2 + \|D^\alpha f\|^2,$$

where α is a multi-index such that $|\alpha| = \alpha_1 + \alpha_2$.

An important result in the general theory of Sobolev spaces [1] is the celebrated *Sobolev lemma*, which states that an element $f \in H^{[n/2]+1+k}(\Omega)$, where n is the dimension of Ω and $[n/2]$ means rounded down to an integer, is also such that $f \in C^k(\Omega)$. Thus as the dimension n of Ω increases, higher-order Sobolev spaces must be taken in order to guarantee the continuity of the elements (functions) that they contain.

For more details of the properties of Sobolev spaces, especially concerning boundary values (traces) and embedding theorems, see [1].

A5: Retarded Potentials

Let f and g be locally integrable functions on $\mathbf{R}^n$ and assume that the function h defined by

$$h(x) = \int_{\mathbf{R}^n} |g(y) f(x-y)| \, dy$$

is also locally integrable on $\mathbf{R}^n$. The function $f * g$ defined by

$$(f*g)(x) = \int_{\mathbf{R}^n} f(y) g(x-y) \, dy$$

$$= \int_{\mathbf{R}^n} g(y) f(x-y) \, dy = (g * f)(x)$$

is called the *convolution* of the functions f and g.

Let $L(D)$ be a differential expression with constant coefficients $a_\alpha(x) \equiv a_\alpha$ that has the typical form

$$L(D) = \sum_{|a_\alpha|=0}^{m} a_\alpha D^\alpha,$$

where α is a multi-index (see appendix A4).

A generalised function $\mathcal{E} \in \mathcal{D}'$ that satisfies in $\mathbf{R}^n$ the equation

$$L(D)\mathcal{E}(x) = \delta(x)$$

is called a *fundamental solution* of the differential expression. It should be noted that in general a fundamental solution is not unique.

Using a fundamental solution $\mathcal{E}(x)$ of the differential expression $L(D)$, it is possible to construct a solution $u(x)$ of the equation

$$L(D)u(x) = f(x),$$

where f is an arbitrary function. In this connection the following theorem is a central result.

Theorem A4 *([43, 102]) Let $f \in \mathcal{D}'$ be such that the convolution $\mathcal{E}*f$ exists in $\mathcal{D}'$. Then the solution $u(x)$ exists in $\mathcal{D}'$and is given by*

$$u(x) = (\mathcal{E}*f)(x).$$

Moreover, this solution is unique in the class of generalised functions in $\mathcal{D}'$ for which a convolution with $\mathcal{E}$ exists.

The above notions are very useful when we deal with wave equation problems. In the particular case of acoustic waves the wave equation has the form

$$\Box_a u(x,t) := \{\partial_t^2 - a^2 \Delta\} u(x,t) = f(x,t),$$

where $\Box_a$ is referred to as the d'Alembert expression (operator).

A fundamental solution of the acoustic wave equation in $\mathbf{R}^n$ is denoted by $\mathcal{E}_n(x,t)$ and satisfies

$$\Box_a \mathcal{E}_n(x,t) = \delta(x,t).$$

Fourier transform techniques provide solutions of this equation in the form [43, 102, 128]

$$\mathcal{E}_1(x,t) = \frac{1}{2a}\theta(at - |x|),$$

$$\mathcal{E}_2(x,t) = \frac{\theta(at-|x|)}{2\pi a\sqrt{a^2t^2 - |x|^2}},$$

$$\mathcal{E}_3(x,t) = \frac{\theta(t)}{2\pi a}\delta(a^2t^2 - |x|^2),$$

where θ denotes the Heaviside unit function defined by

$$\theta(x) = 1 \quad \text{for } x \geq 0, \qquad \theta(x) = 0 \quad \text{for } x < 0.$$

The generalised function V_n defined by

$$V_n(x,t) = \mathcal{E}_n(x,t) * f(x,t),$$

where $\mathcal{E}_n(x,t)$ is a fundamental solution of the d'Alembert equation and f is a generalised function on $\mathbf{R}^{n+1}$ that vanishes on the half-space $t<0$ and is called a *retarded potential* with density f. (With a slight abuse of notation we might write $f(x,t) \in \mathcal{D}'(\mathbf{R}^{n+1})$.)

The retarded potential V_n will, according to the above theorem, satisfy the equation

$$\Box_a V_n(x,t) = f(x,t).$$

It can be shown [43, 128] that if f is a locally integrable function on $\mathbf{R}^{n+1}$, then V_n is a locally integrable function on $\mathbf{R}^{n+1}$ and assumes the following particular forms

$$V_1(x,t) = \frac{1}{2a}\int_0^t \int_{x-a(t-\tau)}^{x+a(t-\tau)} f(\xi,\tau)\, d\xi\, d\tau,$$

$$V_2(x,t) = \frac{1}{2\pi a}\int_0^t \int_{S(x;a(t-\tau))}^{x+a(t+\tau)} \frac{f(\xi,\tau)}{\{a^2(t-\tau)^2 - |x-\xi|^2\}^{1/2}}\, d\xi\, d\tau,$$

$$V_3(x,t) = \frac{1}{4\pi a^2}\int_{U(x;at)} \frac{f(\xi, t - |x-\xi|/a)}{|x-\xi|}\, d\xi,$$

where
$U(x;at) =$ ball centre radius at,
$S(x,at) =$ surface of$U(x;at)$.

As an illustration of the use of retarded potentials, recall [104] that the generalised Cauchy problem for the wave equation is to determine $u \in \mathcal{D}'(\mathbf{R}^{n+1})$, which is zero for $t < 0$ and which satisfies

$$\Box_a u(x,t) := \{\partial_t^2 - a^2\Delta\}u(x,t) = f(x,t), \quad u(x,0) = \varphi(x), \quad u_t(x,0) = \psi(x),$$

where $f \in \mathcal{D}'(\mathbf{R}^{n+1})$ with data $\varphi \in \mathcal{D}'(\mathbf{R}^n)$ and $\psi \in \mathcal{D}'(\mathbf{R}^n)$. It can be shown [155, 104] that this generalised Cauchy problem has a unique solution of the form

$$u(x,t) = V_n(x,t) + V_n^{(0)}(x,t) + V_n^{(1)}(x,t),$$

where

$$V_n(x,t) = \mathcal{E}_n(x,t) * f(x,t), \quad V_n^{(0)}(x,t) = \mathcal{E}_n(x,t) * \psi(x),$$
$$V_n^{(1)}(x,t) = (\mathcal{E}_n(x,t))_t * \psi(x).$$

This leads to the following classical solutions of the Cauchy problem for the wave equation [104].

$n = 3$ (*Kirchhoff's formula*):

$$u(x,t) = \frac{1}{4\pi a^2}\int_{U(x;at)} \frac{f(\xi, t - |x-\xi|\,a)}{|x-\xi|}\,d\xi + \frac{1}{4\pi a^2 t}\int_{S(x;at)} \psi(\xi)\,dS$$
$$+ \frac{1}{4\pi a^2}\frac{\partial}{\partial_t}\left\{\frac{1}{t}\int_{S(x;at)} \varphi(\xi)\,dS\right\}.$$

$n = 2$ (*Poisson's formula*):

$$u(x,t) = \frac{1}{2\pi a}\int_0^t \int_{U(x;a(t-\tau))} \frac{f(\xi,\tau)}{(a^2(t-\tau)^2 - |x-\xi|^2)^{1/2}}\,d\xi\,d\tau$$
$$+ \frac{1}{2\pi a}\int_{U(x;at)} \frac{\psi(\xi)}{(a^2t^2 - |x-\xi|^2)^{1/2}}\,d\xi$$
$$+ \frac{1}{2\pi a}\frac{\partial}{\partial t}\int_{U(x;at)} \frac{\varphi(\xi)}{(a^2t^2 - |x-\xi|^2)^{1/2}}\,d\xi.$$

$n = 1$ (*d'Alembert's formula:*)

$$u(x,t) = \frac{1}{2a}\int_0^t \int_{(x-a(t-\tau))}^{(x+a(t-\tau))} f(\xi,\tau)\,d\xi\,d\tau + \frac{1}{2a}\int_{(x-at)}^{(x+at)} \psi(\xi)\,d\xi$$
$$+ \frac{1}{2a}\int_{(x-at)}^{(x+at)} \varphi(\xi)\,d\xi + \frac{1}{2}\{\varphi(x+at) + \varphi(x-at)\}.$$

Bibliography

[1] R. Adams: *Sobolev Spaces*, Academic Press, New York, 1975.

[2] Z. S. Agranovich and V. A. Marchenko: *The Inverse Problem of Scattering Theory*, Gordon and Breach, New York, 1963.

[3] A. C. Aitken: *Determinants and Matrices*, Oliver and Boyd, Edinburgh, 1951.

[4] N. I. Akheizer and L. M. Glazman: *Theory of Linear Operators in Hilbert Space,* Pitman-Longman, London, 1981.

[5] H. Amann: *Ordinary Differential Equations: An Introduction to Nonlinear Analysis,* W. de Gruyter, Berlin, 1990.

[6] W. O. Amrein, J. M. Jauch and K. B. Sinha: *Scattering Theory in Quantum Mechanics*, Lecture Notes and Supplements in Physics, Benjamin, Reading, MA, 1977.

[7] J. Arsac: *Fourier Transforms and Theory of Distributions*, Prentice Hall, New York, 1966.

[8] C. Athanasiadis, G. F. Roach and I. G. Stratis: A time domain analysis of wave motions in chiral materials, *Math. Nachr.* **250**, 2003, 3–16.

[9] G. N. Balanis: The plasma inverse problem, *J. Math. Phys.* **13**(7), 1972, 1001–1005.

[10] G. N. Balanis: Inverse scattering: Determination of inhomogeneities in sound speed, *J. Math. Phys.* **23**(12), 1982, 2562–2568.

[11] G. R. Baldock and T. Bridgeman: *Mathematical Theory of Wave Motion*, Ellis Horwood, Chichester, U.K., 1981.

[12] H. Baumgärtel and M. Wollenberg: *Mathematical Scattering Theory*, Operator Theory: Advances and Applications, Birkhaüser-Verlag, Stuttgart, 1983.

[13] A. Belleni-Morante: *Applied Semigroups and Evolution Equations*, Clarendon Press, Oxford, 1979.

[14] A. Belleni-Morante and A. C. McBride: *Applied Nonlinear Semigroups*, Mathematical Methods in Practice, Vol. 3, John Wiley, Chichester, U.K., 1998.

[15] A. Belleni-Morante and G. F. Roach: A mathematical model for gamma ray transport in the cardiac region, *J. Math. Anal. Appl.* **244**, 2000, 498–514.

[16] A. Belleni-Morante and G. F. Roach: Gamma ray transport in the cardiac region: An inverse problem, *J. Math. Anal. Appl.* **269**, 2002, 200–2115.

[17] A. M. Berthier: *Spectral Theory and Operators for the Schrödinger Equation*, Pitman Research Notes in Mathematics, No. 71, Pitman, London, 1982.

[18] R. Burridge: The Gel'fand-Levitan, the Marchenko and the Gopinath-Sondhi integral equations of the inverse scattering theory regarded in the context of inverse impulse-response problems, *Wave Motion* **2**, 1980, 305–323.

[19] J. R. Cannon: *The One-Dimensional Heat Equation*, Encyclopedia of Mathematics, Vol. 23, Addison Wesley, Reading, MA, 1984.

[20] K. Chadan and P. Sabatier: *Inverse Problems in Quantum Scattering Theory*, Springer-Verlag, New York, 1977.

[21] D. Colton and R. Kress: *Inverse Acoustic and Electromagnetic Scattering Theory*, Applied Mathematical Sciences, No. 93, Springer-Verlag, Berlin, 1991.

[22] J. C. Cooper: Local decay of solutions of the wave equation in the exterior of a moving body, *J. Math. Anal. Appl.* **49**, 1975, 130–153.

[23] J. C. Cooper: Scattering of plane waves by a moving obstacle, *Arch. Ration. Mech. Anal.* **71**, 1979, 113–141.

[24] J. Cooper, G. Perla Menzala and W. Strauss: On the scattering frequencies of time-dependent potentials, *Math. Methods Appl. Sci.* **8**, 1986, 576–584.

[25] J. Cooper and W. Strauss: Energy boundedness and decay of waves reflecting off a moving obstacle, *Indiana Univ. Math. J.* **25**(7), 1976, 671–690.

[26] J. Cooper and W. Strauss: Scattering of waves by periodically moving bodies, *J. Funct. Anal.* **47**(2), 1982, 180–229.

[27] J. Cooper and W. Strauss: Abstract scattering theory for time periodic systems with applications to electromagnetism, *Indiana Univ. Math. J.* **34**(91), 1985, 33–38.

[28] J. Cooper and W. Strauss: The initial boundary value problems for the Maxwell equations in the presence of a moving body, *SIAM J. Math. Anal.* **16**, 1985, 1165–1179.

[29] R. Courant and D. Hilbert: *Methods of Mathematical Physics*, Vol. II, Wiley-Interscience, New York, 1962.

[30] H. L. Cyon, R. G. Froese, W. Kirsh and B. Simon: *Schrödinger Operators with Applications to Quantum Mechanics and Global Geometry*, Texts and Monographs in Physics, Springer-Verlag, Berlin, 1981.

[31] J. W. Dettman: *Mathematical Methods in Physics and Engineering*, McGraw-Hill, New York, 1962.

[32] N. Dunford and J. T. Schwartz: *Linear Operators,* Vols. 1–3, Wiley-Interscience, New York, 1958.

[33] M. S. P. Eastham: *The Spectral Theory of Periodic Differential Equations,* Scottish Academic Press, Edinburgh, 1973.

[34] D. E. Edmunds and W. D. Evans: *Spectral Theory and Differential Operators*, Clarendon Press, Oxford, 1987.

[35] D. M. Eidus: The principle of limiting absorption, *Sb. Math.* **57**(99), 1962 (in Russian), and *AMS Transl.* **47**(2), 1965, 157–191.

[36] D. M. Eidus: The principle of limiting amplitude, *Uspekhi Mat. Nauk.* **24**(3), 1969, 91–156 (in Russian), and *Russian Math. Surveys* **24**(3), 1969, 97–167.

[37] L. D. Faddeev: On the relation between S-matrix and potential for the one-dimensional Schrödinger operator, *Dokl. Akad. Nauk SSSR, Ser. Mat. Fiz.* **121**(1), 1958, 63–66 (in Russian).

[38] L. D. Faddeev: The inverse problem in the quantum theory of scattering, *Uspekhi Mat. Nauk* **14**, 1959, 57–119 (in Russian), and *J. Math. Phys.* **4**, 1963, 72–104.

[39] J. Fawcett: On the stability of inverse scattering problems, *Wave Motion* **6**, 1984, 489–499.

[40] J. A. Ferreira and G. Perla Menzala: Time dependent approach to the inverse scattering problem for wave equation with time dependent coefficients *Appl. Anal.* **26**, 1988, 223–254.

[41] K. O. Friedrichs: *Spectral Theory of Operators in Hilbert Space*, Springer-Verlag, Berlin, 1973.

[42] I. M. Gel'fand and B. M. Levitan. On the determination of a differential equation from its spectral function, *Izv. Akad. Nauk SSSR Ser. Mat.* **15**, 1951, 309–360 (in Russian), and *Amer. Math. Soc. Transl.* **1**, 1955, 253–304.

[43] M. Gelfand and G. E. Shilov: *Generalised Functions*, Academic Press, New York, 1964.

[44] M. L. Goldberger and K. M. Watson: *Collision Theory*, John Wiley, New York, 1964.

[45] J. A. Goldstein: Semigroups and second order differential equations, *J. Funct. Anal.* **4**, 1969, 50–70.

[46] J. A. Goldstein: Abstract evolution equations, *Trans. Amer. Math. Soc.* **141**, 1969, 159–185.

[47] J. A. Goldstein: Time dependent hyperbolic equations, *J. Funct. Anal.* **4**, 1969, 31–49.

[48] J. A. Goldstein: Variable domain second order evolution equations, *Appl. Anal.* **5**, 1976, 283–291.

[49] J. A. Goldstein: *Semigroups of Linear Operators and Applications*, Oxford University Press, Oxford, 1986.

[50] E. Goursat: *Cours d'analyse Mathématique*, Gauthier-Villars, Paris, 1927.

[51] S. H. Gray: Inverse scattering for the reflectivity function. *J. Math. Phys.* **24**(5), 1983, 1148–1151.

[52] K. E. Gustafson: *An Introduction to Partial Differential Equations and Hilbert Space Methods*, John Wiley, New York, 1980.

[53] G. Helmberg: *Introduction to Spectral Theory in Hilbert Space*, Elsevier, New York, 1969.

[54] T. Ikebe: Eigenfunction expansions associated with the Schrödinger operators and their application to scattering theory, *Arch. Ration. Mech. Anal.* **5**, 1960, 2–33.

[55] D. S. Jones: *Acoustic and Electromagnetic Waves*, Clarendon Press, Oxford, 1986.

[56] T. F. Jordan: *Linear Operators for Quantum Mechanics*, John Wiley, New York, 1969.

[57] L. V. Kantorovich and G. P. Akilov: *Functional Analysis in Normed Spaces*, Pergamon, Oxford, 1964.

[58] T. Kato: On linear differential equations in Banach spaces, *Comm. Pure Appl. Math.* **9**, 1956, 479–486.

[59] T. Kato: *Perturbation Theory for Linear Operators,* Springer, New York, 1966.

[60] I. Kay: The inverse scattering problem when the reflection coefficient is a rational function, *Comm. Pure Appl. Math.* **13**, 1960, 371–393.

[61] R. E. Kleinman and R. B. Mack: Scattering by linearly vibrating objects, *IEEE Trans. Antennas and Propagation* **27**(3), 1979, 344–352.

[62] R. E. Kleinman and G. F. Roach: Boundary integral equations for the three dimensional Helmholtz equation, *SIAM Revi.* **16**(2), 1974, 214–236.

[63] E. Kreyszig: *Introductory Functional Analysis with Applications*, John Wiley, Chichester, U.K., 1978.

[64] R. J. Krueger: An inverse problem for a dissipative hyperbolic equation with discontinuous coefficients, *Quart. Appl. Math.* **34**(2), 1976, 129–147.

[65] A. Kufner and J. Kadelec: *Fourier Series*, Illiffe, London, 1971.

[66] S. T. Kuroda: On the existence and the unitary property of the scattering operator, *Nuovo Cimento*, **12**, 1959, 431–454.

[67] J. A. LaVita, J. R. Schulenberg and C. H. Wilcox: The scattering theory of Lax and Phillips and propagation problems of classical physics, *Appl. Anal.* **3**, 1973, 57–77.

[68] P. D. Lax and R. S. Phillips: *Scattering Theory*, Academic Press, New York, 1967.

[69] H. Levin: *Unidirectional Wave Motions*, North-Holland, Amsterdam, 1978.

[70] B. M. Levitan. *Inverse Sturm-Liouville Problems*, VNU Science Press, Utrecht, 1987.

[71] R. Leis: *Initial Boundary Value Problems of Mathematical Physics*, John Wiley, Chichester, U.K., 1986.

[72] R. Leis and G. F. Roach: A transmission problem for the plate equation, *Proc. Roy. Soc. Edinburgh*, **99A**, 1985, 285–312.

[73] A. E. Lifschitz: *Magnetohydrodynamics and Spectral Theory*, Kluwer, Dordrecht, Netherlands, 1988.

[74] J. Lighthill: *Introduction to Fourier Analysis and Generalised Functions*, Cambridge University Press, Cambridge, 1958.

[75] B. V. Limaye: *Functional Analysis*, Wiley Eastern, New Delhi, 1981.

[76] W. Littman: Fourier transforms of surface-carried measures and differentiability of surface averages, *Bull. Amer. Math. Soc.* **69**, 1963, 766–770.

[77] E. R. Lorch: *Spectral Theory*, Oxford University Press, Oxford, 1962.

[78] M. Matsumura: Uniform estimates of elementary solutions of first order systems of partial differential equations, *Publ. Res. Inst. Math. Sci.* **6**, 1970, 293–305.

[79] M. Mabrouk and Z. Helali: The scattering theory of C. Wilcox in elasticity, *Math. Methods Appl. Sci.* **25**, 2002, 997–1044.

[80] M. Mabrouk and Z. Helali: The elastic echo problem, *Math. Methods Appl. Sci.* **26**, 2003, 119–150.

[81] V. A. Marchenko: Some problems in the theory of one-dimensional linear differential operators of second order. *I. Tr. Mosk. Mat. Ob.* **1**, 1952, 327–420 (in Russian).

[82] V. A. Marchenko: The construction of the potential energy from the phases of the scattered wave, *Dokl. Akad. Nauk SSSR* **104**, 1955, 695–698 (in Russian).

[83] A. C. McBride: *Semigroups of Linear Operators: An Introduction*, Pitman Research Notes in Mathematics, No. 156, Pitman, London, 1987.

[84] P. McCullagh: Scattering by moving penetrable bodies, Ph.D. Thesis, University of Strathclyde, Glasgow, 1996, 1–117.

[85] I. V. Melnikova and A. Filinkov: *Abstract Cauchy Problems: Three Approaches*, Monographs and Surveys in Pure and Applied Mathematics, Vol. 120, Chapman and Hall, London, 2001.

[86] S. G. Mikhlin: *Integral Equations and Their Applications to Certain Problems in Mechanics, Physics and Technology*, Pergamon Press, Oxford, 1957.

[87] P. M. Morse and H. Feshbach: *Methods of Theoretical Physics*, Vols. 1 and 2, McGraw-Hill, New York, 1953.

[88] N. F. Mott and M. S. W. Massey: *The Theory of Atomic Collisions*, Oxford University Press, Oxford, 1949.

[89] A. W. Naylor and G. R. Sell: *Linear Operator Theory in Engineering and Science*, Holt Rinehart and Winston, New York, 1971.

[90] R. G. Newton: *Scattering Theory of Waves and Particles*, McGraw-Hill, New York, 1966.

[91] R. G. Newton: Inverse scattering, *J. Math. Phys.* **23**, 1982, 594–604.

[92] L. P. Niznik: Inverse problem of nonstationary scattering, *SSS Dokl. Akad. Nauk* **196**(5), 1971, 1016–1019 (in Russian).

[93] A. Olek: *Inverse Scattering Problems for Moving, Penetrable Bodies*, Ph.D. Thesis, University of Strathclyde, Glasgow, 1997.

[94] A. Pazy: *Semigroups of Linear Operators and Applications to Partial Differential Equations,* Springer, New York, 1983.

[95] D. B. Pearson: *Quantum Scattering and Spectral Theory,* Academic Press, London, 1988.

[96] G. Perla Menzala: Scattering properties of wave equations with time dependent potentials, *J. Math. Anal. Appl.* **12A**, 1986, 457–475.

[97] G. Perla Menzala: On the inverse problem for three dimensional potential scattering, *J. Differential Equations* **20**, 1976, 233–247.

[98] V. M. Petkov: Scattering theory for mixed problems in the exterior of moving obstacles, Pitman Research Notes in Mathematics, No. 158, Pitman, London, 1987.

[99] V. Petkov: *Scattering Theory for Hyperbolic Operators*, Studies in Mathematics and Its Applications, Vol. 21, North-Holland, New York, 1989.

[100] E. Prugovecki: *Quantum Mechanics in Hilbert Space*, Academic Press, New York, 1981.

[101] M. Reed and B. Simon: *Methods of Mathematical Physics*, Vols. 1–4, Academic Press, New York, 1972–1979.

[102] F. Riesz and B. Sz-Nagy: *Functional Analysis*, Ungar, New York, 1981.

[103] M. A. Rincon and I. Shih Lin: Existence and uniqueness of solutions of elastic string with moving ends, *Math. Methods Appl. Sci.* **27**, 2004.

[104] G. F. Roach: *Green's Functions*, 2nd ed., Cambridge University Press, London 1970/1982.

[105] G. F. Roach: *An Introduction to Linear and Nonlinear Scattering Theory*, Pitman Monographs and Surveys in Pure and Applied Mathematics, Vol. 78, Longman, Essex, U.K., 1995, i–x, 1–254.

[106] G. F. Roach: Wave scattering by time dependent potentials, *Chaos, Solitons Fractals* **12**, 2001, 2833–2847.

[107] G. F. Roach: Wave scattering by time dependent perturbations, *Fract. Calc. Appl. Anal.* **4**(2), 2001, 209–236.

[108] G. F. Roach and B. Zhang: A transmission problem for the reduced wave equation in inhomogeneous media with an infinite interface, *Proc. Roy. Soc. Lond.* **A436,** 1992, 121–140.

[109] G. F. Roach and B. Zhang: The limiting amplitude principle for wave propagation with two unbounded media, *Proc. Camb. Phil. Soc.* **112**, 1993, 207–223.

[110] G. F. Roach and B. Zhang: Spectral representations and scattering theory for the wave equation with two unbounded media, *Proc. Camb. Phil. Soc.* **113**, 1993, 423–447.

[111] G. F. Roach and B. Zhang: On Sommerfeld radiation conditions for wave propagation with two unbounded media, *Proc. Roy. Soc. Edinburgh* 1992, 149–161.

[112] J. Rose, M. Cheney and B. De Facio: Three dimensional inverse plasma and variable velocity wave equations, *J. Math. Phys.* **26**, 1985, 2803–2813.

[113] J. Rose, M. Cheney and B. De Facio: Determination of the wave field from scattering data, *Phys. Rev. Lett.* **57**, 1986, 783–786.

[114] W. Rudin: *Principles of Mathematical Analysis*, 3rd ed., McGraw-Hill, New York, 1976.

[115] D. E. Rutherford: *Vector Methods*, Oliver and Boyd, Edinburgh, 1951.

[116] Y. Saito: An inverse problem in potential theory and the inverse scattering problem, *J. Math. Kyoto Univ.* **22**, 1982, 307–321.

[117] Y. Saito: Some properties of the scattering amplitude and the inverse scattering problem, *Osaka J. Math.* **19**(8), 1982, 527–547.

[118] Y. Saito: An asymptotic behaviour of the S-matrix and the inverse scattering problem, *J. Math. Phys.* **25**(10), 1984, 3105–3111.

[119] Y. Saito: Inverse scattering for the plasm wave equation starting with large-t data, *J. Phys. A* **21**, 1988, 1623–1631.

[120] E. Sanchez-Palencia: *Non-homogeneous Media and Vibration Theory*, Lecture Notes in Physics, Vol. 127, Springer-Verlag, Berlin, 1980.

[121] J. Sanchez Hubert and E. Sanchez-Palencia: *Vibration and Coupling of Continuous Systems: Asymptotic Methods,* Springer-Verlag, Berlin, 1989.

[122] M. Schechter: *Spectra of Partial Differential Operators*, North-Holland, Amsterdam, 1971.

[123] E. J. P. Schmidt: On scattering by time-dependent potentials, *Indiana Univ. Math. J.* **24**(10), 1975, 925–934.

[124] J. R. Schulenberger and C. H. Wilcox: Eigenfunction expansions and scattering theory for wave propagation problems of classical physics. *Arch. Ration. Mech. Anal.* **46**, 280–320, 1972.

[125] L. Schwartz: *Mathematics for the Physical Sciences,* Hermann, Paris, 1966.

[126] N. A. Shenk: Eigenfunction expansions and scattering theory for the wave equation in an exterior domain, *Arch. Ration. Mech.* **21**, 1966, 120–1506.

[127] E. Skudrzyk: *The Foundations of Acoustics*, Springer-Verlag, New York, 1977.

[128] V. I. Smirnov: *Course of Higher Mathematics*, Pergamon, New York, 1965.

[129] I. N. Sneddon: *Fourier Transforms*, McGraw-Hill, New York, 1951.

[130] P. E. Sobolevski: Equations of parabolic type in a Banach space, *Amer. Math. Soc. Transl.* **49**, 1996, 1–62.

[131] B. Spain: *Tensor Calculus*, Oliver and Boyd, Edinburgh, 1956.

[132] P. D. Stefanov: The Newton-Marchenko equation for time-dependent potentials, *Inverse Problems*, **4**, 1988, 921–928.

[133] P. D. Stefanov: Uniqueness of the inverse scattering problem for the wave equation with a potential depending on time, *Inverse Problems* **4**, 1988, 913–920.

[134] P. D. Stefanov: Inverse scattering problem for the wave equation with time-dependent potential, *J. Math. Anal. Appl.* **140**, 1989, 351–362.

[135] P. D. Stefanov: A uniqueness result for the inverse back-scattering problem, *Inverse Problems* **6**, 1990, 1055–1064.

[136] P. D. Stefanov: Inverse scattering problems for the wave equation with time dependent impurities, in *Inverse Methods in Action*, P. Sabatier (ed.), Springer-Verlag, 1990, 1–636.

[137] M. H. Stone: *Linear Transformations in Hilbert Space and Their Appliations to Analysis, Amer. Math. Soc. Colloq. Publ.* **15**, 1932.

[138] W. A. Strauss: *Partial Differential Equations: An Introduction*, John Wiley, New York, 1992.

[139] W. W. Symes: Inverse boundary value problems and a theorem of Gel'fand and Levitan, *J. Math. Anal. Appl.* **71**, 1979, 379–402.

[140] H. Tamura: On the decay of local energy for wave equations with time dependent potentials, *J. Math. Soc. Japan* **33**(4), 1981, 605–618.

[141] H. Tanabe: On the equation of evolution in a Banach space, *Osaka Math. J.* **12**, 1960, 363–376.

[142] H. Tanabe: *Evolution Equations,* Pitman Monographs and Studies in Mathematics, Pitman, London, Vol. 6, 1979.

[143] J. R. Taylor: *Scattering Theory: The Quantum Theory of Non-Relativistic Collisions.* University of Colorado, Boulder.

[144] E. C. Titchmarsh: *Introduction to the Theory of Fourier Integrals,* Oxford University Press, 1937.

[145] F. Tricomi: *Integral Equations*, Interscience, New York, 1957.

[146] B. Vainberg: *Asymptotic Methods in Equations of Mathematical Physics*, Gordon and Breach, New York, 1989.

[147] V. S. Vladymirov: *Equations of Mathematical Physics*, Marcel Dekker, New York, 1971.

[148] J. Van Bladel: Electromagnetic fields in the presence of rotating bodies, *Proc. IEEE* **64**, 1976, 301–318.

[149] H. F. Weinberger: *A First Course in Partial Differential Equations*, Blaisdell, Waltham, MA, 1965.

[150] P. Werner: Regularity properties of the Laplace operator with respect to electric and magnetic boundary conditions, *J. Math. Anal. Appl.* **87**, 1982, 560–602.

[151] P. Werner: Resonances in periodic media, *Math. Methods Appl. Sci.* **14**, 1991, 227–263.

[152] V. H. Weston. On the inverse problem for a hyperbolic dispersive partial differential equation, *J. Math. Phys.* **13**(12), 1972, 1952–1956.

[153] C. H. Wilcox: Scattering states and wave operators in the abstract theory of scattering, *J. Funct. Anal.* **12**, 1973, 257–274.

[154] C. H. Wilcox: *Scattering Theory for the d'Alembert Equation in Exterior Domains*, Lecture Notes in Mathematics, No. 442, Springer, Berlin, 1975.

[155] J. Wloka: *Partial Differential Equations*, Cambridge University Press, Cambridge, 1987.

[156] K. Yosida: *Functional Analysis*, Springer-Verlag, Berlin, 1971.

Index

www.ingramcontent.com/pod-product-compliance
Ingram Content Group UK Ltd.
Pitfield, Milton Keynes, MK11 3LW, UK
UKHW022238130125
453475UK00003B/38